AF520497

ROUGHAGE PROCESSING TECHNOLOGY

ROUGHAGE PROCESSING TECHNOLOGY

Dr. S.S. Kundu
P.S. and Head PAR Division
Indian Grassland and Fodder
Research Institute (ICAR), Jhansi (U.P.)

Dr. S.K. Mahanta
Scientist PAR Division
Indian Grassland and Fodder
Research Institute (ICAR), Jhansi (U.P.)

Dr. Sultan Singh
Scientist PAR Division
Indian Grassland and Fodder
Research Institute (ICAR), Jhansi (U.P.)

Dr. P.S. Pathak
Director
Indian Grassland and Fodder
Research Institute (ICAR), Jhansi (U.P.)

SATISH SERIAL PUBLISHING HOUSE
403, Express Tower, Commercial Complex, Azadpur, Delhi - 110033 (INDIA)
Phone : 011-27672469, 27672852 Fax : 91-11-27672046
E-mail : ssp@ndf.vsnl.net.in, hkjain@satishserial.com
Website : www.satishserial.com

Published by :

SATISH SERIAL PUBLISHING HOUSE
403, Express Tower, Commercial Complex, Azadpur, Delhi - 110033 (INDIA)
Phone : 011-27672469, 27672852 Fax : 91-11-27672046
E-mail : ssp@ndf.vsnl.net.in, hkjain1975@yahoo.com

First Published 2005

ISBN 81-89304-07-0

Printed at
SALASAR IMAGING SYSTEMS, DELHI-35

CONTENTS

PREFACE

Grasses, roughages and crop residues (straw and stovers) along with tree leaves and shrubs constitute the major bulk of available feed pool to subsist the large livestock population of the nation. Presently India is producing 540 million tones of crop residues. The nutritive value of available feed resources is very poor as they are rich in fiber, low in nitrogen, vitamins, minerals and other soluble nutrients. Further the poor palatability; low digestibility and reduced availability of nutrients render them ineligible even to meet the maintenance need of animals. Bulkiness and low density of fibrous feeds further make their handling, transportation and storage operations tedious. In such a state of affair improvement in the nutritive value of these bulky fibrous feeds by tools and techniques of feed technology is the only option to fulfill the nutrients demand of livestock. To meet the objective of sustainable and higher animal production, interaction, coordination, linkage and exchange of scientific inventions amongst the engineers, nutritional scientists, planners and feed manufacturers is the sole viable alternate. In the book entitled " Roughage Processing Technology" editors have attempted to systematically organize, synthesize and supplemented the information contributed by leading experts from the feed processing engineers, animal nutritionists and feed manufacturers. Feed block formation for different livestock species, compaction of fibrous feeds, chemical preservation of forages, densifying machines, roughage particle size, status of densifying technology, straw ammoniation technology, densification for block making, feed block technology, feed and fodder evaluation, leaf meal production technology, forage harvesting machines and codex alimentarius are subjects lie in the scope of book. The editors extend special thanks to contributors for their contribution in form of different chapters.

The encouragement and blessings poured by honorable Director General, ICAR and Secretary DARE, Dr. Mangala Rai ji; Dr. G. Kalloo Deputy Director General (Crop Sciences), Dr. V. K Taneja, Deputy Director General (Animal Sciences) and Dr. S. N. Shukla, ADG are gratefully recorded. It's pleasure in extending thanks to Department of Science and Technology and especially to Dr. V. P. S. Tomar and Dr. J. K. Sharma advisors for their financial support to enable us to arrange this workshop to discuss different aspects of the subject. It is our duty to thank the advisors namely Dr. K Pradhan, Dr. S. K. Ranjhan, Dr. V. K. Singh, Dr. V. D. Mudgal Dr. S. M. Ilyas, Dr. N. N. Patahak for their advice in finalizing the contents of the book.

It is pleasure in extending the thanks to our colleagues namely Dr. N. C. verma, Dr. B. K. Bhadoria, Dr. A. B. Majumdar, Dr. L. K. Karnani, Dr. B. P. Kushwaha, Dr. A. K. Mishra, Dr. S. B. Maity, Dr. N. P. Singh, Dr. K. K. Singh, Dr. A. K. Samanta, Dr. M. M. Das, Dr. G. H. Pailan, Dr. S. K. Nag and students/staff, Dr. Ranjan Kumar, Chanderbhan and Sudheer for heir help at different stages in preparing the book.

S. S. Kundu
S. K. Mahanta
Sultan Singh
P. S. Pathak

LIST OF CONTRIBUTORS

Dr. Amar Singh
Principal Scientist
Division of Agricultural Engineering
Indian Agricultural Research Institute,
New Delhi

Dr. A.K Nagpal
Principal Scientist
National Research Centre on Camel
Bikaner 334001 (Rajasthan)

Dr. Anil Kumar
Scientist
Indian Grassland and Fodder Research Institute, Jhansi-284 003

Dr. A. Pandeya
Scientist
Division of Agricultural Engineering
Indian Agricultural Research Institute,
New Delhi

Dr. A.K. Samanta
Senior Scientist
PAR Division,
Indian Grassland and Fodder Research Institute, Jhansi (U.P.)

Er. Ajay Sharda
Assistant Engineer
Department of FPM
PAU, Ludhiana

Dr. A.B. Mojumder
Principal Scientist
PAR Division,
Indian Grassland and Fodder Research Institute, Jhansi (U.P.)

Dr. B.S. Meena
Scientist
Indian Grassland and
Fodder Research Institute
Jhansi-284 003

Sh. Bharat Sharma
Director
Indian Fodder Care &
Technologies Pvt. Ltd., Delhi

Er. B.D. Shukla
CIAE, Bhopal

Dr. C.S. Sahay
Scientist
FM & PHT Division
Indian Grassland and
Fodder Research Institute,
Jhansi (U.P.)

Dr. D.S. Rajput
Principal Scientist & Head
FM & PHT Division
Indian Grassland and
Fodder Research Institute,
Jhansi (U.P.)

Sh. G.S. Singh
Director
Indian Fodder Care &
Technologies Pvt. Ltd., Delhi

Dr. G P Singh
Principal Scientist
National Research Centre on Camel
Bikaner 334001 (Rajasthan)

Dr. G.H. Pailan
Scientist
PAR Division,
Indian Grassland and
Fodder Research Institute,
Jhansi (U.P.)

Dr. J. S. Pawar
Principal Scientist
Division of Agricultural Engineering
Indian Agricultural Research Institute,
New Delhi

Dr. J.N. Gupta
Principal Scientist
GSM Division
Indian Grassland and
Fodder Research Institute,
Jhansi (U.P.)

Dr. K.K. Singhal
Principal Scientist and Head
Dairy Cattle Nutrition Division
Natonal Dairy Research Institute,
Karnal - 132 001

Dr. K.K. Singh
Senior Scientist
PAR Division,
Indian Grassland and
Fodder Research Institute,
Jhansi (U.P.)

Dr. M. K. Tripathi
Senior Scientist
Division of Animal Nutrition
CSWRI, Avikanagar
Rajasthan-304501, India

Dr. Maharaj Singh
Principal Scientist
Indian Grassland and
Fodder Research Institute
Jhansi-284 003

Dr. M.M. Das
Senior Scientist
PAR Division,
Indian Grassland and Fodder Research Institute, Jhansi (U.P.)

Dr. Nawab Singh
Principal Scientist
CIRG, Makhdoom, Farah - 281122
Mathura (U.P.)

Dr. Nand Kishore
Associate Professor
Department of Animal Nutrition,
CCS Haryana Agricultural University,
Hisar-125004

Dr. Nawab Ali
Director
CIAE, Bhopal - 463038

Dr. O.P. Lohan
Professor
Department of Animal Nutrition
CCS Haryana Agricultural University,
Hisar-125 004

Dr. P. Sharma
Scientist
Indian Grassland and
Fodder Research Institute
Jhansi-284 003

Dr. P.D. Gupta
Principal Scientist
FM & PHT Division
Indian Grassland and
Fodder Research Institute,
Jhansi (U.P.)

Dr. P.N. Dwivedi
Scientist
FM & PHT Division
Indian Grassland and
Fodder Research Institute,
Jhansi (U.P.)

Dr. P.S. Pathak
Director
Indian Grassland and
Fodder Research Institute,
Jhansi (U.P.)

Dr. P.K. Pathak
Senior Scientist
FM & PHT Division
Indian Grassland and
Fodder Research Institute,
Jhansi (U.P.)

Dr. Raman Malik
Senior Scientist
Dairy Cattle Nutrition Division
Natonal Dairy Research Institute,
Karnal - 132 001

Dr. R. C. Jakhmola
Principal Scientist
Division of Animal Nutrition
CSWRI, Avikanagar
Rajasthan-304501, India

Dr. Ram Singh
Scientist
Indian Grassland &
Fodder Research
Institute, Regional Research Centre
CSKHPKV, Campus,
Palampur-176062

Dr. Ranjan Kumar
SRF
PAR Division, Indian Grassland and
Fodder Research Institute,
Jhansi (U.P.)

Dr. R.S. Jaiswal
Associate Professor
Department of Animal Sciences
College of Veterinary Sciences
GBPUA & T
Pantnagar - 263145 (Uttaranchal)

Dr. R.N. Dwivedi
Principal Scientist & Head
Indian Grassland and
Fodder Research Institute
Jhansi-284 003

Dr. S.S.Thakur
Principal Scientist
Dairy Cattle Nutrition Division,
National Dairy Research Institute
Karnal - 132 001, Haryana, India.

Dr. S K Jha
Scientist
Division of Post Harvest Technology
Indian Agricultural Research Institute,
New Delhi

Dr. Sadhna Pandey
Scientist
Indian Grassland and
Fodder Research Institute
Jhansi-284 003

Dr. S. S. Kundu
Principal Scientist & Head
PAR Division,
Indian Grassland and
Fodder Research Institute,
Jhansi (U.P.)

Dr. Sultan Singh
Scientist
PAR Division,
Indian Grassland and Fodder Research
Institute, Jhansi (U.P.)

Dr. S.S. Ahuja
Senior AE & Head
Department of FPM
PAU, Ludhiana

Er. Sanjay Khar
Ph.D Scholar
Department of FPM
PAU, Ludhiana

Er. S. Mangraj
CIAE, Bhopal

Dr. S.A. Karim
Principal Scientist
CSWRI, Avikanagar
Rajasthan - 304501

Dr. T.K. Dutta
Senior Scientist
CIRG, Makhdoom, Farah - 281122
Mathura (U.P.)

Dr. T.C. Thakur
Professor & Head
Department of FM & P
GBPUA & T
Pantnagar - 263145 (Uttaranchal)

CHAPTER - 1

DENSIFICATION FOR BLOCK MAKING : AN INDIAN OVERVIEW

S.S. Kundu, M.M. Das, P.K. Pathak, A.K. Samanta and K.K. Singh

Crop residues such as straws (wheat/Paddy) and stovers (maize, sorghum etc.) are the major component of ruminants diet in developing countries. Economic and balanced feeding is extremely important for optimum livestock production. But in India, lower livestock productivity is mainly due to the scarcity of feeds and unbalanced feeding practices. With the changing scenario of land use, for non-agricultural uses community-grazing lands are shrinking day by day. On the other hand, agricultural operations are mainly diverted towards grain production for human use. Under such circumstances, the animal need to be supplemented with agro-industrial by-products, top feeds etc. Owing to low bulk density (65 - 70 kg / m^3) of most of the crop residues (Yadav et al., 1990), handling, transportation and storage becomes a major problem in the feeding of livestock. Densification of these crop residues to high density (300 - 500 kg/m^3) blocks (Bruhn et al., 1959) containing deficient nutrients by incorporating agro-industrial by-products and top feeds for optimum livestock production may possibly overcomes this difficulty (Preston and Leng, 1984). Besides this, the compressed feed block technology has the potential to provide ready to eat ruminant diet during scarcity period, which are further aggravated by natural calamities like recurrent drought and floods in India.

Machinery for complete feed blocks

Roughage and concentrate mixture when mixed homogenously in a pre determined ratio and formed into pellet, briquette or block to provide a nutritionally balanced diet to animals are called complete feed. The complete feeds are usually formulated on the basis of wheat straw as roughage but other crop residues viz. paddy straw, sugar cane bagasse etc (Singhal and Thakur, 2004) and agro-industrial wastes and by products could also be included. Formation of roughage based complete feed requires unit operations of size reduction, drying, mixing and the subsequent compaction of roughage and concentrate mixture. The compaction could be done for making different end products like pellets, briquettes, wafers or blocks of required size and shape. The complete feed block (CFB) technology saves energy as grinding and steaming are not required. It maintains the integrity of feed ingredients and needs simple press machine to densify the material in block form.

The compaction behaviour of different material as affected by different parameters viz. pressure (Kundu et al., 1998; Kundu and Das, 2004; Singh et al., 2004a, Singh et al., 1986 and Sarsavadia et al., 1991), moisture content (Singh, 2002; Kundu et al., 1998; Dwivedi et al., 2004a; Singh et al., 1986, Sihag et al., 1991 and Sharma et al., 2004a) and molasses addition (Kundu et al., 1998; Sihag et al., 1991; Sharma et al., 2004a; Singh et al., 2004b and Dwivedi et al., 2004a) has already been reported. Clear-cut characteristics of complete feed blocks for commercial manufacturing is yet to be established incorporating the conventional as well as non-conventional feed ingredients including roughages. The desirable characteristics of complete feed blocks for various categories of livestock could be as below (Table 1) and may be altered by the manufacturer according to the demand and convenience.

Table 1 Desirable characteristics of complete feed blocks for various categories of Livestock

Parameter	Type of Animal		
	Growing	Adult	Lactating
Block weight (kg)	5	10	14
Shape	Rectangular	Rectangular	Rectangular
Loose bulk density (kg / m^3)	70 – 80	70 – 80	70 – 80
Compact bulk density (kg / m^3)	200 – 225	200 – 225	200 – 225
Post compression expansion (%)	6 – 12	6 – 12	6 – 12
Durability (%)	< 17	< 17	< 17
CP (% minimum)	10	12	14
TDN (% minimum)	50	55	60

Source: (Singhal and Thakur, 2004)

For the commercial manufacturing of these CFB, two operations are essential and play vital role in selection of the machinery. These are feed mixing and block making. For mixing roughage, concentrates and other feed ingredients the coefficient of variation in the composition remains within the feasible limit. Due to diverse nature of feed ingredients i.e. difference in their physical properties, the satisfactory mix having reasonable coefficient of variation could be achieved by employing the combination of mixing systems (paddle, ribbon or screw type mixer) with proper adjustments of speed. However, care has to be taken to maintain the uniformity in all the batches for desired level of nutrients. Addition of molasses while mixing could also facilitate the homoginity of mix and enhance the durability of blocks. The admixture obtained after mixing is given desirable shape and size by applying the pressure on it. This could be done through hydraulic press or by mechanical press. Singhal and Thakur (2004) reported that the machine developed by Singh et al. (2004b) could manufacture feed blocks of 37 x 37 x 32 cm weighing 18 – 20 kg at a capacity of 1 t/h for the industrial application.

Various types of hydraulic presses have been employed to make the blocks. Kundu et al. (1998) used two pillars type hydraulic press machine, fitted with 3 hp motor, which can also be operated manually with up and down strokes to obtain the circular type of feed blocks having the bulk density of 260 kg / m^3. Verma et al. (1996) used manually operated small capacity animal feed block making machine for making CFB. The machine consists of hydraulic ram, operates with 7.5 hp three-phase motors with a stroke of 750 mm, compaction chamber and frame. Their machine was designed to exert a maximum pressure of 400 kg / cm^2 and produces a block size of 300 x 300 mm at a capacity of 100 kg wheat straw per hour. Ram et al. (1990) used the manually operated screw press consisting of a screw – rotating wheel at one end and a punch on the other end. These machines however, could not be used for commercial manufacturing of CFB due to low capacity. Recently Singh et al. (2004b) has developed horizontal hydraulic type animal feed block formation machine in three models viz. 100 kg/ h, 250 kg/ h and 1000 kg/ h powered by 7.5 hp, 25 hp and 45 hp, respectively. The machine consists of frame, hydraulic cylinder power pack and electric control panel and produces square shape blocks (20 cm x 20 cm) for laboratory model desired thickness and weight with working pressure varied up to 6000 psi and output capacity as 30 – 40 blocks / hour. Singh et al., (2004b) also compared the total cost of feed block formation (Table 2) which indicates that the machine having output capacity of 1000 kg / h has least production cost and machine having medium capacity (250 kg / h) has highest production cost. The reason of variation might be due to variation in the price of the machine, power and labour requirement, which they did not mention. Being, higher machine cost the initial expenditure and fixed cost might also be high in the set – up of the plant. Zombade (2004) reported that cost of feeding of a buffalo, yielding 12 kg milk/ day and cows yielding 25-30 kg milk / day maintained on densified feed block was Rs 65 and Rs 75 per day, respectively.

Table 2 Cost Economics of Feed block formation

	Model I	Model II	Model III
Capacity of the machine (kg/h)	100	250	1000
Price of the machine (Rs)	2.0 lakhs	7.2.0 lakhs	12.0 lakhs
Life of the machine (years)	20	20	20
Interest rate (%)	10	10	10
Working day per year	300	300	300
Working hours per day	16	16	16
Salvage value (%)	20	20	20
Power requirement	7.5hp, 3 phase	25hp, 3 phase	45hp, 3 phase
Labour Requirement (Person / d)	2	4	10
Approximate cost of feed block (Rupees / ton)	473.00	617.00	265.00

Singh et. al., (2004 b)

Mechanical presses also in use for making complete feed blocks, detailed design of these machines are reviewed by Gupta (2000). Indian Grassland and Fodder Research Institute, Jhansi has developed high-density baler (Gupta et al., 1998), which was well employed in making CFB (Dwivedi et al., 2004a). The machine is equipped with crank – piston drive mechanism with rectangular piston cross-section and is powered by 18.5 KW electric motor with V belt and pinion – gear power transmission system. It has a spring tension mechanism at the exit point of compression chamber to adjust the base density and forced feed mechanism for pre compression of loose hay. The bales are tied manually by mild steel wire through the slotted provided in the compression chamber. The cross section area of compression chamber is 305 x 330 mm and the length of bales could be adjusted between 45 to 55 cm for obtaining the bale weight of 20 to 25 kg at the density of 350 – 450 kg / m^3.

The machinery thus available commercially could be selected for making complete set of hard and fast operational procedures for complete feed manufacturing based on all types of roughages and other ingredients. The awareness of the operating characteristics of the machines through trial and error will be helpful for establishing the complete feed block unit.

Factors affecting block formation

A. Pressure

Application of pressure is an critical factor for the formation of block. A pressure of 2.03 to 2.7 MPa / kg feed with one minute dwell time appeared to be sufficient for block formation of wheat straw based complete feed (Kundu et. al., 1998, Kundu and Das, 2004). The higher-pressure application resulted in hard blocks and creates difficulty in biting by the animals. Feed blocks prepared from mustard straw alone or mixed with concentrate mixture and molasses with a pressure of 0.21 - 0.042 kg / cm^2 indicated that bulk density increased with the increase in compaction pressure and resilience of block decreased with increase in compaction pressure (Singh et. al., 2004a). Untreated or 4% urea treated wheat straw; mustard straw and sugarcane bagasse with a pressure application of 350 kg / cm^2 produces more durable blocks. The grass wafers formed at a pressure of 40 MPa were not stable whereas the legume wafers formed at 20 MPa and that of grass legume mixture (1:1) formed at 30 MPa were able to with stand handling (Singh et. al., 1986). Durability test of block of wheat straw and groundnut haulm based complete feed indicated that optimum load at 1 minute holding time should be 10000 kg for producing durable blocks (Sarsavadia et. al., 1991). Application of pressure at three different levels i.e. 3.41 kg / cm^2, 5.69 kg / cm^2 and 7.96 kg / cm^2 for preparation of untreated or 4 % NaOH treated wheat straw based blocks showed that at medium pressure levels there was no significant difference between durability of treated and untreated straw blocks however the durability was higher in treated straw blocks. The durability of blocks formed at 10 minute dwell time found to be better than those formed with 3 and 5 minutes dwell time in both treated and untreated straw based complete feed blocks (Sihag et. al., 1991).

B. Moisture

Moisture content of the feed material used for densification is an important factor influencing the formation of blocks. Densification of untreated or alkali + bleaching powder treated wheat straw based complete feed containing 16 - 18% or 18 - 23% moisture produce blocks i.e. non-breaking during handling and storage (Kundu et .al., 1998). The wheat straw blocks produced at 10, 20 and 30% moisture content were found safe for animal feeding after two and half years of storage (Singh, 2002). Densification of chopped C. ciliaris grass containing 20 - 30% moisture prevent block formation due to very high post operation expansion whereas it was successfully densified at 10% moisture level. However, Subabool leaves can be densified at all moisture levels i.e. 10, 20 and 30% (Dwivedi et. al., 2004a). Resilience of mustard straw block increased with the increased moisture content of feed (Singh et. al., 2004a). Expansion of feed blocks immediately after preparation and after 2 hrs of preparation of block inversely proportional to moisture content. The porosity and compactness of the complete feed block directly proportional to moisture content of the complete feed block (Sharma et. al., 2004a).

The effect of moisture content on stability of wafers of grass, Subabool leaves and grass - Subabool mixtures at 29 to 30 M Pa indicated that the wafers made at moisture content less than 12 and more than 17 percent expanded rapidly after removal from the die and finally dis- integrated during handling (Singh et.al., 1986). The difference in durability is not significantly affected at 15 and 17.5% moisture levels in untreated as well as treated wheat straw blocks, however in both type of straws the durability found to be significantly (P < 0.05%) low at 20% moisture level compared to 15% and 17. 5% moisture levels (Sihag et.al., 1991). Similarly moisture level of 18.5% found optimum for the preparation of durable blocks of paddy and wheat straw than at 15% and slightly non-significantly higher than 20% moisture level (Yadav et. al., 1991). The moisture level of 23% in feed mixture found to be optimum for green fodder based complete feed block formation (Singh et. al., 1998).

C. Molasses

Sugarcane molasses are generally used as a binding agent for preparation of feed block. More than 10% (on fresh basis) or 8% (on DM basis) molasses (66 % DM) resulted sticky surfaces and also cause storage problems (Kundu et. al., 1998). With treated straw, block can be formed even at 5% molasses level but with untreated straw the blocks did not form (Sihag et. al., 1991). Molasses addition (3.5%, 5.5% or 7.5%) to the feed material containing 12, 15 and 20% moisture indicated that with the increase in molasses content in premix, decreases the bulk density and post compression expansion reversely proportional to molasses addition (Sharma et. al., 2004a). Mustard straw based complete feed can be successfully densified at 5% molasses addition (Singh et. al., 2004a). Block of wheat straw and berseem based complete feed can also be prepared by addition of 20% molasses (Dwivedi et. al., 2004a) and from paddy straw based complete feed with 15% molasses. Feed blocks can also be prepared from the groundnut haulm, black gram straw, grass, sorghum stover, stylo straw alone or in combination with concentrate mixture without molasses at 2000 Psi. However, wheat straw and paddy straw need molasses as a binder for durable block formation.

Compaction Behaviour

Bulk density

The bulk density of the block is calculated by dividing sample weight with the measured volume. Various studies (Sarsavadia et. al., 1991, Jha et. al., 2004) indicated that the bulk density of blocks increased with the increase in compression pressure. A similar relationship has been reported for wheat, barley and paddy straw (Singh et. al., 2002, Singh et. al., 2004b). The levels of molasses addition in crop residues for block formation also affect the bulk density. The incorporation of molasses up to 4 % to paddy straw did not make any difference in the bulk density while molasses content higher than 4% caused increase in bulk density. Feed blocks made of urea treated crop residues had significantly ($P < 0.05$) higher bulk density than untreated crop residues (Jaglan and Nand Kishore, 2004). The bulk density of mustard straw increased linearly with the moisture content and the inclusion of supplemental level of concentrate in the feed block reduced significantly ($P < 0.01$) the times increase in bulk density (Nand Kishore and Lohan, 2004). Paddy straw showed more increase in bulk density in comparison to wheat straw and in ground form as compared to that in ungrounded form of the respective straw (Yadav et. al., 1991). The bulk density of complete feed increased 3.19 and 3.89 times for feed blocks based on untreated and alkali treated straw (Sihag et. al., 1994). Bulk density of wafers of grass and legumes decreased after removal from the die. 46 to 70.5% reduction in density recorded in grass and 36 to 66 % in legumes (Singh et al., 1986). The difference in the relaxed densities of different materials indicated that block making process influenced by both the compressive pressure and resilience characteristics of the material used for block making. The bulk density of raw as well as the blocks of some of commonly available feed materials is given in table 3.

Table 3: Bulk density of raw feeds and blocks at compaction pressure of 300 kg / cm^2

Feed/ Crop residue	Bulk density (kg/m^3)		Volume reduction factor
	Feed	Feed block	
Wheat straw	51.20	305.2	5.9
Wheat straw + 5% molasses	57.00	443.5	7.8
Paddy straw	70.0	279.0	3.9
Rice husk	96.0	622.0	6.5
Rice bran	361.0	1197.0	3.3
Maize plant (Chopped)	90.7	520.8	5.7

Feed/ Crop residue	Bulk density (kg/m^3)		Volume reduction factor
	Feed	Feed block	
Maize cob (Ground)	498.0	843.0	1.7
Jowar stover	70.0	437.2	6.2
Bajra stover	119.6	511.1	4.3
Mustard straw	51.48	562.2	10.9
Groundnut haulm	77.1	500.0	6.5
Gram straw	66.5	471.9	7.1
Arhar straw	55.0	689.1	12.5
Sugarcane baggase	61.0	567.5	9.3
Cluster bean	109.0	596.7	5.5
Berseem	55.0	1050.0	19.0
Cotton stalk	140.0	785.0	5.6
Mango leaves	17.0	643.2	37.8
Khejri leaves	67.5	542.0	8.0
Pala leaves	65.5	538.4	8.2
Ardu leaves	51.5	539.1	10.4
Seasam leaves	37.96	691.0	22.6
Jamun leaves	34.26	776.5	24.8
Neem leaves	41.66	1035.9	19.3
Eucalyptus leaves	41.66	806.0	29.3
Pipal leaves	30.55	896.4	22.6
Sewan	91.6	700.0	7.6
Moth chara	84.70	901.0	10.6
Cenchrus grass	67.50	486.8	7.2
Lawn grass	38.00	597.0	15.7
Doob grass	15.00	630.0	42.0
Forest grass	66.33	503.0	7.6

Source: (Singh et. al., 2004b)

Post Compression Expansion

After the removal of load, wafer expanded in axial and radial direction. The expansion in radial direction is only 2 - 3 % and maximum expansion occurred in axial direction. The blocks made from various roughage sources indicated that the maximum expansion in the first 30 - 60 minutes after removal from the die and thereafter it slowed down towards equilibrium state (Singh et al., 1986; Kundu et al., 1998). Maximum axial expansion of 25% of the initial thickness for briquettes having zero level of molasses was recorded in ground wheat straw (Sah et al., 1981), however, for molasses mixed (5, 10 and 15%) straw briquettes, the axial expansion was 12 to 15%. Grass blocks showed maximum axial expansion of 81.25 percent whereas 23.8 percent recorded in legumes. Wheat straw and groundnut haulms based complete feed block expanded rapidly in the first 10 minutes and most of the expansion occurred (70 – 80 %) within 30 minutes after load removal (Sarsavadia et.al., 1991). The post compression expansion of untreated or 4% NaOH treated wheat straw based complete feed block was significantly ($P < 0.05$) lower at low pressure than medium and high pressure (Sihag et.al., 1991). Paddy straw based complete feed block showed significantly ($P < 0.05$) lower post compression expansion than that of wheat straw block and in ground form as compared to ungrounded form (Yadav et. al., 1991). Lower post compression expansion recorded in alkali treated straw than urea treated straw. The increase in the level of concentrate mixture in the feed block also decreased significantly ($P<0.01$) the post compression expansion (Nand Kishore et al., 2004). Grass hay based complete feed block made with different levels of moisture content indicated that expansion of the complete feed block inversely proportional to the moisture content and reversely proportional to pressure and molasses addition (Sharma et.al., 2004a). Wheat straw, mustard straw and sugarcane bagasse produces more durable blocks with low post compression expansion as compared to the blocks of paddy straw. Feed blocks made of urea treated crop residues had significantly ($P < 0.05$) lower post compression expansion as compared to the blocks of untreated crop residues (Jaglan and Nand Kishore, 2004). Wheat straw replaced with paddy straw or sugarcane bagasse caused significantly ($P<0.05$) lower post compression expansion as compared to wheat straw alone. The resiliency of a crop residue indicates the tendency of a compacted block to expand in volume and the moisture content of the material determines the resiliency of the compacted briquette.

Effect of densified block feeding on DM intake

Various studies indicated higher DM intake with densified block however, some studies reported comparable DM intake. The complete feed has been suggested to cause 33 and 72 percent increase in voluntary intake while fed as mash and pellet form respectively over the conventional type of feed in Sheep (Reddy et.al., 1986). Buffaloes fed on wheat straw based compressed complete feed block consumed significantly ($P < 0.05$) more of both DM as well as organic matter (Verma et.al., 1996). However, compression of complete feed prepared with bleaching powder and NaOH treated wheat straw did not significantly affect the DM intake in Murrah buffaloes (Das and Kundu, 1993). Crossbred calves also did not make any significant difference in daily DM intake with dry grass based mash feed or densified feed block (Pachauri et. al., 1999). The complete diet consisting of grass and concentrate mixture at the ratio of 60:40 either in mash form or block form did not reveal any significant difference on the dry matter intake in Barbari goats (Samanta et.al., 2003). Comparable DM intake was noticed in buffaloes fed wheat straw and berseem hay based complete diet in mash or block form (Dwivedi et.al., 2004b). The digestibility coefficients of DM, crude protein, crude fiber and nitrogen free extract were not significantly different for block or mash form of diet in crossbred cows (Berwal et.al.,1997).

Rams fed complete block based on pearl millet kadbi untreated or treated with Sodium hydroxide (2%) or with alkaline hydrogen peroxide showed higher dry matter intake (1369 g vs. 1226g) as compared to those in mash form (Chaturvedi et.al., 2004). DM intake with densified block also affected by the particle size of the roughage. DM intake was significantly ($P < 0.05$) higher in animals consuming with ground straw based feed blocks than the respective block prepared from ungrounded straw. The better intake on feed blocks made up of ground straw may be due to increase rate of passage through the digestive tract (Yadav et.al., 1990). Growing calves consumed more (Sharma et.al., 2004b, Das et al., 2004a) from wheat straw or grass based complete feed block than its loose form. In an on farm trial, lactating buffaloes showed 15% higher intake of DM as compared to conventional feeding adopted by the farmers (Das et.al., 2004b). Higher feed intake was also recorded in the block form of diet (Lohan et.al., 2004) over the conventional feeding system in the villages of Haryana. However, lower average dry matter intake per kg metabolic body size was reported (Singh et al., 1998) in complete feed blocks than those of conventional feeding system in growing buffalo calves which might be due to lower palatability resulted from the presence of urea.

Effect of complete feed block feeding on livestock production

Buffalo calves showed significantly higher average daily gain per day with berseem and wheat straw based complete feed block as compared to mash form (Singh et. al. 1998). The higher growth rate in block fed animals might be due to higher dry matter intake and increase in TDN value of the ration. Camel calves fed chane ki khar, groundnut chara based complete feed block also had better growth rate than the mash fed groups (Nagpal et.al., 2004). Similarly higher growth rate also has been reported (Singh et.al., 2004c) in crossbred calves fed grass based complete feed block. On farm trial study indicated 14% higher milk yield in lactating buffaloes fed grass based complete feed block as compared to the conventional feeding adopted by the farmers of Simardha village of Uttar Pradesh (Das et.al., 2004b). Trial conducted on villages of Haryana also showed 0.85 to 1.2 lit increase in milk yield in lactating animals with complete feed block in comparison to conventional feeding system (Lchan et.al., 2004). Similarly 0.72 lit/ day higher milk yield was recorded in milch buffaloes fed complete feed block consisting of wheat straw, berseem hay and concentrate in Simardha and Budha villages of Jhansi (Dwivedi et.al., 2004b). In crossbred cows, feeding of complete feed block resulted in 0.84 kg/d higher milk yield than its mash form (Berwal et.al., 1997).

Effect of densification on nutrient utilization

Feeding of densified complete feed block in growing calves did not reveal any significant difference in digestibility coefficients of proximate principles and nutritive value (DCP and TDN %) of the ration as compared to its mash form (Singh et al., 1998). Similarly the feeding of wheat straw based complete feed block to Murrah buffaloes had no significant influence on nutrient utilization as well as on the digestibility of various nutrients. However, higher feed intake through complete feed block resulted in significantly higher intake of digestible DM and all other nutrients as compared to the feeding of diets in mash form or concentrate and roughage separately (Verma et. al. 1996). Feeding of densified complete feed block also resulted in higher ($P<0.05$) retention of calcium and phosphorus. Feeding of 2% bleaching powder + 1% NaOH treated wheat straw based complete feed block (350 kg / m^3) to Murrah buffaloes resulted in higher CP digestibility and N balance as compared to its, mash form (Das and Kundu, 1993). In goats, grass based complete feed block did not make significant difference in the digestibility of various nutrients and nutritive value of the ration (Samanta et.al., 2003). Wheat straw and berseem based densified feed block (355 kg / m^3) is well relished by the growing buffalo heifers, however, feeding of densified complete feed block provide nutrient digestibility of various nutrients and nutritive value of the ration similar to its loose form (Dwivedi et.al., 2003). Growing calves consuming wheat straw based complete feed block did not reveal any significant effect on digestibility of nutrients. Digestibility coefficients of various proximate principles of grass based complete feed block also showed similar values to that of its loose form in crossbred calves (Singh et. al., 2004c; Das et. al. 2004a). Intake of DCP, TDN and ME of complete diet consisting of chane ki khar and groundnut chara in mash or block form did not differ significantly in camel calves (Nagpal et. al., 2004). Lactating cross-bred cows did not reveal any significant difference in nutrient utilization from wheat straw based complete diet either in mash or block form (Samanta et. al., 2004). Compaction did not show any significant effect on the digestibility of nutrients in lambs, however, DCP and TDN intake were higher ($P<0.01$) in the block fed animals (Dhuria et. al., 2004). Rams showed increasing trend in intake of digestible OM, DCP and improved nitrogen utilization with untreated or sodium hydroxide or alkaline hydrogen peroxide treated pearl millet kadbi based complete feed block as compared to its mash diet (Chaturvedi et. al., 2004). Milch buffaloes fed wheat straw and berseem based complete feed block (400 kg / m^3) also showed higher DM digestibility (Dwivedi et. al., 2004b). Similarly CP and NFE digestibility of dry grass based complete block were higher in crossbred cows (Prabha et. al., 2004). Paddy straw based complete feed block also improve the OM, EE digestibility and nutritive value of the diet in lactating buffaloes as compared to conventional feeding (Nagalakshmi et.al., 2004).

Agricultural crop residues are the major roughage source to the Indian livestock. These are bulky and difficult to transport from surplus to scarcity regions. Therefore, densification plays vital role to reduce the transportation cost by lowering the bulk density. Moreover, the crop residues are deficient in several nutrients particularly CP and mineral and are not even capable to fulfill the maintenance requirement of adult ruminants. The densification of roughages provides ample scope for incorporation of unconventional or top feeds. Thus, all out efforts are essential to popularize this technology through developmental agencies such as cooperatives, NGO and feed manufacturers etc. so that livestock owners can accrue benefits by feeding the balanced complete feed blocks to their animals.

Bibliography

Berwal, R. S., Lohan, O.P., Yadav, K.K. 1997. Effect of different physical forms of feed on nutrient digestibility, milk yield and composition in crossbred cattle. Indian Journal of Dairy Science 50: 246 - 249

Bruhn, H.D., Zimmerman, Z. and Niedermeir, R.P. 1959. Developments in pelleting forage crops. Journal of Agricultural Engineering 40: 204 - 207

Chaturvedi, O.H., Prasad, R., Tripathi, M.K., Mishra, A,S., Vaithiyanathan, S., Raghuvanshi, S.K.S. and Jackmola, R.C. 2004. Nutrient utilization in sheep fed sodium hydroxide or alkaline hydrogen peroxide treated millet kadbi based complete diets in mash or block. Proceedings of XI Animal Nutrition Conference held at JNKVV, Jabalpur from 5-7th January, 2004. Pp-81

Das, M.M. and Kundu, S.S. 1993. Nutrient utilization in buffaloes fed urea and bleaching powder treated straw based mash and densified feed blocks. Indian Journal of Animal Nutrition 10: 181-183

Das, M.M., Samanta, A.K., Singh, K.K., Kundu, S.S., Sharma, S.D. and Rai, S. 2004a. Effect of different forms of complete diets on nutrient utilization in crossbred calves. Indian Journal of Animal Sciences 74: 969 - 972

Das, M .M., Singh, K.K., Samanta, A.K., Kundu, S.S. and Sharma, S.D. 2004b. On farm trial: Effect of complete feed block on milk yield and its composition in buffaloes. Proceedings of XI Animal Nutrition Conference held at JNKVV, Jabalpur from 5-7th January, 2004. Pp 78.

Dhuria, R.K., Purohit, G.R., Sharma, T., Dhuria, D. and Arya, R.S. 2004. Comparative nutrient utilization of mustard starw based complete feed in mash and block form in lambs. Proceedings of V Biennial Conference of Animal Nutrition Association held at NIANP, Bangalore, from 24-26th November, 2004. Pp 54

Dwivedi, P.N., Goyal, R.K. and Gupta, P.D. 2004a. Effect of moisture levels on densification of forage block. Proceedings of XI Animal Nutrition Conference held at JNKVV, Jabalpur from 5-7th January, 2004. Pp 123

Dwivedi, P.N; Gupta, P.D. and Pathak, P.K. 2004b. On farm evaluation of complete feed blocks on milking buffaloes. In: National Seminar on Economic Feeding and Rearing of Buffaloes Under Different Agro-climatic Regions, March 12 – 13, 2004, held at IGFRI, Jhansi. Pp 22

Dwivedi, P.N., Goyal, R.K. and Singh, K.K. 2003. Preparation and evaluation of densified feed blocks in growing buffaloes. Indian Journal of Animal Nutrition 20: 202 - 205

Gupta, C.P. 2000. Forage baling/ densification equipments, concepts and future. Guest lecture delivered at 1st NATP Workshop on Baling, Densification and Storage of grasses/ fodder/ crop residues". Held at IGFRI, Jhansi from December 14 – 16, 2000

Gupta, P.D, Goyal, R.K. and Chattopadhyay. 1998. Performance evaluation of IGFRI densifying machine for grasses. Journal of Agricultural Engineering 35 : 33 – 38.

Jaglan, B.S. and Nand Kishore. 2004. Effect of untreated or urea treated crop residues on their densification as feed blocks. Proceedings of V Biennial Conference of Animal Nutrition Association held at NIANP, Bangalore, from 24-26th November, 2004. Pp-61

Jha, S.K., Singh, A., Panwar, J.S. and Kumar, A. 2004. Development of feed crusher for forage crops. Proceedings of Workshop on Design and Development of Indigenous Equipments for Feeding, Manufacturing, Preservation and Packaging of Animal Feed held at Vigyan Vishwa vidyalaya Evam Gau Anusandhan Sansthan, Mathura from 16-17th April, 2004. Pp 19-25

Kundu, S.S. and Das, M.M. 2004. Design and development of densification machine for preparation of complete feed. Proceedings of Workshop on Design and Development of Indigenous Equipments for Feeding, Manufacturing, Preservation and Packaging of Animal Feed held at Vigyan Vishwa vidyalaya Evam Gau Anusandhan Sansthan, Mathura from 16-17th April, 2004. Pp 1 - 4

Kundu, S.S., Das, M.M. and Sharma, D.D. 1998. Densification for block making and evaluation of calcium hypochlorite and alkali treated wheat straw based feed blocks in comparison to urea treated straw feed. Buffalo Journal 3: 291 - 302.

Lohan, O.P., Nand Kishore., Sunda, S.K. 2004. Farmers perception for the feeding of complete balanced feed blocks. Proceedings of XI Animal Nutrition Conference held at JNKVV, Jabalpur from 5-7th January, 2004. Pp-84

Nagalakshmi, D., Narsimha Reddy, D. and Rajendra Prasad, M. 2004. Effect of differently processed complete diets containing paddy straw on performance of lactating buffaloes. Proceedings of V Biennial Conference of Animal Nutrition Association held at NIANP, Bangalore, from 24-26th November, 2004. Pp 62.

Nagpal, A.K., Arora, M. and Singh, G.P. 2004. Nutrient utilization in camel calves fed gram straw based complete feed blocks. Proceedings of XI Animal Nutrition Conference held at JNKVV, Jabalpur from 5-7th January, 2004. Pp 84

Nand Kishore and Lohan, O.P. 2004. Effect of replacement of urea and alkali treated paddy straw by concentrate mixture at different levels on its compaction behaviour. Proceedings of XI Animal Nutrition Conference held at JNKVV, Jabalpur from 5-7th January, 2004. Pp 90

Nand Kishore., Lohan, O.P., Sunda, S.K. 2004. Effect of urea treated and alkali treatment of wheat straw and its replacement by concentrate mixture at different levels on its compaction behaviour. Proceedings of XI Animal Nutrition Conference held at JNKVV, Jabalpur from 5-7th January, 2004. Pp 89

Pachauri, V.C., Goyal, R.K., Mahnata, S.K. and Gupta, P.D. 1999. Effect of densification of dry grass on nutrient utilization and growth in crossbred calves. Indian Journal of Animal Nutrition 16: 244 - 247

Prabha, N., Kale. S. and Badve, V.C. 2004. Studies on intake and nutrient utilization pattern of forest grass based complete feed in block form and mash form. Proceedings of V Biennial Conference of Animal Nutrition Association held at NIANP, Bangalore, from 24-26th November, 2004. Pp 69

Preston, T.R. and Leng, R.A. 1984. Supplementation of diets based on the fibrous residues and by products. In: F. Sundstol and E. Owens (Editors). Straws and other fibrous by products as feed. Elsevier, Amsterdam, the Netherlands. Pp 373 - 413

Ram, A.K, Mehra, U.R. and Chalta, J. (1990). Preparation and evalutation of complete feed mix block for goats. Indian Journal of Dairy Science 43 : 458 - 461

Reddy, T.V., Reddy, M.R. and Reddy, G.U.N. 1986. Utilization of wheat straw and agro industrial by products in development of complete feed for sheep. Indian Journal of Animal Sciences 56: 100

Sah, P.C., Singh, B.P.N., Agarwal, U.S. 1981. Compaction behaviour of straw. Journal of Agricultural Engineering XVIII 1 : 89 - 96

Samanta, A.K., Singh, K.K., Das, M.M. and Kuṇdu, S.S. 2004. Effect of complete feed block on dry matter intake, nutrient utilization and milk yield in crossbred cows. Proceedings of V Biennial Conference of Animal Nutrition Association held at NIANP, Bangalore, from 24-26th November, 2004. Pp-49

Samanta, A.K., Singh, K.K., Das, M.M., Maity, S.B., Kundu, S.S. 2003. Effect of complete feed block on nutrient utilization and rumen fermentation in barbari goats. Small Ruminant Research 48: 95 - 102

Sarsavadia, P.N., Patel, N.C., Nayak, N.P. and Joshi, D.C. 1991. Compression of wheat (Triticum aestivum) straw and groundnut haulms (Arachis hypogaea) for low density animal feed block. Indian Journal of Agricultural Engineering 1: 37 - 41.

Sharma, V.K., Chauhan, J.S., Wadhwa, D. and Marwaha, C.L. 2004a. Physical properties of complete feed blocks. Proceedings of V Biennial Conference of Animal Nutrition Association held at NIANP, Bangalore, from 24-26th November, 2004. Pp-48

Sharma, T., Arya, R.S., Dhuria, R.K., Garg, D.D., Harish, M. and Purohit, G.R. 2004b. Comparative nutrient utilization of wheat straw and groundnut fodder based complete feed in loose and block form in calves. Proceedings of XI Animal Nutrition Conference held at JNKVV, Jabalpur from 5-7th January, 2004. Pp 79.

Sihag, Z.S, Rathee, C.S. and Lohan, O.P. 1991. Factors affecting complete feed block formulation of straw based rations. Indian Journal of Animal Sciences 61: 1218 - 1221.

Sihag, Z.S, Rathee, C.S. and Lohan, O.P. 1994. Complete feed blocks for growing buffalo calves. Indian Journal of Animal Sciences 64: 549 - 552.

Singh, A., Panwar, J.S., Kumar, A., Jha, S.K., Pandeya, A. 2002. Management of animal feed materials through densification. Journal of Agricultural Engineering 39: 9

Singh, A., Jha, S.K., Panwar, J.S., Kumar, A., Pandeya, A. and Gautam, R.B. 2004a. Studies on feed block formation using mustard straw. Proceedings of XI Animal Nutrition Conference held at JNKVV, Jabalpur from 5-7th January, 2004. Pp 82

Singh, A., Jha, S.K and Panwar, J.S. 2004b. Development of animal feed block formation machine. Proceedings of Workshop on Design and Development of Indigenous Equipments for Feeding, Manufacturing, Preservation and Packaging of Animal Feed held at Vigyan Vishwa vidyalaya Evam Gau Anusandhan Sansthan, Mathura from 16-17th April, 2004. Pp 26 - 35

Singh, J. 2002. Handling and densification of cereal crop residuea for safe storage and economic transportation: Status and prospects. Proceedings of National Symposium on Grassland and Fodder Research in the New Millennium held at IGFRI, Jhansi from 30th October – 1st November, 2002. Pp 209 - 210

Singh, J., Lohan, O.P. and Rathee, C.S. 1998. Evaluation of berseem based complete feed blocks in growing buffalo calves. Indian Journal of Animal Sciences 68: 480 - 483.

Singh, J., Singh, J.K. and Rajpali, S.K. 1986. Effect of pressure and moisture content on density and stability of wafers of different forages. Journal of Agricultural Engineering 23: 89 - 98.

Singhal, K.K. and Thakur, S.S. 2004. Machinery and equipments for complete feed manufacturing. Proceedings of Workshop on Design and Development of Indigenous Equipments for Feeding, Manufacturing, Preservation and Packaging of Animal Feed held at Vigyan Vishwa vidyalaya Evam Gau Anusandhan Sansthan, Mathura from 16-17th April, 2004. Pp 36 - 41

Singh, K.K., Das, M .M., Samanta, A.K., Kundu, S.S. and Sharma, S.D. 2004c. Effect of complete feed blocks on nutrient utilization and growth in crossbred heifers. Proceedings of XI Animal Nutrition Conference held at JNKVV, Jabalpur from 5-7th January, 2004. Pp 79

Verma, A.K., Mehra, U.R., Dass, R.S., Singh, A. 1996. Nutrient utilization by Murrah buffaloes from compressed complete feed blocks. Animal feed Science technology 59: 255 - 263

Yadav, K.K., Rathee, C.S. and Lohan, O.P. 1990. Effect of compaction of roughage based complete feed on digestibility and rumen parameters. Indian Journal of Animal Nutrition 7: 27 - 30.

Yadav, K.K., Rathee, C.S. and Lohan, O.P. 1991. Compaction behaviour of straw based complete feed. Indian Journal of Animal Sciences 61: 980 - 984.

Zombade, S.S. 2004. Present status of manufacturing densified total mix ration as bheli. Proceedings of XI Animal Nutrition Conference held at JNKVV, Jabalpur from 5-7th January, 2004. Pp 131-132.

CHAPTER - 2

LEAF MEAL PRODUCTION AND UTILIZATION

P.S. Pathak, P. K. Pathak, A.B. Mojumdar and P.N. Dwivedi

Successful animal production requires an adequate supply of feed throughout the year. Increasing demand for cereals, legumes and other food grains for direct human consumption restricts their use in animal production sector. The animal industries must therefore adjust themselves by making changes in their feeding management and production programmes. There exists a big deficit of concentrate in the country to the tune of 60% (Pathak and Mojumdar, 2002). This deficit can be partially bridged by replacing the concentrates in feeds by protein-enriched forages, tree leaves, forage legumes etc. Utilization of such supplements would have a grain-saving effect and would notably increase the currently low feed supply for animals.

The use of tree leaves as component of animal diet is a widespread practice in many tropical countries and the enormous biological diversity within tropical trees provides enormous opportunity for their use for animal feeding, yet only some 20–30 species have been used and out of these Leucaena, Gliricidia, Erythrina and Acacia species have predominated. In India, tree leaves of many species viz. Ficus sp., Neem, Grewia sp., Celtis sp., Sesbania, Prosopis cineraria, Zizyphus sp., Flacourtia indica are used traditionally as fodder. Sesbania sesban, a fast growing legume fodder tree of humid warm climate, is rich in crude protein to the tune of 18-25% (Maity et al. 1997). It can be used as protein supplement to low grade forages for animal production (Upadhyay and Mojumdar, 1991). Flacourtia indica leaves are commonly used as a fodder in the semi arid region of central India and is quite palatable and nutritious for goats and may be fed without any harmful effect on their health (Mojumdar et al., 2002). Zizyphus leaves of different species are also widely used as animal feed particularly in Rajasthan and Central India. Higher feed conversion efficiency with higher gain in live weight has been achieved in Marwari breed on Zizyphus nummularia (Pala) leaf based diet (Ram Ratan et al. 2003). Khejri (Prosopis cineraria) leaves are an important animal feed resource in arid and semi arid regions of India (FAO, 1992). It has been found to contain high CP (12.36%) and total ash (16.00%) and can successfully maintain adult goats

(Khirwar et al., 2003). In fact the use of Leucaena leaves are accepted more widely than others as feed. Systems of biomass production for grazing cattle using Leucaena leucocephala have already been evolved. A recent major development with Leucaena shows great promise both for biomass production and for industrial processing of Leucaena leaf meal (Leng, 1997). Leguminous crops other than fodder tree leaves, which are also equally important for making leaf meal are Stylosanthes (14-16% CP, Chaisang et al. 2004.), Cassava (23.9 to 34.7% CP, Phuc et al., 1996), Lucerne (20.61% CP, Mojumdar et al., 2003) and alphalpha (14.3% crude protein, Guodao et al, 2004), etc. and have commercial potentials for livestock rearing.

This paper attempts to summarize the available information on the aspect of leaf meal production technology, methods, equipments and the effect on feeding different categories of livestock.

Why leaf meal?

Majority of livestock population in India subsists on vegetation of natural grasslands where grasses are cheap source of forage. Crop residues, straw and dry grasses form the basal roughage for feeding to livestock which are poor in quality being deficient in protein, available energy and minerals (Singh et al., 1998). Therefore, supplementation of green fodder and concentrate become essential for sustaining livestock productivity. However, there exists a huge gap between demand and supply of both green fodder as well as concentrate in our country (Pathak and Mojumdar, 2002). Moreover, green fodder is not available throughout the year. Therefore, it becomes imperative to search for alternate source of protein rich forage supplement. Supplementation of leaves of leguminous crops is an important and most practical feeding strategy for improving the feeding value of such poor quality roughages. Leaf meal being rich in CP, carotene and minerals could be an appropriate answer for providing nutritious diet to the livestock throughout the year. Supplementation of leaf meal can meet the requirement of livestock particularly during dry periods when availability of other protein rich forages is scarce. Thus it can bridge the gap between the demand and supply of the green fodder and concentrate. In India, traditionally also tree leaves particularly Zizyphus nummularia (Pala), Khejri (Prosopis cineraria), Ficus sp. and Neem leaves were lopped, dried and fed to different categories of livestock.

Leaf meal production technology

The leaf meal production technology is as simple as a farmer can also produce and sell. In India a leaf meal production plant based on Lucerne was started in 1977 at Udumalpet near Coimbatore, which produced around 12-15 tonnes of leaf meal per month. The setup consists of a chaff cutter feeding sun-dried Lucerne to a powdering mill for grinding and collection of sieved leaf meal (Chakraborty, 2002). Thus for production of leaf meal following operations must be performed.

Harvesting

Harvesting of crop/tree leaves are important in view of time of cutting, which influences the dry matter yield and quality. Cutting too early will give high quality meal but low yields and cutting too late will decrease nutrient content and increase fiber content but give higher dry matter yield. Among the leguminous fodder crops, Stylosanthes seabrana is first cut after 55 days whereas S. scabra and S. guianensis after 60 days and 80 days respectively for leaf meal preparation, depending upon soil fertility and water availability (Guodao et al. 2004). Lucerne has been found to provide 9 t dry matter yield per ha (Singh, 1988) and could also have a great potential for making leaf meal.

The Leucaena, a leguminous fodder tree is cut first at 10–15 cm above soil level when it has grown to a height of 1.2 metres. Thereafter 4–6 cuts can be taken per year when the plant reaches a height of 1 metre. The approximate yield under Cuban farming conditions without irrigation or fertilization is 12 tonnes DM/ha with 22–25% crude protein. In some experimental conditions 20 tonnes DM/ha has been achieved (Leng, 1997). Pathak (1988) compared salient characteristics of three basic types of Leucaena plants (Hawaiians, Peruvian and Salvador) of age 5 years and reported that leaf production varied from 0.15 to 2.73 kg/tree and forage production was 6.0, 8.72 and 9.86 t/ha/year, respectively. Plantation of Leucaena at different densities under energy plantations involving different harvest cycles has indicated its great potentiality. A cycle of 4.5 years was recommended for economic yield of variety K8 (Pathak et al, 1985). Pathak and Patil (1982) while comparing the production at densities varying from 1-40 thousand/ha at 2 years, recorded peak production of 53.5 t/ha by 5000 plants/ha and thus for different cycle duration, different densities were advocated. Rajput et al. (2004) assessed a Leucaena field harvested 15 months back and found the population of the Leucaena plants in the field as 1,32,000 per hectare out of which 30,000 plants/ha having plant base diameter above 2.5cm (mean diameter 3.25cm) was recommended to harvest for production of leaf meal. Yield of fresh Leucaena leaves was estimated as 12.105 t/ha/year by them.

The leguminous crops harvested from the field could be directly used for making leaf meal from the whole plant. However, being different in nature, the tree leaves are detached from the stem/twigs and then the leaf meal is prepared.

The trees, which are having soft twigs viz. Leucaena and Sesbania could also be used for making leaf meal. Herbages/tree leaves can be manually cut or machine harvested and gathered. For harvesting field crops mostly hand operated sickles (simple or cerated) are available having a capacity of about 0.0108 ha/h. However, machineries are already in use for harvesting Stylosanthes guianensis for leaf meal production (Guodao et al. 2004) in China. For harvesting leaves from trees, the branches/twigs are lopped manually using axes or saws or mechanically. Commercial spinning discs are available for mowing trees, which could also be utilized for lopping tree branches/twigs.

Drying

Drying is the single way to reduce moisture from the harvested plants. It could be through many ways including the simplest way of sun drying. In sun drying the materials should be spread in a thin layer on a clean surface preferably concrete and dried for two days, then turned over and dried for another day (Guodao et al. 2004). Alternatively, drying racks can be built using local materials such as bamboo to dry the materials above the soil surface and avoid spoilage. Different kind of racks viz. A frame, boundary fences etc. were evaluated for drying green fodder/grasses by Singh (1984). Indian Grassland and Fodder Research Institute, Jhansi has been developed and evaluated two tier drying structure made up of Subabool (Leucaena) logs for shed/ventilated drying of green fodder and tree leaves. Use of racks can also hasten the drying process and improve quality through increased protein and â-carotene content. Reports are available to sun dry the stylo for three (Guodao et al., 2004) to seven days (Amodu, 2004), the variation in drying time might be due to the local environmental condition and the method (turning in between the process etc.) adopted. Though sun drying is a more acceptable and feasible alternative of many other drying methods, it is difficult to sundry green forage crops/tree leaves, in the rainy season when the production is at large. Also, extending the drying period diminishes the nutritional quality of the product. Hay production from stylo using sun drying is resulted in loss of 15-40% of the dry matter, mostly in the form of leaves. (Amodu, 2004).

Heating air for drying reduces the drying time and maintains the quality of the product and is an essential operation in making leaf meal commercially. Heated air drying can be performed over a wide range of temperatures starting from room temperature as in case of sun drying or at 30°C and 50 percent relative humidity of air (Samual and Anap, 1980) to as high as 600°C as being used for stylo leaf meal production (Chaisang et al, 2004). Table 1 shows the drying temperatures and methods recommended or used for different crops in making leaf meal. Chopped lucerne hay was dried from 65° to 180° C (Leng,

1997). Two different types of heated air-drying methods were reported by Guodao et al. (2004) for preparation of stylo leaf meal. Slow drying at low temperatures (50-60° C) over two to three hours and quick drying in kilns (10-20 seconds) at temperatures well above 60° C. The high temperature drying was reported to be costly to dehydrate leaf meals due to expensive machinery used and the operational cost. The additional cost of drying is however, reported to be compensated for by the high price of the quality product Guodao et al. (2004).

For the hot air-drying operation the forage may be fed fresh, or after part of sun drying. A design approach to a 5 tonnes capacity portable hay drier was dealt with keeping in view of low cost and portability by Samual and Anap (1980) consisting of a fan unit, welded iron mesh floor and slatted floor duct to dry hay from 40% moisture content to a safe limit of 20% within three days. Some times materials are dried in two stages as Stylosanthes are reported to dry in sun up to 45-60% moisture for a short period after cutting and then finally dried using a set of artificial quick drying ovens at high temperatures for three to five minutes to achieve a final moisture content of 9-12% (Guodao et al. 2004). In all the cases however, drying of green fodder crops/tree leaves are accomplished from invariably high moisture content to final moisture content of 9-12%.

Table 1: Drying temperatures and methods recommended or used for different crops in making leaf meal

Crops/Plants	Drying Methods/ Temperature	Referances
Leucaena leucocephala, Calopogonium mucunoids, Centrosema pubescens, Pennisetum purpureum, Manihot utilissima, Albizzia falcataria, Calliandra calothyrsus, Enterolobium cyclocarpum, Gliricidia sepium, Sesbania grandiflora, clidemia hitra and Maesopsis emenyy	Sun drying, Oven drying (40, 60 and 100°C), and Freeze drying	Mahyuddin et al., 1988
Chopped Lucerne hay	65° to 180° C	Leng, 1997
Stylo leaf meal	* Slow drying at low temperatures (50-60° C) over two to three hours * Quick drying in kilns (10-20 seconds) at temperatures well above 60° C	Guodao et al, 2004
	Commercial quick drying technology at higher temperature of 600° C (Kiln drying)	Chaisang et al, 2004
Leucaena leaf meal	Sun drying and drying fresh leaves in an oven at 60°C	D'Mello, and Acamovic, 1989.
Leucaena leaflet meal	Sun drying of branches without damaging the leaflets	Tangendjaja et al. 1990

Size Reduction

In the feed industry the size reduction is usually referred for grinding and producing powdery like material called meal. Commercial products must often meet stringent specification regarding the size and same time the shape of the particle they contain. Reducing the particle size increases the reactivity of solids, reduces the bulk of fibrous material for easier handling (McCabe et al., 1985) and is governed by the ultimate user (animal). For example, smaller particle size is required for small animals and birds whereas larger particle size (straw) may be fed to the larger ruminants. Different mechanism involved in the size reduction process could be compression, impact, rubbing or attrition and cutting. In general, compression is used for coarse reduction of hard solids, to give relatively few fines; impact gives coarse, medium or fine products; attrition yields very fine products from soft, non abrasive materials whereas cutting gives a definite particle size and some times a definite shape with few or no fines (McCabe et al., 1985).

Forages and fodder crops, being fibrous material, could be reduced in their size mostly by using hammer mill type of grinder or beater type of pulveriser with blower connected to a cyclone dust collector as reported for Stylosanthes (Guodao et al. 2004, Amodu, 2004 and Chaisang et al., 2004), twelve Indonesians ruminant feeds including Leucaena leucocephala (Mahyuddin et al., 1988), Gliricidia leaf meal (Osei et al., 1990) and Chaya leaf meal (Cnidoscolus aconitifolius (Mill.) Johnston) (Donkoh et al., 1990) etc. The screen fitted at the bottom controls fineness of grinding in the hammer mills and thus the size of these meals were also different with each other viz. Leucaena leucocephala was milled and passed through a 1 mm screen (Mahyuddin et al., 1988), Chaya leaf meal ground to pass a 100-mesh sieve (Donkoh etal., 1990) whereas Stylosanthes ground through 1-1.5mm sieve in China and 3 mm and 5 mm screen in India, (Table 2) was considered as stylo leaf meal (Guodao et al., 2004)

Table 2: Leaf meal production technology recommended or used for different crops

Crops	Production technology	Reference
Leucaena leucocephala and other eleven Indonesians forage crops	Leaf and edible stem material wilted overnight, chopped, mixed and dried till the moisture was low enough to allow milling through a 1 mm screen.	Mahyuddin et al., 1988
Leucaena leucocephala	Separation of leaves from stalks and twigs after harvesting and drying fresh Leucaena in the sun	Mahanta et al., 1999
	Harvested at 8week intervals to a height of 1.5m, sun dried and leaflets were separated from stem, rachis and rachillae.	Tangendjaja et al., 1990
	Branches are dried on concrete floors in the sun, dried leaves are threshed and incorporated directly into poultry diets. This can be used as whole leaf, ground leaf or as pelletes.	D'Mello, and Acamovic, 1989
Treated Leucaena leaflet meal	Dipping the fresh branches in water at 70°C for 10 min., sun drying and separating the leaflets. Mimosine was reduced by about 80%, mostly by conversion to DHP.	Tangendjaja et al., 1990
Stylosanthes	Harvest, sun dry, grind into a powder using small grinders and crude meal used for feeding chickens, pigs, geese and other animal	Changjun et al., 2004
	Harvesting, drying in sun in a thin layer on a clean hard surface for 2-3 days with turning or machine drying at temperatures 50-60°C or above 60°C, grinding by hammer mill or beater type pulveriser (1-1.5mm sieve in China and 3-5mm screen in India), grading, packaging in 50kg bags with water proof lining and storage at 3-9°C.	Guodao et al., 2004
	Complete stylo meal : Harvesting green crop, transporting to threshing yard, chopping, drying and threshing by commercial thresher and bagging. **Pure stylo leaf meal :** Beating dried crop by stick and collecting the leaves **Stem meal :** The separated Stylo stem could be chopped or densified before feeding.	Pathak and Mojumdar, 2004
Chaya leaf meal (Cnidoscolus aconitifolius (Mill.) Johnston)	Green leaves harvested and air dried, ground to pass a 100-mesh sieve and stored in coloured air-tight containers.	Donkoh et al., 1990
Gliricidia leaf meal	Leaves mixed together and sun cured on a concrete drying surface for 5 days. Then ground in a hammer meal to produce the meal	Osei et al., 1990
Cassava leaves	Spreading the leaves on the ground in the sun and turning them over during a 1-2 day period. The dried leaves were then ground (2mm sieve).	Phuc et al., 1996

Though the size of meal influences the digestibility (Dwivedi and Pathak, 2004), the leaf meal could also be defined as dried whole leaves in some cases as separated from stalks, stem and twigs after harvesting and drying the cut plants/tree branches. Mostly the tree branches (Leucaena) and bushy crops (Stylosanthes) could be used to get whole dried leaf as meal after harvesting, drying and separating the leaves from the twigs/branches due to the self shattering characteristics of leaves and ease in separation at low moisture content (Pathak and Mojumdar, 2004). It waş reported that the leaves of Stylosanthes hamata shatter automatically or by slight jerk when the moisture of leaves reaches up to 13.5% which amount to about 62% of the total leaves. About 15-40% of the dry matter loss, mostly as leaves was reported when stylo is sun dried for hay production (Amodu, 2004). The lost dry leaves could, however, be utilized as whole leaf meal if same is collected without much deterioration in the quality. The whole leaf meal thus separated from dried twigs/branches were well employed in the rations of different categories of livestock such as lambs (Mahanta etal., 1999), poultry (D'Mello, and Acamovic, 1989) etc. Leaf and edible stem material could also be chopped (Mahyuddin et al., 1988) using power operated or manual chaff cutter or threshed using commercial threshers to reduce their size. Chopping green fodder makes more homogenous meal before drying as leaves could be mixed well. Threshing however is possible, provided the moisture content of the crop is low. In a study at IGFRI Jhansi, it was observed that heat is generated when a thresher was operated for threshing Stylosanthes hamata at a moisture content more than 13.5% and a threshing capacity of 500kg/h (Pathak and Mojumdar, 2004). In the process the sun dried stylo crop was threshed by available commercial thresher to reduce its size equivalent to that of wheat straw. The capacity of the thresher was 500 kg/h at average crop moisture content of 7.6%. The process is simple and no special gadgets are required especially for the farmers having the threshers. Benefit of using the thresher for leaf meal production is that it is not busy during the harvest of the stylo crop. Leaf meal production technology as recommended or used for different crops is summarized in the table 2 and is presumed to represent to their respective countries, if otherwise not specified.

Leaf meal Utilization

Leaf meal being made from leaves of legume crops/trees/shrubs rich in protein, could act as a replacer of feed concentrate for livestock to save the valuable feed grain as was reported in case of Stylosanthes (Changjun et al. 2004). In India, however, the legume crops/trees/shrubs were used in many production systems since long, for grazing small ruminants, cut and carry the fodder for domestic and farm animals etc. They were also incorporated in soil conservation and wasteland developmental programmes. Traditionally also, tree leaves particularly Zizyphus nummularia (Pala), Khejri (Prosopis cineraria), Ficus sp. and Neem leaves were used to feed different categories of livestock. Commercialization of leaf meal in India is yet to gain pace, which can be successful through technological achievements of the researchers and the experiences of the farmers. Following are the observations, which were observed in feeding leaf meal to the livestock.

Effect of drying temperature on nutritive value of leaf meal

Various parameters (solubility, digestibility of N and nitrogen balance, IVDMD, carotene and xanthophyll level etc.) were supposed to be affected by the drying temperatures and were well reported. Effect of drying treatments (oven drying at 40, 60 and 100°C; sun drying and freeze drying) on the degree of dry matter disappearance in-vitro (IVDMD) by the rumen fluid – pepsin (RFP) and pepsin – cellulose (PC) procedures on 12 Indonesians ruminant feeds were investigated by Mahyuddin et al (1988). The feed samples were also digested for 24 or 48 h in intra-ruminal nylon bags and the results compared with those from fresh material subjected to 48 h nylon bags in the rumen. The species studied were Calopogonium mucunoids, Centrosema pubescens, Pennisetum purpureum, Manihot utilissima, Albizzia falcataria, Calliandra calothyrsus, Enterolobium cyclocarpum, Gliricidia sepium, Leucaena leucocephala, Sesbania grandiflora, clidemia hitra and Maesopsis emenyy. Gliricidia sepium was the only species in which results were affected by any of the treatments. In cassava leaf, Sesbania grandiflora, Calopogonium mucunoids, Calliandra calothyrsus and Maesopsis emenyy, IVDMD was significantly depressed by drying at elevated temperatures. Even with freeze dried material, the IVDMD figures for Casava, Maesopsis and Calopogonium were 25-30% less than those from the corresponding fresh nylon bag 48-h samples. They also concluded that the feeding value of forages conserved for village animal production may be strongly affected by the drying treatments. Leng (1997) reviewed effect of drying temperature on solubility and digestibility of N and N balance in lambs fed dried, chopped lucerne hay and reported that application of heat might increasingly protect tree leaf meals as apparently occurs when lucerne meal is dehydrated at temperature increasing from 65°C to 160°C (Table 3). Effects of drying fodder tree foliage is shown in table 4, which produced greater live weight gains in goats when supplemented with dried foliages from 5 tree species on a basal diet of poor quality forages than the same quantity of foliage fed fresh (Leng 1997). This response might be related to a change in the solubility of the protein increasing the bypass protein content of the leaf meal or the reduction of anti-nutritive factors or the leaf meal undergoes a mild browning reaction with the reducing sugars that would be invariably present in leaf materials and the protection has little to do with secondary plant components.

Table 3: Effect of drying temperature on solubility and digestibility of N and N balance in lambs fed dried, chopped lucerne hay.

Drying temp (°C)	Solubility of N (%)	N digest (%)	N balance (%)
65	43	49	6.0
130	40	68	7.4
160	40	66	6.9
180	34	52	3.4

Source: Leng (1997)

Table 4: Comparison of the effect of tree foliage supplements fed fresh or after drying to goats.

Supplement	Forage intake (g/kg LWt/d) Tree Foliage				
	Fresh	Dry	Basal forage	Dry digestion (%)	Growth (g/d)
Albizia chinensis	7.5		18.5	51	0
		7.5	16.5	44	54
Calliandra calothyrsus	7.5		18.5	47	24
		7.5	18.5	44	48
Gliricidia sepium	7.5		18.5	56	12
		7.5	15.5	47	42
Leucaena leucocephala	7.5		15.5	49	-18
		7.5	18.5	47	0
Sesbania sesban	7.5		14.5	53	0
		7.5	17.5	52	54

Source: Leng (1997)

Drying was also reported to minimize the anti-nutritional factors present in the leaves. Heating temperature, duration of heating, particle size and moisture conditions are the responsible factors which governs the heat treatments of the Leucaena leaves (D'Mello and Acamovic, 1989). It was reported that moist heating procedures are generally more effective than dry heating methods and there may be further beneficial effects by soaking prior to cooking. Water soaking and washing the Leucaena leaves for one day before sun drying and grinding was reported as an economical way to improve their nutritive value. The best performance was however, obtained by simply sun-drying the leaves prior to use (Table 5). D'Mello and Acamovic (1989) reported that rapid sun drying yielded Leucaena leaf meal with carotene and xanthophylls levels of 484 and 932 mg/kg DM, respectively, but substantial losses occurred during oven drying at 60 and 145°C. They further reported that carotenoids were more stable in sun dried Leucaena leaf meal than in oven-dried samples. It could therefore be inferred that the drying is not only useful for increasing the storage life of the material but also reduces the anti-nutritional factors and makes available the bypass proteins to the ruminants. It also facilitates the size reduction (grinding) operation easier. The process has limitation too viz. excessive temperatures during heat processing and drying can reduce the availability of amino acids. The correct application of heat is therefore crucial to the preparation of leaf meal and could be established before commercializing the leaf meal technology.

Table 5: The effect of including 15% Leucaena leaf meal for pigs (30-60 kg).

Form of administration	Average daily intake (kg/d)	Average daily gain (g)	Average daily feed conversion kg feed/kg weight gain
Dried Leucaena leaves	2.09	770	2.71
Soaked 24 hr, before drying and grinding	1.91	680	2.81
Ground, before soaking for 24 hr	2.00	690	2.90
Fresh, un soaked leaves	1.67	540	3.06

Source: Ruengpaibul (1984)

Effect of particle size on nutritive value of leaf meal

Particle size plays an important role in the digestibility and intake in the livestock. The effect of particle size of Stylosanthes hamata on the digestibility was studied (Table 6 and 7) through an In sacco experiment indicating better digestibility of small particles (Dwivedi and Pathak 2004). Uden, 1988 reported the effect of particle size of grass hay in cows. Digestibility was measured in-vivo and extent of fermentation in-sacco. Grinding the forage caused a reduction in fermentation rate and extent of fermentation, even though the sites for bacterial attack were increased. It decreased total retention time of particles in the digestive tract. Influence of foodstuff particle size on degradation of nitrogen in the rumen of fistulated cows was studied by Michalet-Dreau and Cerneau, 1991. Six feedstuffs (peas, maize, barley, soya-bean meal, Lucerne hay and cocksfoot hay) were ground using three different screen sizes 0.8, 3.0 and 6.0 mm. Feed nitrogen degradability in the rumen increased as the fineness of grinding increases, specially in peas and maize but the degradability of hay did not change for these grinding sizes (0.8 to 6.0 mm). Thus, it is clear that the leaf meal feeding to the livestock is recommended as a protein replacer or suitable for supplementation in the basal diet. The size of various materials including protein supplement as oilcakes, leaf meal could therefore be reduced to facilitate uniform mixing with other feed ingredients, less wastage in form of refusal and improvement in the digestibility.

Table 6: In Sacco DM digestibility of Stylo cut in different sizes

Cut size	**Digestibility (%) at different hours**					
mm	**0 hour**	**12 hour**	**24 hour**	**36 hour**	**48 hour**	**72 hour**
1	14.07	34.23	46.41	57.02	63.24	64.40
2	13.51	32.62	44.96	56.87	63.07	64.32
5	11.97	29.45	41.82	50.77	58.39	59.71
10	11.01	24.73	37.36	41.94	44.63	46.88
15	9.56	23.01	32.28	38.30	41.54	43.02
20	9.11	21.79	30.47	35.82	37.15	39.67
25	8.92	20.58	27.78	32.14	34.58	35.60
30	8.30	20.12	26.35	30.64	33.91	33.06

Source: Dwivedi and Pathak (2004)

Table 7: In Sacco CP digestibility of Stylo cut in different sizes

Cut size	**Digestibility (%) at different hours**					
mm	**0 hour**	**12 hour**	**24 hour**	**36 hour**	**48 hour**	**72 hour**
1	13.81	39.15	51.07	59.22	67.36	67.81
2	13.60	37.06	50.16	57.80	66.75	67.23
5	12.11	33.47	45.02	53.11	62.17	63.36
10	11.23	29.26	39.71	43.87	49.75	50.15
15	09.72	26.60	36.00	40.42	44.06	45.11
20	09.35	24.28	33.58	38.74	41.91	43.20
25	9.07	23.59	30.11	36.02	38.16	39.36
30	8.66	22.37	29.05	35.28	36.00	36.62

Source: Dwivedi and Pathak (2004)

Nutrient composition and level of feeding leaf meal.

Several studies have been conducted to demonstrate the usefulness of leaf meal as a component of feed rations of many farm animals (Changjun et al. 2004). Pigs can be fed a feed ration containing 10-15% stylo leaf meal (SLM) whereas poultry are fed a ration containing 3-5% SLM. Chaisang et al (2004) however summarized the optimum level of feeding of stylo leaf meal to different animals in various feed formulations and are presented in the table 8. Any assessment of the potential of leaf meals in non-ruminant nutrition necessitates a review of nutrient content. A vast number of plants/shrub/trees are available which have great potential in making leaf meal. The outstanding feature of these leguminous leaf meals is their high CP contents (Table 9) and essential amino acid contents (Table 10) in the leaf protein especially in the diet of the non-ruminants. High yielding milch animals need green fodder and concentrate mixtures containing considerable amount of cereal grains (about 40%) and oil cake (30-35%). There exists a deficiency of both green fodder as well as ingredients of concentrate mixture. Tree leaves and leaves of leguminous forages can be blended with roughage as complete feed or transformed into leaf meal for feeding it to livestock.

Table 8: Level of feeding of leaf meal to different animals

Animals	SLM (%) in the ration
Pig	5-10
Broiler	2-5
Duck	8-12
Goose	15-20
Rabbit	30-40
Cattle, goat and deer	40-60

Source: Chaisang et al (2004)

Table 9: CP content (% DM basis) in leaf meals

Source	Crude protein (%)	References
Leucaena leucocephala	18-21	Mojumdar et al. 1987
S. aegyptiaca	20.58	Mojumdar et al. 1987
Prosopis cinereria	12.36	Khirwar et al., 2003
Albizia species	15-16	Mojumdar et al. 1987
Zizyphus numalaria	13-16	Ram Ratan et al. 2003
Stylosanthes species	12-18	Mojumdar et al. 2003
Cassava	23.9 - 34.7	Phuc et al., 1996
Lucerne	20.61	Mojumdar et.al., 2003
alphalpha	14.3	Guodao et al, 2004
Sesbania sesban	18-25	Maity et al. 1997

Table 10: Essential amino acid composition (percent DM basis) of leaf meal

Amino acids	Leucaena	Stylosanthes species (b)		
	leucocephala (a)	scabra	guianensis	seabrana
Threonine	1.21	0.64	0.69	0.61
Glycine	1.33	-	-	-
Valine	1.44	0.73	0.81	0.72
Cysteine	0.20	0.18	0.16	0.19
Methionine	0.46	0.21	0.22	0.20
Isoleusine	1.37	0.56	0.64	0.58
Leucine	2.17	1.06	1.16	1.04
Tyrosine	1.25	-	-	-
Phenylalanine	1.48	-	-	-
Lysine	1.76	0.74	0.80	0.71
Histidine	0.54	0.31	0.35	0.31
Arginine	1.51	0.81	0.93	0.80
Tryptophan	0.38	-	-	-

Source: a. D'Mello and Fraser (1981) b. Changjun et al (2004)

Table 11: Mineral composition of leucaena leaf meal (DM basis)

Minerals	Leucaena leucocephala
Major elements (g/kg)	
Calcium	18.1
Phosphorus	2.5
Sodium	0.0
Potassium	8.0
Magnesium	5.1
Trace elements (mg/kg)	
Copper	9.9
Iron	239.9
Zinc	21.2
Manganese	42.1

Source: a. D'Mello and Fraser (1981)

Leaf meal as feed component in non-ruminant nutrition

Leaf meal has been a great interest for replacement of feed concentrate in the animal diets and performance of leaf meal of different species has been studied in non-ruminant nutrition as well as a ration component for ruminants. The performance of SLM compared with freshly cut Stylosanthes as components of a balanced feed for young rabbits were studied in Hainan, China (Changjun et al., 2004). The results showed that daily weight gain of rabbits was significantly different for three groups. Group I had the highest daily gain, which was more than 28% higher than the group II and more than 98% higher than that of group III (Table 12). In India, two detailed poultry feeding studies using SLM were carried out, one at Acharya NG Ranga Agricultural University at Hyderabad and one at the Tamilnadu Agricultural University (TNAU) at Coimbtore (Changjun et al., 2004). Three levels of SLM (Table 13) incorporated in the diet and compared with a maize – soybean based control diet. A clear weight gain benefit from feeding 3% SLM was observed compared with control and other two levels of SLM (6% and 9%). Findings of these studies, however, revealed that weight gain, feed consumption and feed conversion ratio was not significantly different from control. In the poultry research the dietary treatments too, had significant effects on skin and shank pigmentation, measured as Roche Fan Color Score (Changjun et al., 2004) and could be a consumers choice as in case of Chineas market, who prefers a yellow coloring of the chicken meat and shank and producers often use artificial colouring agent in feed formulations to cater for their demand. Both shank and skin pigmentation was found more intense with SLM than in control and intensity was increased as the SLM level increased from 3 to 9% in the diet (Changjun et al., 2004). This pigmentation could be due to the availability of about 700-1000 mg/kg of ß-carotene in the stylo leaf meal, which is transformed into vitamin A and precipitates under the skin to give it yellow colour. It was observed that 3% stylo leaf meal could completely or partially replaces the artificial staining agent (Changjun et al, 2004).

Table 12: Growth of young rabbits fed on a Stylosathes leaf meal diet for 60 days

Groups	Feed composition	No. rabbits	Initial weight	Final weight	Total weight gain	Daily weight gain
I.	½ SLM + ½ concentrate	20	701	2230	1529	25.48
II.	¼SLM + ?green material + 1/3 concentrate	18	709	1895	1186	19.76
III	½ green material + ½ concentrate	15	702	1474	772	12.86

Source: Changjun et al., 2004

Table 13: Effect of feeding on broilers raised for 42 days on diets containing three different levels of Stylosanthes leaf meal (SLM)*.

Parameters studied	Control (Maize-soybean)	3% SLM	6% SLM	9% SLM
Weight gain (g)	1563	1645	1566	1559
Feed consumption (g)	2924	3117	3002	3104
Feed conversion ratio (FCR)	1.87	1.89	1.91	1.99
Dressed weight (%)	87.5	88.0	88.4	87.5
Skin colour	1.0	1.4	1.8	2.1
Shank colour	2.0	3.1	3.3	3.8

*SLM included at 3 levels by replacing maize and soybean meal.
Source: Changjun et al., 2004

Thakur et al (1985) evaluated ferrous sulphate treated and untreated Leucaena leucocephala leaf meal (LLM) in layer rations and recommended to feed treated and untreated LLM up to 5.0 and 2.5% levels respectively. LLM as a source of pigmenting xanthophylls has also been tried in the poultry ration. Experiment with broilers diet containing two levels (50g and 100 g/kg) of LLM showed that shank pigmentation significantly increased at 50g inclusion level. It was suggested that it would be prudent to restrict L. leucocephala inclusion rates to 50g/kg diet since growth performance was unaffected at this level and broiler pigmentation was only marginally improved by higher inclusion rate of the Leucaena leaf meal (D'Mello et al, 1987). Such restriction would also reduce any mimosine residues in carcass. Egg yolk pigmentation was also improved if leaf meal was included in the diet of hens. Dietary level of 50g/kg leaf meal of Gliricidia sepium (Osei et al, 1990) or Cajanus cajan (D'Mello and Devendra, 1995) significantly improved the egg yolk pigmentation. Sesbania leaf meal can also be used in the rations of poultry (Prasad et al, 1970).

Leaf meal as a ration component for ruminants

For the last two decades, attempts have been made to use the leguminous leaf meal as an animal feed stuff because of high concentration of protein of high biological value and other nutrients such as carotene and various other minerals. Use of some important leaf meal in ruminant ration is being described as below.

Nutrient utilization of Stylo leaf meal (SLM)

Stylo meal prepared from S. seabrana cultivars has been found rich in amino acids particularly lysine and sulfur containing amino acids (cysteine and methionine) as well as essential amino acids and can thus be incorporated in the ration of livestocks including poultry (Mojumdar et al., 2004) as quality protein supplement. The leaf meals have also been found a good replacer of the concentrate mixture including oilseed cakes. Stylo meal prepared from S. hamata harvested at 100% flowering stage was found to replace concentrate mixture when low-grade dry grass based ration was fed to sheep. There was no adverse impact upon dry matter intake (2.73-2.97%) and digestibility co-efficient of other nutrients. However dry matter digestibility (69.29%) was found significantly higher ($p < 0.05$) when 50% of concentrate ration was replaced with stylo meal (Majumdar et al., 2004). Similar observations were also observed when stylo leaf meal was fed to ewes where NFE digestibility was found significantly ($p < 0.05$) higher (76.36%) when 50% concentrate was replaced by leaf meal. Total protozoal population (4.08×10^4 per ml) was highest in the diet of sheep where 50% of concentrate mixture has been replaced with stylo leaf meal and thus providing a better rumen eco-system in the animals (Mojumdar et al., 2004).

Nutrient utilization of Leucaena leaf meal (LLM)

Leucaena leaves incorporation (20%) by replacing 30% crude protein of mustard cake in concentrate mixture and its subsequent usage during block making with natural grass did not interfere with nutrient utilization in Barbari goats and can be incorporated by replacing mustard cake safely without affecting nutrient utilization in Barbari goats. (Samanta et al., 2003). The LLM can be safely incorporated in diets of growing lambs by replacing mustard cake to meet up to 45% to the protein requirement without affecting nutrient and energy utilization and feed cost per kg of weight gain was 19% less than the control diet (Mahanta et al, 1999).

Dry matter intake of Leucaena and other leguminous leaf meal by the ruminants were found comparable like other protein sources as oil cakes. Thus, dietary inclusion of leaf meal did not elicit any palatability problem. Replacement of groundnut cake (GNC) -sun flower cake mixture with Leucaena leaf meal in the diet of lactating buffaloes (Garg and Kumar, 1994) and goats (Rai et al, 1994) did not influence dry matter (DM) intake. However, growing lambs fed isonitrogenous complete diets in which LLM replaced 15% of GNC exhibited highest DM intake (Vahidulla and Reddy, 1987). Thus, intake of diets containing Leucaena appears variable depending on experimental conditions such as other dietary constituents (Garcia et al., 1996). In a comparative evaluation of cotton seed cake and LLM, Barman and Rai (2003) observed higher effective degradability of both DM and OM as well as un-degradable protein (144.62 vs 76.89 g/kg DM) in LLM than that on cotton seed cake. Higher UDP values of LLM has also been reported by Garcia et al., 1996. Total tannin content in the LLM has been found low (2.83%) which was below the threshold level in ruminant ration (>4% on DM). LLM was also found to contain beneficial level of condensed tannin (1.44%) from the point of increasing the by-pass protein values in ruminants (Barman and Rai, 2003). Less than 4% condensed tannin in ruminant diet increases the by-pass protein, which increases both milk yield and protein contents in milk (Woodward et al., 1999). In another studies, Abdulrajak et al (2001) found that LLM contained 1.8% total tannins and 0.5% condensed tannins. Tangendjaja and Wina (2001) reported higher in-vivo CP digestibility (77.70%) in sheep fed on LLM. Total GNC protein when replaced by LLM in the concentrate mixture of lactating goats was found to increase ($p<0.01$) milk yield (Rai et al., 1994). Economics of leaf meal supplementation and live weight gain depends on feed intake, cost of feed and efficiency of feed utilization. It was economically advantageous to replace GNC or sunflower meal with Leucaena leaf meal for feeding lactating goats (Rai et al., 1994) and buffaloes (Garg and Kumar, 1994) where economics were expressed as feed cost per kg of milk produced, because feed intake was similar among treatments.

It could therefore be said that leaf meal provides ample scope for replacement of concentrate mixture in the ration of ruminants, without any adverse impact upon nutrient utilization and blood profile.

Sesbania leaves as a protein supplement

Sesbania sesban, a fast growing legume fodder tree is rich in CP (16.80%) can be used as protein source in lieu of conventional concentrate feed in growing calves in barley straw based ration without any ill effect on the digestibility and body weight gain of animals (Rekib and Shukla, 1994). Sesbania sesban leaves with soft twigs can replace one third of CP from conventional concentrate mixture in sorghum stover based diet without affecting nutrient utilization and milk production in crossbred cows. (Maity et al.1997). Sesbania grandiflora leaf meal was found to contain 31.71 %CP, 4.95% EE, 10.00% TA, 39.00% NDF, 18.87% ADF and 20.13% hemicellulose which could be used to completely replace GNC-nitrogen in complete diets for Nellore rams without any adverse effect on N retention and nutritive value (Bala Peeraiah et al, 1993).

Teak leaves as a supplement in animal ration

The complete feed prepared using fallen teak leaves (Techtona grandis) 40%, maize 15%, groundnut cake 20%, de-oiled rice polish 23%, mineral mixture 1% and common salt 1% was fed to male cross-bred calves which attained 591g LWG/head/day (Tiwari and Baghel, 1987). Gupta and Patle (1992) also recorded the chemical composition of teak leaves (7.00% CP, 2.63% EE, 17.35% CF, 55.92% NFE and 21.7% total ash) and observed that such dry fallen leaves can be fed as a sole feed to cross bred calves if fed as complete feed.

Leaf meal being rich in protein and essential amino acids has a wide scope for its commercialization and acceptance. A recent major development with Leucaena shows great promise both for biomass production and for industrial processing of Leucaena leaf meal. In China the acceptability of Stylosanthes leaf meal has been successfully adopted in a commercial way. Farmers started SLM production at rural level and presently two factories have also been established to make SLM a commercial commodity at a sale price of 1.6 yuan/ kg (approximately Rs. 8.70/ kg). In India also, a reduced feed cost can be achieved from the use of stylo leaf meal. It can be recommended to annex the leaf meal production programme with the crop husbandry practices of a seed production farming system where stylo is grown as a crop. There is potential of stylo leaf meal due to a fledgling Lucerne leaf meal industry developing in India. Success of leaf meal industry could however be dependent upon the nutritional requirement of livestock and its type, processing requirements to be accomplished (sole or complete feed, powder/pellets/blocks/threshed/chopped), additional nutritional advantages (carotenes, xanthophylls, by-pass protein, minerals etc.) to enhance the production quality of meat, milk and wool etc.

Bibliography

Abdulrazak SA, Nayangaga Jand Fujihara T. 2001. Relative palatability to sheep of some browse species, their in sacco degradability and gas production. Asian- Aust. J. Animal Sci. 14:1580-1584.

Amodu, 2004. Stylosanthes : a promising legume in Africa Chapter 22. In: High yielding anthracnose-resistant stylosanthes for agricultural system. Edited by: Sukumar Chakraborty. Australian Centre for International Agricultural Research, Canberra. Pp 225-233.

Bala peeraiah K, Prabhakara Rao Z and Anjaneya Prasad D. 1993. Sesbania grandiflora as areplacement for groundnut cake in complete diets for sheep. Indian J. Anim. Nutr. 10(2):95-99.

Barman K and Rai SN. 2003. Comparative evaluation of cotton seed cake and Leucaena leaf meal on profiles of amino acids, tannins and their influence on digestion kinetics. Indian J. Anim. Nutri. 20(4):378-388.

Chaisang Phailkaew, Ramesh CR, Kexian Yi and Stur, W. 2004. Utilization of stylosanthes as a forage crop in Asia. In: High yielding anthracnose-resistant stylosanthes for agricultural system. Edited by: Sukumar Chakraborty. Australian Centre for International Agricultural Research, Canberra. Pp 65-76.

Chakraborty Sukumar 2002. Stylo leaf meal shows potential for India, Internet: Stylo News Letter. CSIRO Plant Industry, Brisbane, Australia

Changjun Bai, Guodao Liu, Wang Dongjun, Krishna Daida, Qudratullah S., Prasad VLK, Rama Rao SV, Rao P. Parthasarthy, Ramesh CR, Balagopal R and Gopalan A. 2004. Stylosanthes leaf meal for animal industries in China and India. In: High yielding anthracnose-resistant stylosanthes for agricultural system. Edited by: Sukumar Chakraborty. Australian Centre for International Agricultural Research, Canberra. Pp 243-252.

D'Mellow JPF and Fraser KW 1981. The composition of leaf meal from Leucaena leucocephala. Tropical Science. 23: 75-78.

D'Mello, JPF and Acamovic T. 1989. Leucaena Leucocephala in poultry nutrition – A review. Animal Feed Science and Technology. 26: 1-28.

D'Mello, JPF and Devendra C. 1995. Tropical Legumes in Animal Nutrition. CAB International. Wallingforde, UK. Pp 247-282.

Donkoh A., Kese AG and Atuahene CC 1990. Chemical composition of chaya leaf meal (Cnidoscolus aconitifolius (Mill.) Johnston) and availability of its amino acids to chicks. Short Communication. Animal Feed Science and Technology. 30: 155-162.

Dwivedi PN and Pathak PK. 2004. Insacco evaluation of stylosanthes hamata meal threshed in different sizes. V[th] Biennial Conference of animal Nutrition Association. Banglore. 24-26 November, 2004. pp 67-68

FAO. 1992. Legume trees and other fodder trees as protein source for livestock. Animal Production and Health Paper No. 102. FAO, Rome.

Garcia, GW, Ferguson, TU, Neckles FA, Arlchibsald KAE 1996. The nutritive value and forage productivity of Leucaena leucocephala. Animal Feed Science and Technology. 60: 29-41.

Garg MC and Kumar Sanjiv. 1994. Effect of replacement of oil cake protein in concentrate mixture by Leucaena leucocephala leaf meal on the nutrient utilization and milk yield in Murrah buffaloes. Indian J. of Anim. Nutr. 11: 43-46.

Guodao Liu, Changjun Bai, Dongjun Wang, Ramesh CR and Rao P. Parthasarthy. 2004 . Leaf meal production from stylosanthes in China and India. In: High yielding anthracnose-resistant stylosanthes for agricultural system. Edited by: Sukumar Chakraborty. Australian Centre for International Agricultural Research, Canberra. Pp 253-256.

Gupta RS and Patle. 1992. Availability and chemical composition of dry fallen teak leaves and possibility of its feeding to ruminants. Indian J. Dairy Sci. 14(7): 383-384.

Khirwar SS, Tewatia BS and Hooda MS. 2003. Comparative nutritional evaluation of different species of Prosopis loppings in goats. Indian J. Anim. Nutr. 20(4): 400-404.

Leng RA. 1997. Bypass proteins from tree foliages (Chapter 5). In : Tree foliage in ruminant nutrition FAO Animal Production and health Paper 139). http://www.fao.org/docrep/003/w7448e/w7448E00.htm#TOC.

Mahanta SK, Singh S., Kumar A. and Pachauri VC. 1999. Subabul leaf meal as a replacement of mustard cake in lamb diets. Small Ruminant Research. 32: 37-42

Mahyuddin Prapti, Little D.A. and Lowry J.B. 1988. Drying treatments drastically affects feed evaluation and feed quality with certan tropical forage species. Animal Feed Science and Technology. 22: 69-78.

Maity SB, Rekib A and Shukla NP. 1997. Sesbania sesban as replacement for conventional concentrate mixture in silage based ration for crossbred lactating cows. Indian J. Anim. Sci. 67 (2): 158-159.

McCabe WL, Smith JC and Harriott P 1983. Unit operations of chemical engineering. Fourth Edition McGraw Hill Book Company, New York pp 767-781.

Michalet-Doreau B and Cerneau P. 1991. Influence of foodstuff particle size on in situ degradation of nitrogen in the rumen. Animal Feed Science and Technology. 35:69-81.

Mojumdar AB, Mishra AK, Das MM and Bhadoria BK. 2002. Nutrient utilization and blood profile in goats fed Flacourtia indica leaves. Anim. Nutr. Fd. Technol. 2:177-181.

Mojumdar, AB, Pathak PS, Mahanta, SK and Das, MM. 2003. Exploitation of stylo meal as quality protein supplement in the ration of buffaloes. Paper presented at National Seminar on "Economic Feeding of Buffaloes under different Agro-climatic Regions" held at IGFRI, Jhansi. 12-13 March, 2003.

Mojumdar, AB, Pathak PS, Das, MM, Mahanta, SK and Pathak PK. 2003. Effect of replacing concentrate mixture with stylo meal in nutritional status in ewes. Proceeding IX Animal Nutrition Conference, Jabalpur (3-5 January, 2004). pp 41.

Mojumdar, AB, Pathak PS, Das, MM, Mahanta, SK and Bhatt RK. 2004. Effect of replacing concentrate mixture with stylo leaf meal in ruminent utilization and blood metabolites in ewes. Proceedings. ANA conference on New Dimensions of Animal Feeding (Editors Sharma et al). pp 50.

Mojumdar, AB, Pathak PS, Mahanta, SK, Das, MM and Chandra A. 2003. Evaluation of promising Stylosanthes accessions for their nutritional values. Proc. National Symposium on Agroforestry and Sustainable Production (November 7-9, 2003). IGFRI, Jhansi.

Mojumdar, AB, Rekib A. and Debroy R. 1987. Seasonal effects on the nutritive quality of fodder tree leaves. Agroforestry for rural needs. (ed. Khurana and Khosla). Pp 312-318.

Mojumdar, AB, Samanta, AK, Pathak PS, Das, MM, Mahanta, SK, Pathak PK and Bhatt RK. 2004. Influence of feeding stylo leaf meal on the ruman eco-system of sheep. Paper presented at V[th] Biennial Conference of animal Nutrition Association. Banglore. 24-26 November, 2004.

Mojumdar AB, Shukla GP, Bhatt Vishnu and Kohli KS. 2003. Variability for quality traits in forage alfalfa (Medicago sativa L.). Range Managt. Agroforestry, 24(2):131-136.

Osei, S.A., Opoku RS and Atuahene, Comfort C. 1990. Gliricidia leaf meal as an ingradient in layer diets. Animal Feed Science and Technology. 29: 303-308.

Pathak PK and Mojumdar AB. 2004. Stylo leaf meal production and utilization. A lecture delivered on the occasion of Technology Day at IGFRI, Jhansi. 14th May 2004.

Pathak PS. 1988. Ecology and potentials of Su-babool in agroforestry systems. In : Pasture and Forage Crop Research (Ed. Panjab Singh). RMSI, Jhansi. Pp 332-341.

Pathak PS, Gupta SK and Debroy R. 1985. Biomass accumulation pattern and the productivity of a Leucaena mini rotation on moist lands. J. Trop. For. I (II):112-123.

Pathak PS and Mojumdar AB, 2002. Current status of feed and forages: prospects and constraints. Base Paper presented at UPCAR, Lucknow. 23rd October, 2002.

Pathak PS and Patil BD. 1982. Leucaena research at IGFRI. Proc Leucaena Workshop, Singapore. 129 p.

Parthsarthy Rao, Ramesh CR, Pathak PS, Mohan Rao, Y. and 2004. Recent trends in stylosanthes seed production by small holders in India. In: High yielding anthracnose-resistant stylosanthes for agricultural system. Edited by: Sukumar Chakraborty. Australian Centre for International Agricultural Research, Canberra. Pp 235-241.

Phuc Bui Huy Nhu, Ogle R B, Lindberg J E and Preston T R.1996.The nutritive value of sun-dried and ensiled cassava leaves for growing pigs. Livestock Research for Rural Development Volume 8, Number 3, September 1996

Prasad DA, Joji Reddy K and Subba Reddy C. 1970. Agathi leaf meal (S. grandiflora) in the ration of growing chicken. Indian Vet. J. 47: 164.

Ram Ratan, Ayub M, Bapna DL and Sawal RK. 2003. Effect of feeding Pala, (Zizyphus, nummularia) leaves in the diets of bucks on their growth perfomance in arid regions. Indian J. Anim. Nutr., 20(4):449-453.

Ravindran V 1991 Preparation of cassava leaf products and their use as animal feed. In; Roots, tubers, plantains and bananas in animal feeding (Editors: D Machin and Solveig Nyvold). FAOAnimal Production and Health Paper No 95: 111-122 seen in Phuc Bui Huy Nhu, Ogle R B, Lindberg J E and Preston T R The nutritive value of sun-dried and ensiled cassava leaves for growing pigs. Livestock Research for Rural Development Volume 8, Number 3, September 1996 University of Agriculture and Forestry, Ho Chi Minh City, Vietnam

Rai SN, Walli TK, Srivastava A, Verma GS. 1994. Effect of replacement of groundnut cake protein by leucaena leaf meal on milk production performance in goats during early lactation. Indian Jornal of Animal Nutrition. 11:149-154.

Rajput D.S., Dwivedi P.N., Tamhankar M.B. and Pathak P.K. 2004 Prediction of production potential of leucaena for energy and animal nutrition. Paper presented in 38th ISAE Annual Convention and Symposium held at DR. B.S. Sawant Konkan Krishi Vidyapeeth, Dapoli from 16-18 January, 2004.

Rekib A and Shukla NP. 1994. Evaluation of Sesbania sesban as a protein supplement to low-grade roughage (barley bhusa) for growing calves. Indian Jornal of Animal Sciences. 65(1):113-116.

Samanta AK, Singh KK, Das MM, Maity SB and Kundu SS. 2003. Effect of complete feed block on nutrient utilization and ruman fermentation in Barbari goats. Small Ruminant Research. 48: 95-102.

Samual DVK and Anap GR. 1980. Design criteria for portable hay drier. Forage Research. 6:209-214.

Singh Jai 1984. Mechanical contrivances in hay quality amelioration. PhD (mechanical Engineering), Thesis submitted to the faculty of Engineering, Jiwaji University, Gwalior. Division of Agricultural Engineering, Indian Grassland and Fodder Research Institute, Jhansi

Singh LN. 1988. Forage production in temperate and sub alpine regions. In : Pasture and Forage Crop Research (Ed. Panjab Singh). RMSI, Jhansi. Pp 66-62.

Singh AK, Upadhyay VS, Singh KK and Mishra AK. 1998. Effect of dietary levels of Abezia lebbek leaves on the utilization of matured dry grass in goats. Indian J. Anim. Nutr. 15:290-292.

Sonaiya E.B. 1995. Feed resources for smallholder poultry in Nigeria. In: Better feed for animals: more food for people. Edited by D. Chupin FAO, Rome

Tangendjaja Budi, Rahardjo Yono C. and Lowry J.B. 1990. Leucaena leaf meal in the diet of growing rabbits: evaluation and effect of low-mimosine treatment. Animal Feed Science and Technology. 29: 63-72.

Tangendjaja B and Wina E. 2001. Tannins and ruminant production in Indonesia. ACIAR Proc. No. 92. Australlia. Pp 40-43.

Thakur RS, Gupta PC, Prasad D and Yadava IS. 1985 evaluation of subabool (Leucaena leucocephala) leaf meal in layer rations. Forage Research 11(1): 63-68.

Tiwari DP and Baghel RPS. 1987. use of dry fallen tead leaves as roughage in complete feed for cross breed calves. Indian J. Anim. Nutr. 4: 214-216.

Uden P. 1988. The effect of grinding and pelleting hay on digestibility, fermentation rate, digesta passage and rumen and faecal particle size in cows. Animal Feed Science and Technology. 19: 145-157.

Upadhyay VS and Mojumdar AB. 1991. Studies on quality, intake and digestibility of sorghum silage as influenced by addition of Leucaena and Sesbania leaves. Forage Research. 17(1): 69-72.

Vahidulla S, Reddy KR 1987. Feeding value of Subabool leaf meal in complete rations for growing Nellore brown lambs. Cherion 16: 199-204.

Woodward Sl, Auldist MJ, Labuyrie PJ and Jansan ERL, 1999. Effect of Lotus corniculatus and condensed tannin in milk yield and milk composition of dairy cows. Newzealand Soc. Anim. Prod. 59: 152-155.

CHAPTER - 3
COMPACTION OF FIBROUS FEEDS

O.P. Lohan

Dry roughages, forages and agricultural byproducts containing more than 18 percent crude fibre or 35 percent cell-wall constituents on dry matter bases, are the staple feeds of Indian livestock, which are poorer in quality and are poorly utilized due to high fibre content. They include green or dry forages, conserved forages, fibrous crop residues and agro-industrial byproducts. These are the primary feed resources for feeding ruminants in majority of the developing countries including India in the tropics. Under global context wheat straw (25.3%) is the most abundant crop residue followed by rice straw (22.8%), corn stover (15.8%), barley straw (8.3%) sugarcane tops (8.1%) and sugarcane byproducts (6.3%) as reported by Han and Garrett (1986). Asia is producing 46 percent of world supply of crop residues, therefore, Asian Animal Nutritionist paid more attention to improve the nutritive value and utilization of crop residues as animal feed in relation to other regions. Ninety two percent of rice straw and thirty four percent of the wheat straw are produced in Asia.

The major constraints associated with the use of crop residues as ruminant feed are low nutrient density and poor digestibility. The low nitrogen and mineral content, along with high lignin and silica levels associated with roughages result in their low nutrient digestibility and consequently resulting in hardly meeting the maintenance needs of even non-productive animals.

The results of many studies confirm that untreated or treated crop residues supplemented with concentrates, legume forages, protein supplements can significantly improve the dry matter (DM) digestibility, feed intake and ultimately the performance of animals. Supplementation will also help in overcoming the low palatability of certain crop residues. Hence, the mixing of concentrates/ green forages with crop residues and then compressing it in the form of complete feed is a practical approach to improve their utilization.

Interest in complete feed system is increasing in the world because it offers a mean of controlling the ratio of concentrate to roughage and the nutrients consumed. Also it reduces the cost and helps in the utilization of locally available agro industrial products efficiency. Now a days, the complete feed is produced in the form of pellets. The major limitation in the production of complete feed pellets, it requires grinding of roughage portion of feed, which is costly and also reduces the digestibility of crude fibre and dry matter. However, efforts have been made by several workers (Yadav, 1988; Sihag, 1998 and Mann, 1992) to prepare complete feed in the form of blocks in which grinding of roughage portion is not needed.

DESIGN AND DEVELOPMENT OF BLOCK MAKING MACHINE

A power operated vertical type straw densification machine (Fig. 1) was developed to exert a compressive force of 300-400 kg/cm^2. The densifying machine was fabricated for the production of complete feed blocks. The basis of the design was to move a piston plate vertically upward against the strength of straws in two jackets provided in the machine. The feed mixture in the jackets was compressed to make stable feed blocks strong enough to withstand jerks during transportation. The rotary motion of the prime mover (electric motor) was converted to give linear vertical motion to cross rods provided on the top of the piston. A hydraulic pump to generate pressure upto 400 kg/cm^2 was provided in the machine. To store the hydraulic oil, a tank of 8 liters capacity made from 10 gauge MS sheet was also provided at the right side of the pump. A pressure pipe fitted with a pressure regulating valve and a pressure gauge was also provided to regulate and measure the oil pressure exerted by the hydraulic pump. A three-phase electric motor 5 HP, 7.5 Amp. 440 V, 1.440 RPM, 50 Hz and 3.7 watt was provided in the machine as prime mover. The electric motor was fitted on an extended frame of 43 cm and 35 cm wide made from angle iron. The drive to hydraulic pump from motor was provided by belt and pulley arrangement. An online three phase starter suitable for 5 HP motor was provided on the panel. A phase reversing switch was provided to control the movement of cross rods and piston plates within the desired distance. The machine was supported on a platform fabricated from 16 gauge MS sheet. The power from three-phase electric motor was suitably transmitted by means of belt and pulley arrangement mounted on the side of the bottom of the main frame. A matching motor starter a phase reversing switch and a clutch pedal were provided to control the movement of the piston within the designed range. Two die boxes (jackets) were designed to accommodate the required quantity of straw mixture and to withstand lateral pressure of feed during compression. The prototype of the machine was fabricated using mild steel channels, rods, shafts, sheets and angles of appropriate strength and sizes. Mainframe of size 160 x 56 x 25 cm was fabricated from 65 x 65 x 10 mm mild steel iron. Two feed jackets were supported on top portion of the main frame and hydraulic press and pump were fitted at the bottom of the frame. Two rectangular jackets of the size 23 x 23 x

46 cms fabricated from 10 gauge MS steel were fitted at upper portion of the main frame. Two square iron plates each of size 22.5 x 22.5 cm of the jackets were fabricated from 10 gauge MS steel and are connected with cross rods of piston with the help of two squares iron rods. The cross rods were welded with hydraulic piston. To bring the piston back to its original position after applying pressure for required time cross rod were connected with two springs. Other end of the spring was attached to the foundation of the machine. To transport the machine at desired place of working rubberized cast iron wheels were provided for each movement of the machine.

DESIGN AND DEVELOPMENT OF A MIXER

The feed mixing system is the heart of any feed manufacturing unit/plant. Each ingredient has its own identity unit, it is mixed with one or more other ingredients to make a formula feed. For an efficient growth, reproduction and lactation and to remain in good health, animal should receive a balanced ration that supplies nutrients at proper concentration.

There are several factors which affect uniform mixing of ingredients viz. type of mixer, type of flights on the shaft in the mixer, mixing time and the physical properties of ingredients. Wilcox and Balding (1976) listed the ingredient characteristics that have the greatest impact on mix uniformity. These factors include:

- Particle size and shape
- Density
- Electrostatic charge
- Hygroscopicity
- Flow ability

The mixing of molasses with wheat straw is very difficult due to their physical states and bulk density. Therefore, to facilitate the mixing of molasses with wheat bhoosa a round bottom rectangular batch type feed mixer having 50 kg capacity per batch was designed and fabricated for. It comprises of four vertical 65 x 65 x 8 mm angle bars to hold a bottom rectangular bowl type mixer. The mixer bowl was fabricated from 14 G sheet. A bottom hole with a sliding gate of 20 cm x 14 cm was provided in the bottom for unloading. The rectangular shaft was fabricated from 50 x 50 mm square shaft 175 cm long. The shaft was supported on two bearing housing. Four 'T' shaped mixer blade fitted on 3 x 3 mm square shafts were bolted on these shafts for mixing of molasses with wheat straw. The clearance between blades and bowl can be adjusted. To decrease the speed of rotating mixing shafts a reduction gear set consisting of two gears having 52 and 13 teeth were used to decrease the speed to one fourth time as taken from the pulley. A 10 HP 3 phase induction motor was fitted as a prime mover for the working of all the moving parts of the machine. To spray the molasses effectively on the wheat straw a stator rotor type pump capable of delivering pressure upto 5 kg/cm^2 was provided. This pump took drive from the driver shafts through chain and sprocket system. G.I. Pipe of "H" shape with outer diameter of 3.5 cm was provided to spray molasses. Holes of 3 mm dia at the regular interval of 10 cm to spray the molasses uniformly over the entire material in the mixing bowl were provided in the spray pipe. An overhead tank of 100-litre capacity was provided to store the molasses. The tank was fitted on 156 cm. High tripod stand made from angle iron 40 x 40 x 5 mm with 3 rods welded acrossed to work as a ladder. To heat the molasses the tank was fitted with 1 KW heater. Heating of molasses was done to increase in flow ability. A depth gauge was provided to monitor the quantity of the molasses in the tank. Gate valve was provided to regulate the flow of molasses. One pulley having 'B' type groove can be fitted to take the designed RPM (20, 35 and 40). Three different pulleys of diameter (117, 67, 52 cm) were used with their respective belt of B-112, B-84 and B-72 type. These pulleys were used as driven pulley and take drive from prime mover (Fig. 2).

Development and evaluation of feed blocks

The material to be used for densification was thoroughly mixed and required quantity was filled in the jackets of the press machine. The weight and dimensions of the blocks were taken just after ejection from the block-making machine. The blocks so formed were stored and then tested for quality on the basis of their durability, post compression expansion and increase in density.

FACTORS AFFECTING COMPACTION BEHAVIOUR

I) Moisture

Yadav (1988) conducted an experiment with wheat straw and paddy straw in unground and ground forms at 15, 17.5 and 20% moisture levels and 5, 10 and 15 minutes dwell time. Seventeen and half per cent of moisture level (Table 1) and 5 minutes dwell time (Table 2) produced more compact and durable blocks. The average increase in bulk density by compaction was 2.5, 2.5, 2.7 and 2.4 times for unground wheat straw, ground wheat straw, unground paddy straw and ground paddy straw based blocks, respectively (Table 3). The unground form of both straws was better than ground form of respective straws, as the former resulted in more durable blocks.

Table 1. Effect of moisture levels on post compression expansion and durability of Blocks

Parameters	Wheat straw Moisture (%)			Paddy straw Moisture (%)		
	15.0	**17.5**	**20.0**	**15**	**17.5**	**20.0**
Post compression Expansion	70.11	62.56	63.83	53.88	44.67	47.00
Durability	44.46	57.82	56.19	75.52	82.25	80.25

Each figure is an angular transformed value of percent durability.

II) Dwell time

Table 2. Effect of dwell time on post compression expansion and durability of blocks

Parameters	Wheat straw			Paddy straw		
	Dwell time (minutes)					
	5	**10**	**15**	**5**	**10**	**15**
Post compression expansion	68.33	63.66	65.11	50.33	49.00	46.43
Durability	52.12	51.31	54.24	78.15	78.57	80.79

Each figure is an angular transformed value of percent durability

III) Grinding

Table 3. Effect of grinding roughage on the bulk density of feed blocks

Parameters	Bulk density (kg/m³)			
	Wheat straw		Paddy straw	
	Unground	**Ground**	**Unground**	**Ground**
Loose bulk density	130	164	125	160
Compact density	325	369	337.5	384
Increase in bulk density (%)	250	225	270	240

Each figure is an angular transformed value of percent durability.

Sihag (1989) found that the post compression expansion was lower and the durability was higher at 15 and 17.5% moisture levels than at 20% moisture level in wheat straw based complete feed blocks. The durability of the blocks was significantly higher at 20% molasses level than at 15 and 10%.

IV) Green forage incorporation

Durable blocks could be formed from wilted berseem. (Mann, 1992) post compression expansion reduced significantly only after 96h of wilting. The berseem that was not wilted formed less durable blocks than the one which was wilted (Table 4). A higher durability was observed at 70% as compared to 30 and 40% berseem levels. As the berseem level increased post compression expansion, decreased and durability and density increased (Table 5).

Table 4. Effect of wilting time of berseem on compaction of feed blocks

Parameters	Wilting time (hrs)				
	0	**24**	**48**	**72**	**96**
Post compression expansion (%)	39.71	41.36	46.43	40.10	33.27
Durability (%)	24.50	37.80	38.11	32.21	57.78
Increase in density (time)	2.9	3.41	3.49	3.78	4.01

Each figure is an angular transformed value of percent durability.

Table 5: Effect of level of berseem on compaction behavior of feed blocks

Parameters	Level of berseem (%)				
	30	**40**	**50**	**60**	**70**
Post compression expansion (%)	45.38	39.36	38.36	38.42	39.37
Durability (%)	22.70	31.07	42.30	43.55	48.78
Increase in density (time)	3.10	3.27	3.40	3.71	4.14

Each figure is an angular transformed value of percent durability.

V) Type of crop residues

The different crop residues had been compared (Jaglan, 2002) for their densification behaviour. Paddy straw based feed blocks had maximum ($P<0.01$) increased in density followed by the blocks having sugarcane bagasse, mustard straw and wheat straw in decreasing order. Post compression expansion was minimum in feed blocks made from sugarcane bagasse and maximum in paddy straw blocks. Whereas, paddy straw based blocks were least durable, while sugarcane bagasse blocks were most durable (Table 6).

Table 6. Effect of different crop residues on their densification

Crop residues	Loose bulk density (kg/m3)	Compact bulk (kg/m3)	Increase density (kg/m3) %	Post compression expansion	Durability (%)
Wheat straw	71.97^{b}	199.92^{a}	177.75^{a}	9.30^{b}	52.65^{b}
Paddy straw	66.86^{a}	219.49^{b}	231.72^{c}	12.62^{c}	42.54^{a}
Mustard straw	77.89^{c}	222.21^{b}	184.80^{b}	7.81^{a}	59.58^{b}
Sugarcane bagasse	77.43^{c}	222.50^{b}	186.83^{b}	6.89^{a}	64.37^{b}

a, b, c values bearing different superscript in a column differ significantly ($P= 0.01$)

VI). Urea treatment

Wheat straw, paddy straw, mustard straw and sugarcane bagasse were treated with 4 percent urea as per the standard procedure (Kundu *et al.* 19). The treated crop residues were compressed in the form of blocks (Jaglan 2002). Urea treatment of crop residues increased bulk density and durability and reduced post compression expansion of densified blocks in comparison to untreated crop residues (Table 7).

Table 7. Effect of urea treatment of crop residues on their densification as feed blocks

Treatment	Loose Bulk density (kg/m3)	Compact Bulk density (kg/m3)	Increase in Bulk density (%)	Post Compression expansion (%)	Durability (%)
Untreated	74.07[a] ±0.86	211.62[a] ±3.33	182.22[a] ±1.26	9.11[b] ±0.40	55.50[a] ±1.49
Urea Treated	78.59[b] ±0.74	227.22[b] ±3.15	190.38[b] ±1.11	7.15[a] ±0.24	74.38[b] ±1.33

a, b values bearing different superscript in a column differ significantly (P= 0.01)

VII) Sodium hydroxide and steam treatment

Prior treatment of crop residues using sodium hydroxide improved the quality of feed blocks. It decreased post compression expansion and increased durability of blocks, irrespective of crop residues. Similarly, steam treatment followed by densification increased the bulk density of wheat bhoosa to two folds as compared to untreated wheat bhoosa (Table 8). The durability increased and the post compression expansion decreased due to the treatment (Sunda, 2004).

Table 8. Effect of NaOH and steam treatment.

Parameters	Untreated	Steam treatment		steam plus NaOH treatment	
		Before molasses	After molasses	Before steam	After steam
Post compression expansion (%)	10.01	16.39	14.43	51.84	33.27
Durability (%)	74.65	68.09	52.60	88.40	83.11
Initial bulk density (kg/m³)	69.14	78.16	78.26	74.34	74.34
Block's bulk density (kg/m³)	139.00	125.63	115.16	218.10	181.33
Increase in bulk density	2.01	1.61	1.47	2.93	2.44

Cost of densification.

The economics of densification of crop residues was calculated taking into account the labour charges and the cost of electricity consumed in the process of densification. The total cost of densification was Rs.36.03, 35.95, 36.24 and 37.88 per qtl. for wheat straw, paddy straw, mustard straw and sugarcane bagasse based blocks, respectively (Jaglan, 2002).

Storage of complete feed blocks

Complete feed blocks (Fig. 3) were manufactured and stored in conventional godown under natural climatic conditions for one year. The temperature ranged from 10 to 42°C and relative humidity ranged from 20 to 100%. The samples were taken every month and were analysed for CP, DM, total ash and organic matter contents. The blocks were regularly inspected for growth of fungus.

The changes became prominent after 3 months of storage, the crude protein content started decreasing slowly and gradually and decreased by 1.10 per cent at the end of the year, the total ash content of feed started increasing and increased by 2.48 percent, whereas the organic matter contents started decreasing and decreased by 2.49 percent at the end of the year (Table-9).

Table 9. Effect of storage period on the chemical composition of complete feed blocks

Parameters (%)	Period of storage (Months) 1	2	3	4	5	6	8	10	12
Total ash	8.22	8.25	8.40	8.48	8.75	9.00	9.46	10.10	10.70
Organic matter	91.78	91.75	91.60	91.52	91.25	91.00	90.25	89.90	89.30
DM	88.83	90.59	91.57	91.45	91.60	91.00	91.00	91.50	90.00
CP	12.10	12.11	12.08	12.00	11.90	11.70	11.56	11.27	11.00

Overall physical appearance of the feed blocks showed no change upto one year of storage. The smell was also normal. There appeared no visible growth of fungus during the storage period.

In vivo evaluation of feed blocks

In an experiment (Yadav, 1988) the adult buffaloes were fed complete feed blocks made of four similar rations except the type and nature of straw (Table 10). The DM intake was more in paddy straw based blocks as compared to wheat straw based block. Grinding of straws improved its DM intake. The digestibility co-efficients of all the nutrients were higher in paddy straw than wheat straw based blocks and the co-efficients decreased due to grinding of straws.

Table 10. DM intake and digestibility co-efficients in buffaloes

Parameters	Complete feed blocks			
	Wheat straw unground	**Wheat straw ground**	**Paddy straw unground**	**Paddy straw ground**
DM Intake/100 Kg body wt.	1.90	2.26	2.36	2.46
Digestibility co-efficients (%)				
DM	50.69	49.49	58.25	55.02
CP	66.59	64.52	69.10	64.99
CF	56.00	54.02	65.54	63.89
EE	53.74	49.00	59.52	54.56
NFE	54.14	52.10	60.00	58.96

In an another experiment, Singh (1992) fed growing buffalo calves on green forage based complete feed blocks (wheat straw, berseem, molasses, urea, mineral mixture and salt) and conventional feed (wheat straw, green fodder and concentrate mixture). The dry matter, nutrient digestibility, nutritive value and body weight gain was significantly higher in second than the first group, but, the cost per kg weight gain was Rs. 3.58 more in group fed on conventional feed. The results indicated that green fodder based complete feed blocks can be developed after wilting (20-25% moisture) and can be fed to the animals safely.

Jaglan (2002) conducted an experiment on crossbred heifers by feeding complete feed blocks containing urea treated wheat straw (T_1). Wheat straw was replaced by urea treated paddy straw (T_2), mustard straw (T_3) and sugarcane bagasse (T_4) at 50 percent level, respectively (Table 11). In each treatment roughage to concentrate ratio was kept 60:40. The DM intake and daily body weight gain were higher, while the cost per kg live weight gain was minimum in T_1 than other treatments. The results showed that urea treated crop residues can successfully replace urea treated wheat straw at 50% level in complete feed blocks with a daily gain over 530 g/day.

Table 11. Dry matter intake, nutritive value, daily gain and cost of feeding in heifers

Parameters	Treatments			
	T_1	T_2	T_3	T_4
DM Intake/100 Kg body wt. (kg)	2.81	2.90	2.79	2.51
Nutritive value				
DCP	9.58	9.38	9.20	9.59
TDN	62.05	60.17	59.50	58.38
Body weight gain (g/day)	727.78	590.74	585.19	531.48
Cost/kg live weight gain (Rs)	29.56	35.14	33.78	35.75

Fistulated adult crossbred bulls were fed complete feed blocks, isonitrogenous and with same composition except the type of crop residue (T1), wheat straw; (T2), paddy straw; (T3), sugarcane bagasse (T4) and mustard straw (Sunda, 2004). The DM intake was similar in all the treatments; however, the digestibilities of DM, CP and EE were higher in T_1 and T_2 than the T_3 and T_4 treatments. The rumen metabolites and than the pH of the SRL were not affected by the type of crop residues.

In another experiment (Sunda, 2004) on buffalo calves fed complete feed blocks having concentrate and wheat bhoosa (T_1), 20 percent (T_2) and 30 percent (T_3) of wheat bhoosa was replaced by mustard straw. Roughage to concentrate ratio was kept 60:40 and all the diets were isonitrogenous. Neither the DM intake and digestibility co-efficients nor the TDN content differed significantly among the treatments. The results (Table 12) indicate that replacement of wheat bhoosa with mustard straw had adverse effect on the growth performance.

Table 12. Effect of replacement of wheat straw by mustard straw on buffalo calves performance

Parameters	**Treatments**		
	T_1	**T_2**	**T_3**
DM Intake/100 Kg body wt. (kg)	1.92	1.91	1.69
Nutritive value			
DCP	7.93	7.93	7.19
TDN	51.71	51.41	48.80
Body weight gain (g/day)	509[b]	423[a]	355[a]
Feed consumed/kg gain (kg)	11.9	14.31	14.44
Cost/kg live weight gain (Rs)	38.37	44.29	47.86

In another experiment on lactating crossbred cattle, Sunda, (2004) studied the effect of 50 percent replacement of wheat straw (T_1) with paddy straw (T_2), sugarcane bagasse (T_3) and mustard straw (T_4) on milk yield and its composition. The roughage to concentrate ratio was maintained 60:40. The digestibilities for all the nutrients were higher in T_1 and T_2 as compared to T_3 and T_4 treatments. The milk yield significantly ($P<0.05$) reduced when 50 percent of the wheat straw was replaced by sugarcane bagasse or mustard straw, however, the treatments did not affect the milk composition (Table 13).

Table. 13. Effect of different sources of crop residues on milk yield and its composition in cattle

Parameters	**Treatments**			
	T_1	**T_2**	**T_3**	**T_4**
Digestibility co-efficients (%)				
DM	63.36[b]	63.98[b]	57.45[a]	58.08[a]
CF	60.21[b]	68.84[b]	56.36[a]	56.50[a]
NFE	65.96	64.65	63.00	62.80
Milk yield (kg/animal/day)	10.80	10.87	10.15	10.02
Milk composition (%)				
Fat	4.18	4.0	4.49	4.36
Total solids	12.36	11.84	12.56	11.68
SNF	8.54	8.24	8.52	7.86
CP	3.57	3.54	3.48	3.46

Probable composition (%) and cost as per prevailing market rates of complete feed blocks

Ingredients	Cost (Rs./ qutl.)	Growth	Lactation	Maintenance
Wheat/paddy straw	160	60	30	71
Berseem wilted	300	-	27	-
Mustard cake	600	19	15	-
Deoiled Rice bran	225	2	-	9
Barley	625	-	10	-
Urea	500	1	-	2
Molasses	300	15	15	15
Mineral mixture	2100	2	2	2
Common salt	70	1	1	1
Cost (Rs/kg)		3.08	3.65	2.32
DCP %		8.0	7.7	5.5
TDN %		61.5	62.6	58.5

In an experiment Berwal et al. (1997) compared feed block, pellets and mesh form of feeds in lactating crossbred cows. They reported that milk yield kg/animal/day was slightly higher in blocks fed group as compared to other physical form of feeds. The fat, total solids and SNF were significantly ($P<0.05$) higher in complete feed block group than those of pellet and mesh form of feed.

Table 14. Effect of physical form of feed on milk yield and its composition

Parameters	Physical form of feed			
	Block	Pellet	Mesh	C.D.
Milk yield, kg/animal/day	7.96	7.28	7.12	NS
4% FCM kg/animal/day	8.09	7.09	6.77	NS
Fat (%)	4.13	3.68	3.83	0.31
TS (%)	12.59	11.52	12.07	0.71
SNF (%)	8.46	7.74	8.24	0.59
CP (%)	3.7	3.56	3.76	NS

Benefits of densified feed block

i) Reduces the cost of transportation of cereal straws and will also minimize the risks of road accidents, because of pregnant trucks.

ii) Minimizes the cost of storage of cereal straws and thus will help in solving the problem of acute shortage of storage space, particularly in urban and peri urban areas.

iii) The feeding of balanced complete feed blocks saves time, labour and reduces the problem of nutrients deficiency and wastage of feeds.

iv) The feed blocks are cheap and more palatable and digestible as compared to pelleted feeds as well as conventional feeding of straws and concentrates, fed separately. Thus the technology developed increases animal performance and will curtail the cost of livestock production.

v) The technology will be particularly helpful in solving the problem of feeding livestock in floods/droughts affected areas.

vi) In case of complete feed blocks manufactured by this technology, there is no need to grind the roughage portion as required in feed pelleting and thus the digestibility of the feed is not affected.

vii) The machinery and equipment required is not so costly as in case of pelleting.

Complete feed blocks can be manufactured using all types of conventional crop residues with particle size 2.5 to 3 cm. For preparing good quality blocks, molasses to be added @ 15% and moisture @ 15 to 17% of the material on dry matter basis. Concentrate mixture can be included up to any level and green fodder as berseem upto 70% of the total material after wilting for 96 hrs (17-20% moisture). The material can be densified to 2.5 to 4 times, when pressed at 350 to 400 kg/cm^2 for 5 minutes as dwell time. Densification reduces the cost of transportation and storage. The product is exclusively important in flood/drought conditions. During scarcity, unconventional feed ingredients can be easily incorporated into the feed blocks. The feeding of densified feed results in to improvement in animal performance.

Limitations/constraints in commercialization

1) 'Ready to use' plant is not available in the market, it has to be got fabricated and standardized.

2) Commercial production of complete feed blocks at large scale will require systematic large-scale commercial plant.

3) Huge infrastructure facilities are required to handle large quantities of feed ingredients, particularly low-density roughages and complete feed blocks.

Bibliography

Berwal, R.S.; Lohan, O.P. and Yadav, K.K. 1997. Effect of feeding different physical forms of feed on nutrients digestibility milk yield and composition in crossbred cattle. Indian J. Dairy Sci. 50: 246.

Han, I.K; Garrette, W.N. (1986). Improving the dry matter digestibility and voluntary intake of low quality roughages by various treatment. A review, Korean Journal of Animal Sciences 28 (4). 199 – 236.

Jaglan, B.S. 2002. Urea treated crop residues based complete feed blocks for growing animal. M.Sc. Thesis submitted to CCS Haryana Agricultural University, Hisar.

Mann, J.S. 1992. Development of green roughage based complete feed blocks for growing calves. M.V.Sc. Thesis submitted to CCS Haryana Agricultural University, Hisar.

Sajjan, Sihag, Yadav, K.R; Vidya Sagar (1998). Effect of feeding sunflower cake on nutrient utilization in cattle. Indian Journal of Animal Nutrition 15 (2) 118 – 121.

Sihag, Z.S. 1989. Development of balanced and cheap roughage based complete feed blocks for growing buffalo calves. M.Sc. Thesis submitted to CCS Haryana Agricultural University, Hisar

Sunda, S.K. 2004. Evaluation of crop Residue Based Complete Feed Blocks in Ruminants. Ph.D. Thesis submitted to CCS Haryana Agricultural University, Hisar, India.

Wilcox, R.A. and Balding, J.L. (1976). Incomplete mixing and segregation. Kansas State University, Co-operative Extension Bulletin C – 555.

Yadav, K.K; Rathee C.S. and Lohan, O.P. (1991). Compaction behavior of straw based complete feed. Indian J. Animal Science 61: 980.

Yadav, K.K. 1988. Development of complete famine feed blocks for adult cattle. M.Sc. Thesis submitted to CCS Haryana Agricultural University, Hisar.

CHAPTER - 4

PROCESSING AND UTILIZATION OF SOYBEAN IN LIVESTOCK FEED

Nawab Ali

Global production of soybean is about 203 million tonnes and the share of India is about 7.0 million tonnes (3.4 %). The major soybean producing countries of the world are USA, Brazil, Argentina, China and India (Table -1). Processing of soybeans results into products that can be used in livestock feed formulations. Whole soybean can also be used in animal feed. Soybean is mainly processed for its oil, protein and lecithin. Soybean processing for oil results in soybean meal/flakes which contain protein and fibre as a major components. Soyhull is another by products of soyfood industry. Whole soybean contain 43% protein compared to 50% in soybean meal, 54% in dehulled soymeal and 14% in hulls (Table 2).

Table 1. Estimated world production of soybean during 2003-04

Country	Area	Productivity t/ha	Production Mt.	% of world production
USA	46.0	1.8	83.0	41.0
Brazil	19.0	2.8	53.0	26.0
Argentina	13.0	2.7	35.0	17.0
China	9.0	1.8	16.0	8.0
India	7.0	1.0	7.0	3.0
Others	6.0	1.5	9.0	5.0
Total	**100.00**	**2.03**	**203**	**100.0**

Combined soybean production of Brazil and Argentina has surpassed that of USA.

Table 2. Dry matter, protein and oil contents of soybean and its parts.

Soybean and its parts	Major constituents, %				
	Dry matter	**Crude Protein**	**FAT**	**NDF**	**NFC**
Whole soybean	91	43	19	22	17
Soybean hull	91	14	3	60	22
SE soymeal	89	50	2	15	27
SE dehulled SBM	90	54	1	10	30
ME soy-meal	90	46	8	22	28

NDF - Neutral detergent fibre SE - Solvent extracted SBM - Soybean meal
NF - Nonfibre carbohydrates ME- Mechanically expelled.

Hulls are high in fibre containing 60% neutral detergent fibre (NDF) compared to 15% for regular soybean meal and 10% for dehulled soybean meal. Soybean hulls are a source of rumen digestible fibre and low in starch. Because of this characteristics, hulls can replace some of the feed grains such as corn and barley in dairy cattle diets. Although high in fibre, it is not effective in stimulating rumination and wold not substitute for forage. No processing of hull is needed before feeding. However, its protein value and digestibility could be enhanced through fermentation.

Feed Ingredients

India's population which is over one billion now and continue to rise, will create a greater demand for milk, meat, eggs and fish. In order to meet the demand, the quantity and quality of affordable feed ingredients must be made available tot the poultry, aquaculture, and livestock sectors through domestic and or global production. Since poultry, cattle, and aqua are the main Indian feed industries, feed ingredients utilized by these sectors are of primary concern. Common feed sources for poultry include maize, sorghum, wheat, rice, millets, soybean meal, sunflower meal, peanut meal, cotton seed meal and to a lesser extent rapeseed meal, coconut meal, sesame meal, palm kernel meal, safflower meal, fish meal, meat meal, including byproducts such as rice bran, wheat bran, molasses, etc. According to industry sources, maize and soybean meal are the dominant energy and protein sources, respectively, in broiler diets. Common feed ingredients for cattle include deoiled rice bran, cottonseed meal, groundnut meal, and bagasse with the primary feed coming form grazing. Aqua feed commonly contains wheat, soybean meal, groundnut meal, deoiled rice bran and fish meal.

Nearly all of India's current feed demand is met from indigenously produced feeds; however, with the ever increasing food demand and rapid growth of poultry production, India's domestic capabilities to keep up with this feed ingredient demand will be challenged. India faces many agricultural challenges such as water, insects, weeds, environmental concerns, cost efficiency, sustainability, declining agriculture land base etc. The adoption of biotechnology has provided a tool to help accelerate the genetic progress needed in meeting the demands for food and feed.

Maize and Soybean

Maize and soybean meal are very important feed ingredients. Their production and uses show that a greater proportions of maize and soybean meal are going to feed sector (Fig. 1 & 2). It shows that maize feed use has grown much faster than nonfeed uses and the 14 per cent annual growth in feed maize use in the last half of the 1990s is comparable to the growth in broiler production during this same time period. Domestic maize Production in India has the potential to expand with average yield (1.8 tons per hectare) still well below other countries, including the United States (8.0 tons/ha), China (4.4 tonnes/ha), and Thailand (3.2 tonnes/ha). Soybean production has rapidly expanded in India (Table 3). However, improvement in yields have not made a significant progress. Though the production of soybeans has slowed and declined since late 1990s, feed use has significantly increased (Fig. 2). The increasing demand for soybean meal by the feed sector could in the future create pressure for imports; where as, currently all feed requirements can be met by the domestic supply. For both maize and soybeans, the adoption of newer technologies and the use of biotechnology can result in increased yields helping to provide an economical source of these feed ingredients.

Table 3. Production of soybean in India

Year	Production, Mt	Remarks
1981-82	0.47	
1999-00	7.10	AGR of soybean during 1981-82 to 2002-03 was 14.42%
2000-01	5.28	
2001-02	5.86	
2003-03	4.57	
2003-04	7.00	There were droughts during 2000-01 to 2002-03.

World soybean production during 2002-03 was 195 Mt and the estimate for 2003-04 is 203 Mt.

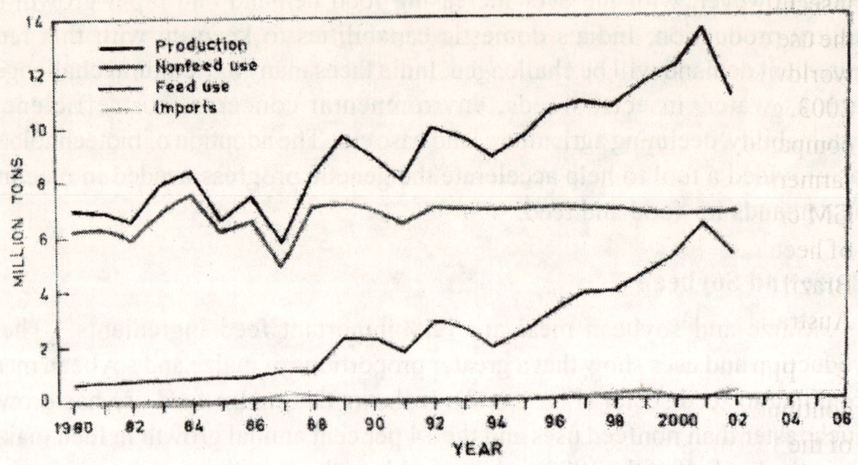

Fig. 1. Supply and use of corn in India

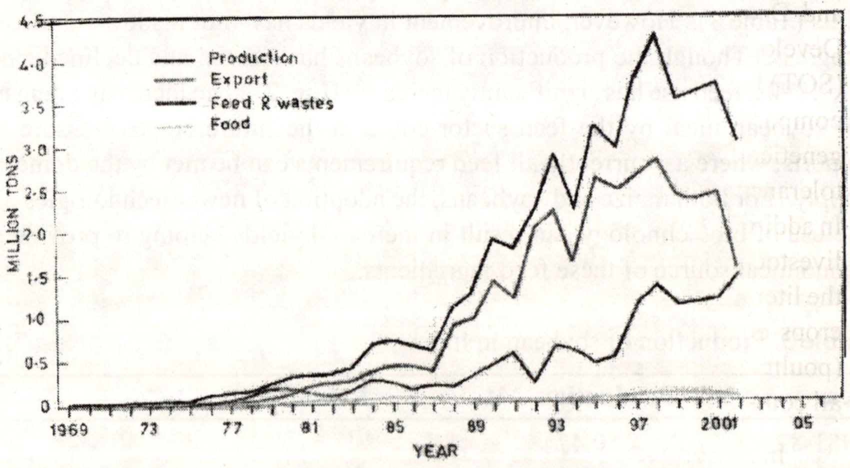

Fig. 2. Supply and use of soybean meal in India

Globally, efforts have been made to enhance yields, herbicide tolerance, insect resistance and virus protection through both conventional breeding and the use of biotechnology. The fruits of these efforts have been reaped by farmers worldwide with the rapid adoption of products derived from biotechnology. In 2003, global area of GM crops continued to grow at the rate of 15% per year, as compared to 12% in 2002, with 67.7 million hectares planted by seven million farmers (one million more than in 2002) within 18 countries. Since 1996 the GM crop area has grown over 40 fold. In 2003, growers of GM crops in order of hectarage include the United States (63%), Argentina (21%), Canada (6%), Brazil (4%) and China (4%), and South Africa (1%) with the balance grown by Australia, India, Romania, Spain, Uruguay, Mexico, Bulgaria, Indonesia,

Philippines, Columbia, Honduras and Germany. The main GM crops continued to be soybean (55% of the 76 million soybean hectares), maize (11% of the 140 million hectares of maize), cotton (21% of the 34 million hectares of cotton) and canola (16% of the 25 million hectares of canola).

Biotech feed ingredients are as safe as their conventional counterparts. Government and international scientific organizations including the Food and Agricultural Organization/World Health Organization of United Nations, Food and Drug Administration (FDA), Organization for Economic Cooperation Development (OECD), French Academy of Science and Society of Toxicology (SOT) have concluded that plant biotechnology does not pose any unique risk compared with other production methods. All commercially approved genetically modified (GM) products containing insect protection or herbicide tolerance traits are substantially equivalent to their conventional counterparts. In addition, nutritional equivalence has been further demonstrated in numerous livestock feeding studies conducted globally. There are over 100 references in the literature reporting results and/or reviewing studies covering the feeding of crops containing insect-protected and/or herbicide-tolerant traits to animals (poultry, swine, beef cattle, dairy cattle, sheep, water buffalo, fish, and rabbits) all concluding nutritional equivalence.

Insect-protected plants that have been commercialized are genetically enhanced to produce in plant insect control proteins similar to those produced by the soil bacterium Bacillus thuringiensis (Bt). Research has consistently demonstrated significant reductions in total fumonisin (B1,B2 and B3) concentrations in maize kernels containing the Cry1Ab protein when the hybrids were infested with European Corn Borer. This benefit provides the poultry and livestock user with an even safer source of maize.

Genetically Modified (GM) Crop Based Feed

The future outlook for GM crops is optimistic. Within the next five years, global planting of GM crops may grow to about 100 million hectares in 25 countries grown by up to 10 million farmers. The number and proportion of small farmers growing GM crops in developing countries will continue to increase. A more diversified portfolio of GM products will help drive the adoption and use of GM crops in both industrialized and developing countries. These include agronomic properties in crops with drought, salinity, and cold tolerance, increased yield, more efficient use of nitrogen, virus protection, and broader insect protection which would result in even greater insecticide reduction and further reduction in mycotoxins including aflatoxin.

Through biotechnology, the livestock industry will have the opportunity to select crops as feed ingredients that optimize performance and profits for their specific enterprise. Precise gene insertion through recombinant technology now enables one to up-regulate or down-regulate specific metabolic pathways or insert new pathways in plants resulting in enhanced protein quality and/or quantity, modified amino acid levels (i.e. high lysine maize), modified carbohydrate composition and content (i.e. starch, oil, cellulose, hemicellulose, pectin), reduced lignin, reduced indigestible oligosaccharide content, increased levels of oligofructans, enhanced vitamin content, and incorporation of natural compounds for improved feed efficiency and growth performance. High oleic acid soybean will provide high-energy concentrations. Grains with oligofructans may reduce the need for antibiotics. The transformation of active components found in botanicals into crops may offer a future substitute for antimicrobials.

Livestock Feed producers may have the ability to select specific grains or protein sources that were designed for optimizing performance and profitability of their particular livestock enterprise. It is reported that reported an eight percent increase in wool growth and seven percent increase in live weight gain in sheep fed modified lupin. Lupin was genetically modified to contain a sunflower gene into subterranean clover resulting in an accumulation of this protein representing 0.3 to 1.3% of the total leaf protein.

Future opportunities associated with biotechnology could lead to transfer of traits to meat, milk and or eggs which produce products such as anti-cancer agents, anti-oxidants, anti-aging agents, health enhancing agents, natural meat tenderizers, modified fatty acid products, and disease preventive agents.

Quality Enhanced and Safe Feed

Development of crops having low or no anti-nutritional factors will benefit animals and producers through improved digestibility and nutrient utilization.

Biotechnology is a valuable tool to plant scientists to reduce or eliminate undesirable components of feed ingredients such as erucic acid and glucosinolates in rapeseeds; trypsin inhibitor, lectin, raffinose and stachyose in soybean; tannin in sorghum; gloucosinolates in cotton; and protease inhibitors, phytohemagglutinins, and cyanogens and bloat compounds in legumes in a timely fashion. Barley, wheat, maize and soybeans contain beta 1, 4 glucan and xylan in their hulls that result in increased viscosity of intestinal contents resulting in decreased digestibility and reduced animal performance. In the future these crops may contain enzymes like beta glucanase, and xylanase to increase digestibility of fibrous ingredients contained in monogastric animal feeds.

Plants have considerable potential for Production of biopharmaceutical proteins and peptides including edible vaccines that can be stored and distributed as seeds, tubers, or fruits. In particular, the chicken interferon gene is being incorporated into maize so a particular protein can be produced in the seed as a vaccine to help combat avian influenza and help keep chickens healthy. The market (estimated $ 1.7 billion USD market for chicken vaccines) is high because farmers who raise the world's 18 billion chickens must be careful to protect against the outbreak of virulent avian influenza strains. The vaccines produced via this technology can be marketed either in edible form, made from parts of the fruit, vegetable or grain plant or in injectable form. Hepatitis B in humans and transmissible gastroenteritis virus (TGV) in swine are two diseases being targeted.

Livestock producers will continue to face issues regarding livestock waste management including waste disposal, excessive phosphorus and nitrogen excretion, methane production and odor. Increasing digestibility of nutrients in animal feeds will be a major factor in reducing livestock waste problems. In the future, forages will be bioengineered to contain more digestible fibre components and/or fibre digestive enzymes. The plant may contain lignin, celllulose and hemicellulose that are more easily degraded or cellulase, hemicellulase and lignase enzymes may be bioengineered into plant or rumen microbes that enhance the utilization of these energy sources. Low lignin corn is already here via plant breeding. The brown mid rib mutant gene was identified and bred into certain corn hybrids. This product contains about 40% less lignin than its parental variety. Enzymes could be genetically engineered to convert roughages to substrates now utilizable by monogastrics. Microbes (silage microbial inoculants) could be genetically modified to enhance the fermentation process of forages in silos. Feedstuffs will have more balanced amino acid patterns, more digestible carbohydrate components and optimum fatty acid profiles.

Mineral bio-availability could be improved to make more efficient use of the minerals in plans and reduce the contribution to the environment where leaching can occur in streams and lakes. Phosphorus is the main mineral of concern at this time. Monogastric animals do not have the necessary enzyme machinery to access the phosphorus so it ends up in animal waste. The waste is applied to the fields where leaching and run-off can occur. To prevent this problem phytase enzyme is added to poultry and swine feed to unlock this valuable mineral. Through biotechnology, plants could contain low phytate (high available phosphorus) or added phytase activity thus allowing for more efficient use of phosphorus with reduced supplementation of inorganic sources.

Bio-engineered crops may be developed resulting in reduced animal wastes, greenhouse gases and less offensive manure odors. It may be concluded that the Recombinant DNA technology is producing revolutionary changes in agriculture. With the adoption of biotechnology, food and feeds will be more abundant, more nutritious, more stable in storage, safer, and healthier than current foodstuffs. The fruits of agricultural biotechnology will also include an improved environment for humans and livestock producers as foods are produced using less pesticides and herbicides. Improved animal efficiency and feed digestibility will result in less animal wastes. Biotechnology must be considered as an important tool for sustaining agriculture to meet the demands of the world's rapidly increasing population for now and the future.

Soybean Meal Quality

Soybean meal is the most important source of dietary protein for swine and poultry in most of the world. It would be an uncommon poultry feed that did not contain 10% soybean meal, and some poultry feeds may contain as much as 35% of soybean meal. Soybean meal is also one of the highest quality protein sources with the least variability. Some variation, however, does occur, in both the nutrient quantity (analysis) and nutrient quality (Digestibility) between samples and sources of soybean meal. Variation may occur because of differences in varieties of soybeans, growing conditions, storage conditions, and processing variation. Soybean meal produced in the United States is usually from closely related varieties of soybeans, processed in large plants, with carefully controlled precessing conditions. At the other extreme, In some countries the varieties of soybeans grown are markedly different and processing plants may be small and poorly managed. Maintaining the quality of the soybean meal that are used in feed is essential because:

- With high quality soybean meal you get the most nutrients for your money
- High quality soybean meal supports better growth, better feed efficiency, more eggs production.
- Better performance generally results in lower cost per unit of production and greater profits to the industry.

The basic quality of soymeal is related to its protein content (the higher the better); fibre content (the lower the better); ash content (the lower the better); fat content (usually 12%, certainly less than 15%); urea and mycotoxin. High protein soybean meal (476.5 to 48.5% protein) is usually the most economic source of nutrients. The hulls are removed form low protein soybean meal (44 to 45% protein) to produce high protein soybean meal. The hulls are low in protein and low in digestibility. There are some feeds where the low protein soybean meal will fit nicely because a lower nutrient density is desirable. Following quality points maybe noted :

- Optimum heat processing for maximum available amino acid content;
- Soybeans contain anti-nutritional factors (Trypsin inhibitors, lectins and others). Trypsin inhibitors interfere with trypsin enzymes and reduce amino acid digestion,
- Soybean meal and FF-soy must be heated to destroy anti-nutritional factors;
- Heating improves the digestibility of ALL amino acids and improves performance; and
- Over-heating destroys amino acids and decreases amino acid digestibility (most critical for lysine and cystine).

The objective in soybean meal processing is to apply the optimum amount of heat to produce the most nutritious product. Insufficient heating, or under processing, will negatively impact amino acid digestibility because the anti-nutritional factors are not adequately destroyed. Excessive heating, or over processing, will negatively impact amino acid digestibility because a portion of the amino acids have either been destroyed or tied-up as indigestible bound compounds.

Table 4 shows that the digestibility of four amino acids was greatly increased as the result of autoclaving raw soybeans at 121 °C and 1 kg/cm^2 pressure for 0-18 minutes. There is a dramatic improvement in the digestibility of all of the amino acids with increasing heat treatment. It is important to understand that the improvement in amino acid digestibility that occurs as a result of the initial heat treatment of both soybeans and raw soybean meal occurs for all of the amino acids, not just some of the amino acids. This improvement in digestibility is the result of the destruction of anti-nutritional factors by the heat treatment.

Table 4. Effect of under-processing on amino acid digestibility of raw soybeans, %

Autoclave time, min	Lysine	Methionine	Cystine	Threonine
0	73	65	67	64
9	78	70	70	68
18	87	86	83	82

When too much heat is applied to soybeans or soybean meal there is a reduction in the concentration of some of the amino acids as well as a decrease in the digestibility of some of the amino acids. Table 5 shows the effect of overheating (over processing) on a commercial soybean meal. A commercial soybean meal was placed in an autoclave for upto 40 minutes. The result was a significant reduction in the analytical concentration of lysine and cystine as well as a reduction in the digestibility of lysine and cystine. There was little or no effect on methionine and threonine. Most of the other amino acids are also not affected by excessive heat treatment or over processing.

Methods of Monitoring Soybean Meal Quality

Under heated soybeans or soybean meals have reduced amino acid digestibility which reduces growth performance. Over heated soybean or soybean meals have some of the amino acids

Table 5. Effect of over-processing on amino acid concentrations and digestibility of commercial soybean meal

	Digestibility coefficient, %			
Autoclave time, min	**Lysine**	**Methionine**	**Cystine**	**Threonine**
0	91	82	86	84
20	78	69	86	86
40	69	62	83	80
	Analytical concentration, %			
Autoclave time, min	**Lysine**	**Methionine**	**Cystine**	**Threonine**
0	3.27	0.70	0.71	1.89
20	2.95	0.66	0.71	1.92
40	2.76	0.63	0.71	1.87
	Available concentration, %			
Autoclave time, min	**Lysine**	**Methionine**	**Cystine**	**Threonine**
0	2.98	0.57	0.61	1.59
20	2.30	0.46	0.61	1.65
40	1.90	0.39	0.59	1.50

Pearson et al., 1992.

destroyed and also a reduction in amino acid digestibility. It is important to be able to identify those products that have been heated sufficiently to destroy the anti-nutritional factors but not over heated. A brief descriptions of three different assays that are used to evaluate soybean meal quality are as under:

The Urease Assay

Soybeans and soybean meals contain urease, an enzyme that hydrolyzes urea to produce carbon dioxide and ammonia. The production of ammonia causes the pH of a solution to increase. Urease is not detrimental. However, urease is destroyed by heating. The destruction of urease is highly correlated with the destruction of trypsin inhibitors and other anti-nutritional factors. The urease assay is based on the pH increase form ammonia released urea by residual urease enzyme in a soybean meal. The primary purpose of the urease assay is to determine if soybean meal has been sufficiently heated to destroy most of the anti-nutritional factors. The optimum pH increase has generally been considered to be 0.20 to 0.05.

The KOH Protein Solubility Test

The solubility of soybean proteins decreases with heating. The KOH solubility test is based on the solubility of soybean proteins in a dilute solution of potassium hydroxide. The soybean protein decrease in solubility as the result of heat processing. The acceptable or satisfactory range for KOH values is generally considered to be 70 to 85%.

The Protein Dispersibility Index (PDI)

The Protein Dispersibility Index assay is also based on the solubility of soybean protein. For this test the solubility is in water. A sample of ground soybean meal is combined with water and blended in a blender for 10 minutes. The sample is then centrifuged or filtered and the soluble nitrogen measured by a standard method and converted to soluble crude protein.

Research generally indicates that a soybean meal with a PDI of 45% or lower is a adequately heat processed. All of these tests are relatively simple to conduct and can be performed in most feed quality control laboratories. All of these tests have application to soybeans, soybean meal and full fat soybeans. All of these tests are relevant to both poultry and swine.

This research shows that PDI gives a more consistent, predictable and linear response to the application of heat treatment. The PDI may be more useful in identifying soybean meal that has been processed just right. The PDI is also the simplest measurement tool of those that were evaluated.

Samples for Quality Evaluation

Every analysis starts with the sample. A representative sample is essential. Representative probe samples should be combined from each lot of material to be evaluated (a truck load or railcar). Each method of analysis requires a ground sample. Particle size does make a difference. A grinder or grinding method should be used that will produce a relatively uniform particle. Changing the moisture content will also change the crude protein, crude fat, crude fibre and ash determinations. Any comparisons between samples of soybeans, soybean meal or FF-Soy should be made using the same method of sample preparation, including grinding method. In summary it may be said that it is better to purchase and use the highest quality soybeans, soybean meal and FF-Soy that is available. The Protein Dispersibility Index (PDI) is the best method of evaluating these soybean ingredients for both under heating and overheating.

On a global scale soybeans are the primary vegetable protein source in animal feed. The use of soybean in animal feed has grown steadily, reaching

125 million metric tons in 2001. Despite some regional and seasonal fluctuations, the increase in soy protein use in animal diets accelerated over the past 3 years averaging 5% per year. Consequently, soybeans account for most of the increase in world production of protein meals.

The increased use of vegetable protein meals in animal feeds has been especially pronounced in W Europe where legal restrictions have virtually eliminated the use of animal proteins form feeds. Countries outside the EEC, especially in Central and East Europe, have implemented similar legislation while others are adapting feed formulations to increase the use of plant proteins. With the increased use of vegetable protein sources, the importance of protein quality and the methods used to control quality have increased proportionally. This is evident in the case of young-animal nutrition where protein quality is a major factor in early growth and performance, but it is equally important in the feeding of older animals where feed cost and feed conversion are major economic parameters.

The selection of soybean meal for specific purpose and the application of different quality criteria as a selection tools have thus become increasingly important steps in diet formulation and feed production. The correct and detailed description of the soybean product and the strict application of quality criteria are therefore, essential pre-requisites to maximizing the efficient use of soy products under modern least cost formulation settings.

Soy-meal Differentiation

Many different types of soy products are marketed and used in the animal feed industry. Classification of the various types of soybean products depends largely on the specific association or organization dealing with the product. Traders and raw material suppliers will generally prefer a limited number of classifications. For the classical soybean meals typical specifications have therefore been established and these are often used as reference values in trading or classification. Table 6 gives the specifications for solvent extracted soybean meal and for dehulled soybean meal often referred to as Lo-and Hi-Pro soybean meal. These minimum quality standards allow the establishment of trading rules and adjustments for quality in case of disputes involving nutrient concentrations. They are often completed with additional quality recommendations related to colour, odour, texture, contaminants, ash levels and protein quality characteristics.

Table 6. Specifications for solvent extracted and dehulled soybean meal, %

	Min/Max	Solvent extracted SBM	Dehulled Soybean meal
Moisture	Max	12.0	12.0
Protein	Min	44.0	47.5-49.0
Fat	Min	0.5	0.50
Crude fibre	Max	7.0	3.3-3.5
Anticacking agent	Max	0.5	0.5

Specifications such as these may vary among organizations and regions and although they provide a solid basis for commercial transactions they are inadequate and of limited value when formulating diets. Formulators, seeking to maximize the value of a raw material and minimize formula cost often prefer to separate raw materials beyond trader classifications. Common criteria of segregation are origin, supplier and especially quality characteristics often combined with chemical composition. This hold true for all raw materials including soybean meal. Under these conditions, storage capacity and the number of silos or bins at the production facility become the major limiting factor. An additional, important criterion for the separation of soy meals is the use of the ingredient in specific diets. Fro instance, dehulled SBM will be used preferentially in poultry diets and solvent extracted SBM in swine diets.

Inclusion of the different protein sources in animal diets depends much on the type of diets that needs to be formulated. In the case of full fat soya the level of unsaturated fatty acids is a major limiting factor for inclusion. For soybean meal incorporation in diets is mainly determined by the nutrient specifications of the diet and price.

Nutritional Aspects

Compared to other sources of plant protein, soy products stand out by their large diversity and their advantageous levels of many critical nutrients. This is especially true where it concerns the use in concentrate feeds. At the same protein levels, fibre concentrations are generally low and metabolizable energy relatively high. However, since they are predominantly used as protein sources, the most important criteria is the level of digestible amino acids. The lysine concentration of all soy products is high and the concentrations of other amino acids relative to lysine are fairly constant across soy products indicating that the specific treatments to which the soybeans were subjected did not affect the amino acid profile.

The special elaborations and the heat treatments needed to permit inclusion of soybean products in complex diets or in normal diets at the above indicated maximum levels in normal performance diets are related to the overall digestibility of bio-availability of nutrients. Modern diets are largely formulated on the basis of digestible amino acids. In fast growing, intensively raised animals such as hogs this is normally expressed in terms of ideal digestibility. Expression of nutrients on this basis allows formulation and production of diets more closely related to real physiological requirements. Digestibility of protein lysine agree closely and this holds true for all essential amino acids. Values are generally higher for the soybean meals especially in the case of lysine. This combined with the higher levels of protein and lysine results in significantly higher concentrations of digestible protein and lysine. Soybean meals are thus clearly a more concentrated source of digestible amino acids which is especially advantages when formulating high performance diets. Digestibility and quality of soybean meal depends to a large degree on the processing applied to soybeans. Under practical conditions, these are differences of significant importance in diet. formulation and they can make the difference between in high performing diet and a poor or medium performing diet that is otherwise correctly formulated.

Quality Variation and Control Methods

While soybeans and soybean meals are one of the most homogenous ingredients used in feed manufacturing quality is variable and a large number of factors will affect quality especially in terms of protein and amino acid digestibility. Among them variety, origin and growing conditions are important. A number of comparisons have been carried out among countries all showing significant differences in nutrient composition and general protein quality. From the review of the available tests, it is evident that PDI with its good accuracy and low skill requirements is probably the most suited to effectively and rapidly measure protein quality in an industrial setting. It should be pointed out however, that like the urease index, PDI is a test that is most suited to detect under-heating while KOH solubility of protein is more suited to detect overheating. A combination of both tests is recommended as the most effective quality control measure.

Commercial specifications of soybean products provide limited information on soy products for diet formulation. Detailed nutritional specifications are required. Those specifications have to be on a species-specific basis and should provide measures of bio-availability or digestibility. Table values are inadequate since measures of bio-availability will vary with origin, variety and production condition. However, within a specific soybean product major variation in digestibility can and will occur with degree and quality of processing. Tests have been developed to assess variation in quality due to heat processing. Those tests need to be applied in order to assure the consistent formulation and production of high performing diets.

Feed Uses of Soybean Meal

Soybean meal, because of its high content of nutritionally valuable proteins, became an indispensable ingredient for animal feed. It can be said that the economic Production of animals is impossible without properly processed soybean meal. One of the major processing requirements of a good quality soybean meal is proper heat treatment: on one hand it is necessary to deactivate all anti-nutritional factors which are naturally present in the soybean but on the other hand, over-heating decreases the digestibility of the protein. This heat treatment process is normally called toasting. The various feed uses of soybean meal are given below.

Aquafeed

Industrial fishing is depleting the world's reserves of fish in the seas. Industrial production of fish is becoming very quickly an absolute necessity to satisfy the needs of an ever-growing population. Soybean proteins play an important role in the production of fish feed. The production of feed products for aquaculture is a specialized area where a lot of know-how is needed. Indeed, fish require a feed product adapted to the species and the age. In addition, for some fish the feed should sink whereas for other the feed needs to remain in suspension in the water.

Bee Feeds

Soybean meal can be added to the mixture of pollen and honey that is provided to the drones for feeding the bee larvae.

Calf milk replacers

Both soy proteins and lecithin are common ingredients for the production of calf milk replacers.

Fox and mink Feeds

Fur bearing animals require special attention. Soybean meal used in their rations should be properly heat-treated and over-cooking, as well as under-cooking, needs to be avoided as the delicate animals are extremely sensitive to anti-nutritional factors. High-protein soybean meal and also full fat soybean meal might represent a better feed ingredient for these animals. The linoleic and linolenic acids present in full fat soybean meal may even improve the quality of the fur.

Livestock feeds

Soybean meal represents one of the major feed ingredients for cattle, especially in winter time. The protein content of the soybean meal (mostly 44% in livestock feed) make it an indispensable ingredient and without soybean meal, economic Production seems to be very impossible.

Poultry feeds

An economic production of poultry meat and eggs is impossible without the use of properly toasted, high protein level (mostly around 48%) soybean meal. The presence of soybean meal in the rations of poultry allow, optimum growth of the animals. Lest cost formulation to fit the nutritional needs of the animal at its particular stage of the life cycle will always insist one the presence of soy protein in the feed.

Protein concentrates

After leaching out the soluble carbohydrates form the soybean white flakes, a protein content of minimum 65% is obtained. This product is offered as a flour or in texturized form and became a popular ingredient in processed meat products. It is also used fin pet foods, basically because the leaching out of the soluble carbohydrates, overcomes the flatulence problems.

Pet foods

Pet foods prefer to use scy protein concentrates because the absence of the water-soluble carbohydrates not only increased the protein content but more so because the flatulence problem is under control.

Dairy feed

Ruminants need in their rations a higher content of fibre. Soybean hulls are extensively and practically exclusively sued for roughage in feedings livestock.

The Feed Industry is charged with the responsibility of producing a high quality, consistent feed from a number of variable ingredients and doing it economically, quickly and efficiently. It must put in front of the target animal a feed of a quality and the conditions specified by the nutritionist. In fact, the formulation of a feed by the nutritionist is only the beginning of a process that ends when the feed is consumed.

There are two main reasons to implement a Quality Assurance Programme. The most important is feed quality, but we need also to mention Government Regulations. There are several reasons why we need regulations in an industry such as ours. The customer must be confident that the feed he buys is safe, is free of toxic elements, is properly medicated and generally meets nutritional guidelines. Regulations differ from country to country.

Quality Assurance in the Feed Mill

Quality assurance is a program that is composed of many parts, all of which are important. In general, the program consists of three parts;

- Setting of nutritional stands
  in ingredients
  in finished feeds
- Technology of Manufacturing
- Method s of Quality Control

Nutritional standards

The Nutritionist is central to the program in that he or she sets the nutritional standards for incoming ingredients as well as finished feeds and ensures that they are met. Also, the nutritionist ensures that the manufacturing process is set up in such a way that it can produce and deliver a feed that is uniform, palatable and efficacious for the class of animal that it is being fed to. As part of this role, the nutritionist is involved in ;

- Designing the Quality Control Programme
- Training mill and laboratory personnel
- Updating the Quality Assurance/Control programme
- Overseeing the Quality assurance
- Control Reports
- Keeping feed formulae up to date
- Control risks associated with novel feed ingredients.

Technology and feed manufacturing

The process of feed manufacturing produces several critical Control Points at which the Qaulity Assurance Programme takes on special importance and at which we have to pay special attention. These areas of special interest are as follows :

- Receiving
- Raw product storage
- Raw product processing
- Weighting and mixing
- Pelleting
- Delivery of feeds
- Cross contamination

Receiving

The receiving area is the last line of defense in preventing poor quality ingredients form entering the mill. Once those products are in a bin, we lose track of them and cannot differentiate the poor ingredients form the good ones. The Receiver must know what quality of ingredient has been ordered, have enough information from the QA program to be able to recognize that quality of product, and must have the authority to reject an of-spec product. Samples of all incoming ingredients should be taken for analysis, if required, and for retention in case of a dispute over the ingredient. A recording system should be set up for each ingredient with at least the following information :

- Receiving date
- Manufacturers name
- Shippers name
- Product name
- Number of packages
- Package size
- Lot Number
- Receivers signature
- Comments with regard to quality

Combining efficient receiving procedures with standards of quality set by the Nutritionist should minimize problems with poor quality ingredients.

Raw products storage :

Storage of bulk ingredients is fairly straight forward. Basically, one has to ensure that the product gets into the correct bin on receiving and is removed from the correct bin during use. Bagged ingredients are a little more difficult because extra attention must be paid to ensure that bags or containers are properly labeled. It is very easy to tear off a label when opening a bag or switch lids of containers, and if only the lid is labeled, products can be confused. This is particularly important in the case of medications.

Raw feed processing

Processing of raw materials includes such things as grinding, rolling and flaking. Attention paid to these areas of the manufacturing process will result in an attractive, uniform feed but, more importantly it will lead to better acceptance by the animal, less ingredient separation and will improve the nutritive value of the feed.

To be consistent in processing requires an equipment maintenance program to ensure that the equipment is capable of performing to its potential. Equipment such as roller mills and hammer mills need to be adjusted and maintained continually to ensure a consistent product.

Weighing and Mixing

The weighing and mixing process is the heart of feed manufacturing. It is absolutely essential that ingredients going into a feed by accurately weighed and then be mixed into a homogenous feed. The important aspect of weighing scales to remember is that their accuracy decreases with increasing size of the scale. As a general rule, a scale is accurate to no more than 0.1% of its total capacity. So scales need to be sized according tot heir use in the mill. Feed uniformity is also important because no matter how good the formula, it is not possible to get maximum results if feed is poorly mixed. Poor mixing results in animals getting less nutrients than they need on some days and more than they need on other days. Either way, performance is affected. It is essential that mixers be checked periodically to ensure that they are producing a uniform product. There are several ways of checking mixers for uniformity and they all involve analyzing the feed for a nutrient or additive that is added at low levels and comes primarily form one source.

Pelleting

The pelleting process affects the nutritional as well as the handling characteristics of the feed. The quality of the pellet can also be easily assessed by the customer who often equates a poor pellet with a poor quality feed. So it is important to produce the best pellet possible. The proper type and amount of steam, amount of fat addition and rate of product flow that is required for the production of good pellets is something that must be derived for each type of feed and each pellet mill. Wear on dies and rolls can, of course, greatly affect pellet quality, as can the general standards of maintenance of the equipment.

Delivery of feeds

Once a high quality feed has been produced, we do not want to reduce the quality by poor handling of the feed during delivery. Even more important, we do not want to put the feed in the wrong bin. So it is important that truck bins and the feed handling equipment on the trucks be properly maintained and operated in such a way as to minimize a reduction in pellet quality and to reduce separation of feed ingredients in the truck. It is also important that feed bins on the farm be clearly marked so no mistakes are made when delivering feed to the farm.

Cross-contamination

Cross contamination of drugs in feeds is an ongoing problem that the feed industry must solve. The actual contamination, when a drug gets into a feed unintentionally, can occur almost any where in the system, but the main areas are in the mixer, the elevators and the load-out bins. Worn paddles or ribbons on a mixer have been shown to be one reason for failure, while poor electrical grounding can lead to a build up of electrostatic drugs in the mixer. Also, improperly cleaned load-out bins can lead to mixing of feeds. While a small amount of a drug in the wrong feed may not seem like much, drugs commonly used for some species can be harmful to others. There are two main methods of attempting to reduce cross-contamination of medicated feeds.

Flushing

This is a technique whereby an ingredient such as a ground grain is run through the system after a batch of medicated feed is made. This ingredient will pick up much of the contamination in the system and then must be bagged up and sued later when feed containing the same medication is made. The disadvantage of this method is the handling of the bagged ingredient and keeping track of the various products.

Sequencing

This technique involves the production of feeds containing the same medication at the same time. This reduces the number of times contamination may occur. A run of the medicated feed would then be followed by a run of a type of feed that would not be adversely affected by a low level of a contamination; for example a grower feed being fed to the same species. This technique would do away with the need for flushing material unless the proper feed for sequencing was not being made.

Methods of quality control

A quality control program is a procedure that is set up to evaluate the effectiveness of the Quality assurance program. QC usually entails analytical work to assess the incoming ingredients as well as the finished feed.

Incoming Ingredients

In order to properly and accurately use many of the feed ingredients available to the feed industry, we must have an accurate assessment of the nutrient content of the ingredient and the variation between samples of that ingredient. Only by having that information can a nutritionist accurately and economically formulate feeds. Each of the raw materials to be purchased must have definite purchasing specifications set up by the nutritionist so that the purchasing agent will not have any difficulties in buying exactly what is required. A standard internal purchasing specification indicating analytical tolerances should be used. These specifications should include :

- Definition of product to be purchased
- Normal physical characteristics (Colour, odour, texture)
- Chemical analyses (minimums, maximums, and tolerances)
- Particle size (mesh sizes)

Finished feed

The final step in the Quality Assurance programme is the evaluation of the finished feed. If the program is working properly this analysis will confirm that fact. If it is not working properly then it will show up on the analysis. Samples should be taken on every load of feed that is delivered. These samples should be kept for at least three months. Regular analyses of these feeds should be conducted to evaluate the quality of feeds produced.

Feed Safety

It is estimated that by 2050 the number of people on earth will be about nine billion and Asia will have a major contribution in this growth. Therefore, the challenges faced today by meat, milk and egg producers will become hot topics. To meet the demand, they must be able to produce large volumes of food. At the same time, the consumer expects low prices and a good quality. Food safety is also a clear expectation. It includes different issues such as food hygiene; healthy animals, low level of antibiotics and other medicinal products, and reduced impact of animal production on environment. The major cost of production in livestock husbandry is the feed. To ensure that the farm generates profits, feed must provide an adequate supply of nutrients at the lowest cost. A good knowledge of the different raw materials and their nutritional characteristics allows optimizing the feed formula. The digestibility of nutrients can be enhanced by the addition of different enzymes. Feed additives such as mould inhibitors and antioxidants can be used to preserve the feed quality. Feed must also be free from pathogens and control the flora of the digestive tract to optimize digestion and prevent enteric diseases. Also, feed must help in maintaining an efficient immune system to prevent loss of production.

Antioxidants are very important under Asian conditions. Oxidation will damage the quality and reduce the digestibility of fats and oils used to bring energy in the diet. This naturally-occurring process is promoted by climates with high temperature and humidify. Antioxidants protect the feed from oxidation processes and ensure a good conservation of raw materials or feed. They prevent loss of performance resulting from the consumption of feeds containing oxidized fats. They also limit the degradation of vitamins in feed and avoid a number of diseases related with vitamin deficiencies. Ultimately, by preventing off-odours in meat or reduced pigmentation in eggs and meat, they avoid loss of market acceptance, resulting in higher revenue for the producer. Best results are obtained when using combinations of antioxidants, because they can ensure a synergistic effect. This means that the effectiveness of the combination is much higher than when using each ingredient separately. Homogeneity of the combination is crucial in that case. The formation of moulds is another major concern in hot and humid climates. This will lead to a loss in the nutritive value and palatability of the feed. When animals are fed with mould-contaminated diets, feed intake is reduced and feed waste increases, resulting in poor performances and an increased feed conversion ratio. The growth is slower and the time to market is longer, which significantly increases costs for the producer. Moulds can also produce mycotoxins and flavour. When dealing with moulds, prevention is the keyword. Different mould inhibitors are available today. Depending on the objective to achieve, they can be in liquid or powder forms. A correct selection of the ingredients ensures that the formula effectively protects feeds from the formation of moulds and their toxins, but also yeasts and a variety of bacteria.

Mycotoxins are toxins produced by fungi before or after harvest, or during storage. Most of the mycotoxins are very stable and can resist to changes in temperatures or to storage and processing conditions. Therefore, they can be found in feeds or foods produced with contaminated raw materials. Some mycotoxins ingested by animals, or their derivatives, can even be transferred to meat, milk or eggs and could contaminate the people eating these products. In Asia, due to unfavourable conditions, the growth of fungi is very difficult to control and many commodities or feeds are contaminated. In general, mycotoxins lower the resistance to diseases, increase the sensitivity of animals to stress and affect organs such as liver (e.g. aflatoxin), kidney (e.g. ochratoxins), lungs or even brain (e.g. fumonisins). In the case of zearalenone, they can also seriously affect reproductive performance of animals. When mycotoxins are combined, as it is usually the case, they will have a much more damaging effect than if they are taken separately. Different methods have been tested to get rid of mycotoxins. When possible, prevention is the most effective strategy. But once the raw material is contaminated, it is very difficult to remove the mycotoxin. Mycotoxin binders have been developed to bring an answer. Unfortunately, not all mycotoxins can be removed with these products which are efficient against polar mycotoxins. The solution lies in the combination of different strategies. For example, products consisting of toxin-degrading enzymes, toxin-binding natural biopolymers, and toxin binders of mineral origin are already efficiently used in various countries. Additionally some natural extracts and vitamins can be used to reduce the stress-related symptoms.

Enzymes are among the solutions for the future. Enzymes are biocatalysts in which means that they speed up chemical reactions. The feed is better digested, reducing the cost of production. Enzymes also allow the use of greater quantities of cheaper raw materials in the feed formula, which improves the profitability of the farm. They also contribute to food safety. For example, when improving digestion they contribute to reducing the wastes in the environment. This is particularly true in the case of phytase, which improves digestibility of phosphorous in diets. Another successful use of enzymes is the case of Non-Starch Polysaccharides degrading enzymes. These formulas are widely used in markets such as Europe where cereals (wheat, barley) are the major energy source. They will help in reducing not only the volume, but also the stickiness of manure. This is very important in poultry production, to reduce the problems associated with contaminated eggs or carcasses. Enzymes specially designed to increase the digestibility of corn-soya diets are now developed as answer to Asia conditions. They certainly have a bright future because they can significantly improve the digestibility of the diet, reducing cost but also pollution.

Achieving food safety must be the target for producers. Consumers are more and more concerned about health when it comes to food. This subject becomes crucial when it comes to international trade, and in many countries requirements become more and more strict. When looking for successful solutions to achieve food safety without compromising on the economical aspects, the producer should consider using different feed additives based on their efficacy, quality, and reliability of the supplier.

Soybean in Dairy and Fish Feed

Soybeans and its products can be used in animals feeds. Soybean hulls are a source of rumen digestible fibre and low in starch. Because of this characteristic, hulls can replace some of the feed grains such as corn and barley ind airy cattle diets. Although high in fibre it is not effective in stimulating rumination and would not substitute for forage. No processing of hulls is needed before feeding.

Soybean meal is a major source of plant protein fed to dairy cattle. It is usually economical and blends well with other feeds. It can be purchased in large quantities at lower prices. Also no further processing is needed before feeding. Soybean meal that is produced as a result of solvent extraction of oil is relatively high in rumen degradable protein, a weakness, but if oil is mechanically extracted (expeller) or specially treated it has less degradable protein. Soybean meal produced by mechanical extraction is usually sold at a higher price. Also soybean meal is a poor source of methionine, a limiting amino acid. Whole soybeans or full fat soy can be fed cracked or rolled to dairy cattle. They contain fat, a concentrated energy source. Raw, unheated soybeans can become rancid because this fat is unstable. Also, raw soybeans have a high amount of rumen degradable protein. Roasting or heating makes the fat more stable, reduces rumen degradability, and increases palatability. Therefore, heated soybeans are preferred over raw soybeans for dairy cattle. Roasting and extrusion are the two most common methods of heating.

Rumen protein undegradibiltiy (% of total protein) of solvent extracted soybean meal is estimated to be 33% and 68% for expeller extracted soybean meal, 29% for raw soybeans, and 38% for roasted. Also there are maximums for inclusion in dairy cattle rations. Soybean meal has no maximum and can be fed to meet protein requirements. Soybean hulls should be limited to 20-0% in concentrate mixes and 2-3 kg/cow/day. Raw soybeans can be fed in up to 2 kg/cow/day and roasted soybeans should be limited to 3 kg. Inclusion of roasted soybeans at 20-30% of the concentrate is acceptable (Table 7).

Table 7 Use of soybean and its products in animal feed

Maximum	% of supplement	kg/cow/day
Soybean hulls	20-30	2-3
Soybean meal	as needed	as needed
Soybeans, full fat heated	20-30	2-3
Soybeans, full fat unheated	not recommended	1-2

There are advantages in feeding whole soybeans versus whole cottonseeds. Soybeans contain 43% protein versus 23% for cotton seeds. Roasted or heated soybeans have more rumen undegradable protein. Also roasted soybeans are very palatable and can be mixed with other concentrate ingredients before feeding. Linted cottonseeds can't be mixed into concentrate formulas very easily. Rations that substitute some soybean hulls for corn or barley will have reduced levels of non-fibre carbohydrates such as starch. This can be a benefit because excess non-fibre carbohydrates can cause an acid rumen, laminitis, and erratic feed intakes, especially on high corn silage rations. However, low levels of non-fibre carbohydrates should be avoided because microbial yields form the rumen will be reduced. The recommended level of non-fibre carbohydrates in rations for dairy cows is 36-40% of the dry matter. Also, high protein degradabilities in the rumen of cattle results in wastage of excess nitrogen. Rations should contain 60–65% of the protein as rumen degradable and 35-40% undegradable protein under most circumstances.

The far-off dry cow is defined as being more than 3 weeks before calving and the close-up dry cow is within the final 3 weeks of pregnancy. Notice that we recommend that the nutrition program be increased for the close-up dry cow in order to make a smooth transition form being non-lactating to lactating. After calving the nutrient requirements increase further to reflect milk production. Mid to late lactation animals will need fewer nutrients than early lactation because they produce less milk. In summary, soybean hulls, soybean meals, and full fat soybeans can all be included in the diet of a lactating dairy cow as long as maximum are not exceeded. Soybean products are generally economical sources of high quality protein, energy, and digestible fibre. Since methionine is limited in soy products some consideration should be made for higher milk producing animals. Use of other feeds high in methionine such as fish meal and special treated soybean meal is possible. Rumen protected methionine is also commercially available but is an expensive feed additive and should be used only selectively.

The use of feeds in aquaculture systems has increased production and profits considerably. However, fish feeds are expensive and can amount to over two-thirds of the variable co to a fish culture operation. Fish feeds usually contain 25% to 50% protein. To obtain these protein levels in fish feeds, protein feedstuffs are utilized extensively and comprise 50% to 75% of the feed formula. Fish feeds were developed on the concept of fish eat fish, therefore fish meal is a major and traditional component of fish feeds. But limited world supplies and the high price of ish meal have forced aquatic nutritionists to consider alternative sources of protein. Plant protein feedstuffs are generally cheaper than animal protein feedstuffs. Of these plant protein feedstuffs, soybean and soybean meal are increasingly being utilized in fish feeds due to their nutritional quality, lower cost, and availability.

The quality standard specifications for soy products is given in Table 8.

Table 8. Quality standard specifications of soy products

Nutrients	**soybean cake, %**	**Soybean meal, %**	**Dehulled soybean meal, %**
Protein, minimum	41.0	44.0	43.0
Fat, minimum	3.5	0.5	0.5
fire, maximum	6.5	7.0	3.5
Moisture, maximum	12.0	12.0	12.0
Urease activity	0.05-0.20	0.05-0.20	0.05-0.20

(Range L increase in pH)

Fish feeds are the highest variable cost of a fish culture system. The utilization of cost effective feeds would improve profitability of a fish farm. Commercial fish feeds have been traditionally based on fish meal. Soybean meal is a consistently available, cost effective, and palatable to most species of fish. It has been demonstrated that soybean meal will replace a considerable amount of fish meal with no loss in production and fish performance. Energy, phosphorus, and methionine levels must be considered when utilizing high levels of soybean meal in fish feeds. Soybean meal is not nutritionally better than fish meal, it is more cost effective.

Changing Scenario in Livestock Feed Industry

Livestock production is being practiced in India since ages and a vast knowledge is available in ancient Indian scriptures about the feeding, breeding and management of animals including those of cattle, horses, elephants, etc. however, we now follow modern science in India from 1964 with the establishment of a cattle feed plant in Gujarat. Since then the feed processing is spreading in all parts of the country and contributed in White Revolution, which made India proud by becoming the largest milk producer in the world.

India has large livestock population with low productivity, which is mainly attributed to the poor quality of available feedstuffs that too in insufficient quantity. This is due to the tropical climate, poor purchasing power of farmers and lack of scientific temperament among livestock owners. As per an estimate, the total availability of DM is about 600 million tonnes and because of poor nutrient content of available feeds, protein and energy deficiency has been reported to the tune of 60 and 40%, respectively. The available feed includes corp residues, agro-industrial by-products and other feed ingredient unfit for human consumption. Usually quantity of grains and oil cakes, vital sources of energy and protein, in the animal ration is meager. Therefore, feed manufacturing technologies being used in developed countries can not be employed as such and need indigenization improvisation and blending with our traditional knowledge to make the optimum benefit. India has large infrastructure, expertise and experience for manufacturing and marketing of compound cattle feed clubbed with scientific know how. Therefore, we have to make rapid strides in manufacturing of compound cattle feed to make available nutritionally balanced feeds to our livestock so as to provide the nutritional security to its population. This can be achieved by incorporating the modern knowledge about feed processing to make the best use of so called poor nutritional quality feeds. This approach may help in enhancing the productivity of our livestock by greater availability of good quality feeds so that animal products can match international standards.

International Feed Code

Codex Alimentarious Commission (CAD) was established in 1963 by the United Nations with the objective to develop international food safety standards to protect human health, while facilitating trade in food, feed and agricultural products and resolving international disputes among member countries. Recently, CAC ad hoc international task force on animal feeding met in Copenhagen to finalize the international code of animal feeds with an objective of safe feed for safe food. Delegates including those from India attempted to resolve the pending issues pertaining to the labeling of genetically modified organisms (GMO), feed additives, traceability of ingredients and feed products in a bid for quality assurance and risk analysis. In the aforesaid meeting, the feed additives definition was modified slightly to accommodate the views of Indian delegates as well as those form Japan and Argentina to include the herbal preparations for allowing a broader sense of regulatory discretion. The GMO labeling of feed was modified to 'modern biotechnology to pattern in after the CAD biotechnology task force language as the earlier provision of labeling was difficult to implement in countries like India where GM cotton was recently introduced and the byproduct like cottonseed are being processed in ginning mills and oil extraction industry along with the byproduct of traditional varieties of cotton crop. The language of cod about traceability was adopted from the Codex Committee on General Principles (CCGP) definition with minor details concerning to feed ingredients and feed manufacturing because the traceability in countries as large as India having diverse mode of Production and transportation from procurement of ingredients to consumption of different type of feeds is difficult, if not impossible. In spite of modification about the code on traceability, country has to gear up its system to follow the international code. Being a member of WTO, India has to enforce Codex code of practice in developing domestic feed and food safety standards. Indian government is fully aware about its responsibility and she has brought the feed manufacturers and the scientists to a common platform to facilitate and formulate the strategy for the enforcement of code. Efforts are on to fill the gaps in information about Indian feedstuffs such as levels of incriminating factors (pesticide residues, mycotoxins, heavy metals etc. for this purpose, National Dairy Development Board, Compounded Feed Manufacturers Association of India and scientific organization such as National Dairy Research Institute are working together relentlessly.

Codex code outlines good manufacturing practices and where applicable hazard analysis and critical control points (HACCP) principles to control the hazards to the reasonably achievable extent in a bid for the safety of food producing animals. HACCP approach is being applied at NDRI and it should be applied in larger perspective to control the hazards at least on 'hot spots' after their identification. In addition, code of GMP may be popularized so that feed manufacturers can understand and practice it at their plants. In the begin inning these may appear bit difficult to adopt, but it will be beneficial for the country in the long run.

The code has been developed to ensure the safety of food for human consumption through adherence to food feeding practice a the farm level and GMP during the procurement, handling, storage, processing and distribution of animal feed and feed ingredients for food producing animals. It distinctly mentions that animal products, which could be a source of Bovine spongiform Encephalopathy (BSE), should not be used for feeding directly to or feed manufacturing for ruminants. Cod also elaborates actions to be taken in order to prevent and reduce the heavy metals including that of lead contamination in foods from fam till consumers.

The finally approved code will be used by 130 member countries as a minimal standard for safe use of feed and feed ingredients for consumer safety and fair trade practice. To enforce the cod e in India, w e have to prepare ourselves with new infrastructure for improved machinery, quality control equipments and storage facility as the days of grinding and mixing of ingredients for manufacturing the feed are gone and there is realization among Indian manufacturers about these aspects, which is collaborated by the recent survey by Feed International. Besides the manufacturing of cattle and poultry feed, the demand for pet foods, fish feed and other specialized feeds will grow in Indian market and we have to update our feed manufacturing system to cater the need.

Post-Pellet Liquid Application (PPCA) System

The objective of the feed manufacturing is not only to manufacture the feed but also to provide the nutrition. As the manufacturers are increasingly realizing about their objective in clear terms, there are in search of technologies to achieve their goal. PPLA technology has been perfected during last 5 years, which demonstrated how feed manufacturing could upgrade to a fat plus enzyme application system at a reasonable cost and improve process control accuracy. Ingredients that are resistant to heat and need heat proofing are expensive for lower cost alternative s and these resistant products are liquids. Using PPLA technology to apply heat sensitive ingredients (Enzymes such as phytase) allows the feed manufacturers to use the lowest cost ingredient possible. There are several application technologies for PPLA systems, including centrifugal disk, pressurized nozzle, air assisted atomization. All of these application methods will apply the micro ingredients. Choice of technology is determined by type of liquid and or micro powder ingredient (Aicher, 2004).

Herbal/ Botanical Additives to Feed

Herbal or botanical additives are being used in human diets in India and other several countries since ages as a traditional mean to keep the human fit and healthy. Such botanical ingredients or their extracts have also many applications in animals and these are being used for this purpose extensively through out the length and breadth of the country in spite of the fact that these are not considered strictly nutritional feed ingredients or pre-mixtures. The documentation of knowledge about such herbal products is limited; however, their potential as feed additive has been recognized at International level. Therefore, such additives have been taken into consideration while defining the additives in animal diet by the International community while formulating the Codex code. Newly established European Food safety Authority (EFSA) is going to put in place a new range of feed additives regulation, which will include most botanicals or plant extracts and plant essential oils. The main functions of botanicals cover :

- Pathogen control or resistance including antimicrobial and antifungal activity.
- Toxin tolerance, including mycotoxins tolerance and liver activity

- Digestion aid, including stimulation of endogenous enzyme activity and N_2 absorption as well as beneficial physiological and morphological effects
- Pollution control, including manure odour and ammonia control and N_2 binding activity
- Antioxidant activity as in control of metabolic auto oxidative stress caused by blood borne fee radicals.
- Other effects such as stimulation or modification of immune response, increased feed intake, animal product pigmentation etc.
- Enhanced animal production in terms of growth and milk.

Manipulation of Minerals

Role of minerals and vitamins in animal production is clear in their maintenance and production. Certain trace elements besides playing their useful role in metabolism, also create deficiency or toxicity conditions due to interaction with other minerals and/or nutrients and both the conditions affect the animal productivity adversely. Mostly, minerals are blended in ionizable form in mineral mixture and their availability to animals according to the requirements remains uncertain. Ceratin reports suggests that manipulation of minerals in the prechelated form in the mineral mixture improves their availability, therefore, chelated minerals supplements are being manufactured in developed countries. Minerals in chelated form are better utilized even at low minerals concentration in animals including swine, poultry and diary cows. The chelating ligands most commonly used are EDTA and amino acids which functions as mineral sequestering agent depending up on their stability constant in different environments in the gut. In such situations, the chelated mineral remains in soluble form but do not permit the reaction of mineral with other components or chemical substance to form insoluble salt. It carries the mineral ions at the absorption site and makes them more absorbable at suitable pH.

High yielding cows are normally unable to maintain the adequate reserves of certain minerals, which are necessary to sustain the milk Production. Therefore, the deficiency or toxicity of such minerals due to their interactions with other minerals and vitamins reduce milk Production. The field trials showed better performance of animals in terms o milk Production, butter fat production and reproduction. Manipulation of minerals in the form of amino acid chelate in the ration of high yielding lactating cows have positively significant effect. A technology has been developed to prepare Cu, Mn and Zn chelates using milk casein as an organic legend and efforts are on to find the alternative source of amino acids. We hope that chelated minerals will be available in the country by employing the indigenous technology.

Complete Feed Block

As stated earlier, crop residues are the integral part of livestock ration and the situation is not going to change in near future. If it is so inevitable that we have to tackle the problem by indigenous way. Complete feed system is the system by which we can provide the nutritionally balanced feed to the cows and buffaloes and this approach offers an avenue to incorporate the scientific know how in feeds such as incorporation of deficient minerals in ionic or chelated form, feed additives, protected protein and fat etc. which a common farmer can't do himself. By manufacturing such nutritionally balanced complete feeds Codex code for animal feeds can be implemented in true spirit smoothly. This system reduces the wastage and provides the opportunity of optimum rumen fermentation environment. A beginning has already been made and the concept is spreading rapidly with an encouraging response form feed industry and farmers. Natural calamities such as recurring floods, drought in one par t or other necessitated the bulk manufacturing of such feed and it is proposed to Government of India to create Fee d Banks using the concept of complete feed to tackle such calamities. Government has been sensitized on this development and committees have been constituted to finalize the specifications and manufacturing practices as well as quality control of complete feeds.

Feed Processing equipment

Various machinery and equipments are required from the point of feed ingredients procurement to the point of consumption of finished feed. These equipments are being manufactured indigenously and country turned from importer of such equipments to exporter. However, the Indian feed industry needs better machinery and equipments to cater the needs of consumers as being felt in other countries. However, capacity of feed manufacturers depends upon the location of plant and surrounding consumers of feed. As in past, significant expenditure will be in developed countries on key capital equipment and production line components but recent survey of Feed International indicates that there will be a phenomenal growth on expenditure in Asia countries as well on complete production lines, storage and conveying equipments. As per the survey, about 40% of the Asia respondents indicated interest in upgrading conveying and handling equipments as well as process control, while nearly 18% intended to install or replace new equipments as new process control technology in the feed plant management can achieve strong productive gains, especially in regions with higher labour costs. Intentions of Asia countries for the upgradation of equipments required for laboratory and quality control was far higher (50%) than North America (22%) and European countries (38%), which further show the awakening about the quality feed in Asian countries including India.

The optimism of Asian feed manufacturers reflects the state of economic growth and their willingness for competition with advanced countries. In India, awakening has come partly due to consumer's emphasis on quality of animal product and partly due to the liberalization policy of government, which emphasis on quality and safety of animal product as per code recommended by CAC. The most positive aspect of Indian feed industry is that it does not depend upon imports of feed ingredients such as grains and oilmeals, hence it is immune to the international fluctuations in prices, however, Indian feedstuffs besides being poor in nutritional quality have one or other incriminating factors, which is a major concern. Unfortunately, information about these incriminating factors such as mycotoxins, pesticide residues, have metals, pathogenic bacteria etc. and infrastructure to control them is not adequate in the country. More storage capacity has to be created in India for the proper protection f feed ingredients and also make possible for their traceability and inventory control.

Dry Extruded Soybean Meal

It is produced using dry extruders which are capable of cooking, expanding, sterilizing, dehydrating and texturizing soybean by creating heat through friction in a short-time maintaining feed quality. Bulk density of soybean extrudates is about 570 kg/m^3. The proximate analysis of dry extruded soybean meal is given in Table 9.

Table 9. Proximate analysis of dry extruded

Item	**Value, %**
Dry matter	94.0
Crude protein	46.0
Ether extract	7.0
Acid hydrolysis	7.7
Crude fibre	5.5
Ash	5.3
Undegradable protein	50.0
Free fatty acids	0.5
Linoleic acid	4.5
Linolenic acid	0.9
Lecithen	0.2

Feed Pellets

Pelleting costs money, and better pellets cost a little more. Payback comes in the form of convenience in handling the feed and more efficient use of the feed by the animal. Improved efficiency is due in part to heat processing which reduces pathogens and makes starches more digestible. A significant portion of the improvement is related to the physical form of the pellet. Durable pellets reduce waste, reduce segregation, improve palatability, and allow larger meals to be eaten in less time. All these factors contribute to optimized feed efficiency. Pellet quality does have economic importance; it should not be left to chance. Fines are generally formed by mechanical action on the pellets during transport. These forces may be classified as impact, compression, and shear. Impact shatters the pellet surface and any natural cleavage planes in the pellet; compression forces crush the pellet; shear forces cause abrasion of the edges and surface of the pellet.

Pellet length is a simple way to monitor quality. When samples are collected at the same sampling point, longer pellets indicate greater durability and less fines. When collected at multiple sampling points pellet length can indicate points of destruction in the handling system. Whenever feed is too expensive to waste, pellet quality begins to have an economic value. Expect your customers to give the quality of your pelleted feed at least a subjective measurement. There are a number of good ways that can be used to objectively measure and record the quality of pellets during the manufacture process. Objective measurement and recording is the first step toward correcting destructive conditions and improving quality. It may also allow opportunity for cost savings when quality is found to be above standard specification.

Recommendations

Soybean, a grain legume, is one of the most versatile plants known to people, science and industry. It is, as of now, grown through the world because of its high quality protein, oil, complex carbohydrates, minerals and phytochemicals. Soybean cake/meal that remains after oil extraction, is a valuable high-protein product for feeding livestock. it increased milk, meat, fish and egg production. Soybean leaves, stems and pod residues are also good feedstuffs. Use of soybean meal and plants residues in livestock feeds is very cost-effective. The rising income of people in Asia and Africa would demand more animal products which in turn require low-cost feed that could be used to produce more animal based food products. Soybean and its products being a rich source of protein, energy and phytochemicals at an affordable cost, would continue to be the major ingredient in livestock feed, the world over.

Bibliography

Akiyama DM and Tan Ronnie KH. 1991. Proceedings of the Aquaculture Feed Processing and Nutrition Workshop. 19-25 September, ASA, Singapore.

Ali Nawab. 2003. Soybean production and utilization in India-status, prospects and the strategy. Lecture delivered to the participants of Winter School on Value Addition and Entrepreneurship Development based on Soybean. Held at CIAE, Bhopal during 1-30 September, 2003.

Ali Nawab. 2004. Soybean oil milling technology and modified usage of oil and its byproducts. Proceedings of the 4th ISPUC-IV held at Foz do Iguassu, Brazil during 29 Feb-5 Mar. Pages 852-862.

Ali Nawab. 2004. Global scenario of soybean production and utilization. Lecture delivered to the participants of Winter School on Quality Assurance in Foods with Special Reference to Soybean and Soy-products. Held at CIAE, Bhopal during 16 Nov-06 Dec., 2004.

Cortyl Mathieu. 2004. Achieving food Safety - an ongoing challenge. Soya Update. Vo. 1(11):26-27.

Dudley-Cash WA. 2004. Soybean meal quality. Soya Update. Vol 1(3):16-21.

Hartnell Gary F. 2004. Feed ingredients situation in retrospect and prospects. Soya Update. Vo. 1(11):30-34.

Leslie AJ. 2004. Quality control in feed milling-procedure for an effective programme. Vol. 1(12):24-26.

Patil RT, Kulkarni SD and Ali Nawab. 2004. Strategic issues to expand soybean utilization in India. Proceedings of the 4th ISPUC-IV held at Foz do Iguassu, Brazil, during 29 Feb-5 March, 2004. Pages 800-807.

Singhal KK and Thakur SS. 2004. Changing scenario in feed processing industry in India. Soya Update. Vol 1(2):27-31.

Stallings CC. 2004. Soy in dairy feeds- hulls, roasted soybeans, full fat soybean meal and dehulled soybean meal. Soya Update. Vol. 1(9):36-38.

Van Eys JE. 2004. Soy and soybean meal use in the feed industry-formulation and quality considerations. Soya Update. Vol. 1(3):10-14.

Winowiski TS. 2004. Pellet quality in animal feeds. Soya Update. Viol. 1(12):21-23.

CHAPTER - 5
STRAW AMMONIATION TECHNOLOGY

T.C. Thakur and R.S. Jaiswal

Crop residues like wheat and rice straws, and stovers of maize and sorghum are the staple cattle feeds in India and other countries of South East Asia. However, due to their low nitrogen and mineral contents along with high lignin content and poor digestibility, they cannot meet even the maintenance requirement of animals. Because of increasing livestock population density and decreasing grazing lands, the straw feeding is becoming more and more important in livestock diet. About 50-60% of dry matter intake in large ruminants in India is through the crop residues and therefore, improved utilization of these crop residues is very essential. Doyle et al. (1987) indicated that low quality of roughage has both physical and chemical constraints to obtain optimum animal production. The ruminants on straw diet alone lose body weight but they respond to additional nitrogen and energy.

The crop residues may be processed by a number of methods to improve their nutritional value viz. (i) Chemical treatment i.e. alkali treatment, ammonia treatment either with anhydrous ammonia or aqueous ammonia or urea, (ii) Physical treatment i.e. chopping, grinding, soaking, pelleting, steaming and cooking, and (iii) Biological treatment using organism/fungi. The chemical treatments have been comprehensively reviewed by Jackson (1978), Sundstøl et al., (1978) and Jaiswal, (1990). The conclusions drawn by these workers show that the use of sodium hydroxide is neither technically feasible nor economical to farmers. However, the technique applicable for treatment of low quality roughages with ammonia has been found technically and economically superior which may vary considerably with the form of ammonia, level of technology, type of material to be treated and weather conditions. The ammonia used for treatment could be produced by using ammonium hydroxide, anhydrous ammonia, aqueous ammonia and urea.

The use of ammonia either in anhydrous or aqueous forms is rather more effective but in view of the problems associated with its transport, storage and application, it may not be feasible on small farm conditions of Asia. The use of urea as a source of ammonia has been, therefore, tried by several workers (Gadre and Jackson, 1980; and Saadullah et al., 1981 and Jaiswal et al., 1988). Ammonia (urea) treated straw has been reported to be superior to untreated straw in terms of its digestibility (Rashiq, 1980; Saadullah et al., 1981 and Jaiswal et al., 1988), intake of nutrients (Gadre, 1979 and Jaiswal et al., 1988) and animal performance in terms of growth (Perdok et al., 1983; Hamid et al., 1983 and Jaiswal et al., 1988) and in terms of milk production (Perdok et al., 1983 and Mazumdar et al., 2003).

The topic of improving the nutritive value of crop residues has been debated at a number of forums and many recommendations have been made but none of the methods recommended so far has found acceptability among the livestock farmers of India and other South-East Asian countries. However, the progress made in adoption of ammonia (urea) treatment technology in China has been encouraging (Dolberg, 1992). This calls for introspection among the scientific community and a fresh look on various practical aspects of the methods recommended so far. This paper, therefore, describes the technologies developed for farm scale ammoniation of straw with particular reference to Indian conditions.

Methods for Ammonia Treatment of Straw

Treatment with anhydrous ammonia

Anhydrous ammonia is dry or pure undiluted ammonia that is often used as nitrogenous fertilizer in Western countries and is required in small amount for treatment of straw. Another advantage is its gaseous nature which ensures rapid penetration into the straw. However, due to numerous disadvantages, such as handling in pressurized container (8 kg/cm^2 at 20°C), improper distribution network to ensure regular and cheap supply and its toxic nature, this method of ammonia treatment may not be practically feasible under Indian conditions at present. However, it has been used extensively for straw treatment in European countries and therefore, only a brief account of this technology is being presented.

The treatment of straw with anhydrous ammonia is mainly done in two different ways. The first method is the farm scale treatment in which low quality crop residues are treated on the farm of their origin to avoid the need of transportation. Another method is the industrial scale treatment in which crop residues are turned by chemical treatment into an energy rich feed for ruminants.

Farm scale treatment of baled straw

The stack method of treatment was developed in Norway in which a stack of straw bales is made of size varying from 2 to 4 tonnes in the open on a sheet of polyethylene spread on the ground. The stack is covered with another similar sheet. These two sheets are then rolled up at the edges to make an airtight container. Before finally closing the stack, anhydrous ammonia from a tanker truck is introduced by means of a pipe inserted into the stack (Fig.1).

Fig.1. Injection of anhydrous ammonia into baled straw stack covered with air tight plastic sheets

The ammonia is added at the rate of 3.5 kg per 100 kg of straw and treatment period ranges from 4 to 8 weeks. The straw is exposed to air for a day or so after opening the stack and then fed to the animals. This method became popular with farmers in Northern Europe. In 1976, more than a thousand farmers treated wheat straw by this method in Norway (Sundstøl et al., 1978).

Industrial scale treatment of baled straw

Lucht (1990) reported the non-traditional feed mills installed in Germany and suggested the possibilities available to feed mill owners for improving the nutritive value of straw by appropriate conditioning and production of a complete feed for ruminants by adding other raw feed minerals. The flow chart of the process for treatment with ammonia is shown in Fig. 2. The process of decomposing straw by ammonia is characterized by the treatment of unshredded straw with ammonia, the shredding of straw, adding molasses and other compounds to the mixture and then compacting and cooling the straw. The straw bales are treated continuously. A horizontal reactor is filled with straw bales and then ammonia gas is injected. The reaction time of ammonia is about 12 to 16 hours at 90°C. The unit is then aerated i.e. excess ammonia which has not reacted is withdrawn and conditioned for the next production cycle by a recycling system. The entire batch cycle, including filling and emptying takes about 24 hours. One reactor is filled with approximately 4 to 5 tonnes of straw, therefore, depending upon the output, several reactors may be needed. The quantity of ammonia added is approximately 2.5 to 3% of straw DM. The excess of NH_3 after conditioning is siphoned-off by the recycling system, therefore, the actual consumption only amounts to 2% of NH_3.

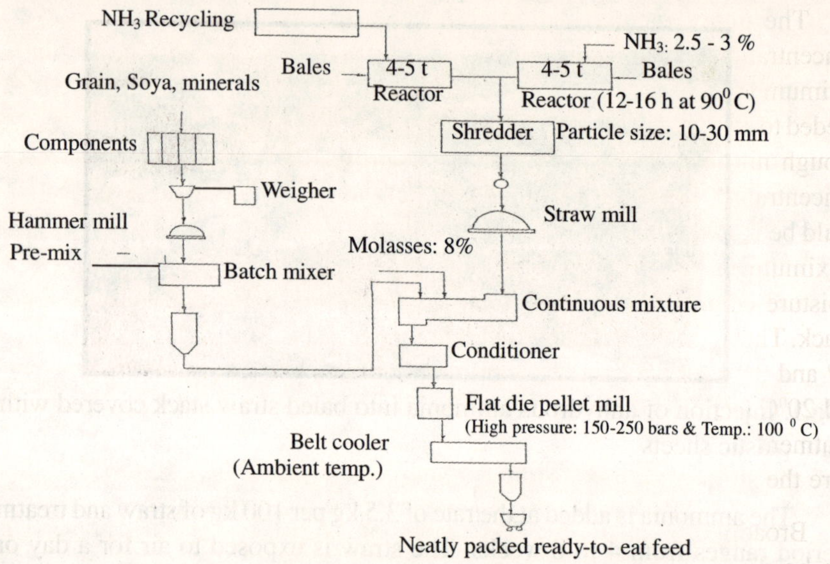

Fig.2. Flow chart for non-traditional feed mill treatment with NH_3

The production costs of the treatments are relatively high in relation to the improvement of the products. Conditioning will only be profitable when high straw quantities are available at a low price and when the straw must not be mechanically dried. This system would surely be profitable in those cases where other feedstuffs are not available (apart from concentrate feed) as ruminants also need a basic ration that is rich in raw fibers for covering their basic nutritive requirements. The production cost itself depends on the energy costs which differ from region to region and from the costs to be paid for the alkali substances. Therefore, the equipment of the non-traditional feed mill should be designed so that not only straw based quality feed can be produced but also other roughage feeds and 'traditional' feed in the form of meal and pellets for poultry etc. are produced. This flexibility would allow for immediate adjustment of the production to the market place and to the raw materials that are available.

Treatment with ammonia (urea)

Urea as a source of ammonia might be rated as one of the best alternatives for treating straw in Asia because of its ready availability, conversance of farmers with its transport, storage and application, in addition to its superior effects on DMD, intake, growth and production. Urea has also the advantage compared to anhydrous or aqueous ammonia as no health risks are involved in handling and using this chemical.

The major factors, which affect the urea treatment of straw are urea concentration, moisture content, temperature and curing period. The economical optimum level of urea has been suggested as 5% (wt/wt) of straw DM which is needed to obtain the higher value of DMD and crude protein and also for ensuring enough nitrogen in treated roughage to provide desirable level of ammonia concentration in rumen for microbial activities. The moisture level of straw could be increased up to 40% (w.b.) while treating straw with urea to obtain the maximum value of DMD and crude protein in treated straw. However, the moisture content above 50 % (w.b.) is not recommended because of fungus attack. The optimum temperature for urea treatment has been suggested between 30^0 and 35^0C. However, the minimum temperature should vary between 15^0 and 20^0C. The curing period required for reaction to take place is affected by treatment temperature. However, a minimum of 3 to 4 weeks time is needed to store the straw after treatment under Indian conditions.

Broadly speaking the methods of ammonia (urea) treatment of crop residues can be divided in two categories viz. the on-farm and commercial scale treatments.

On-farm ammonia (urea) treatment

A technique for farm scale treatment of loose roughages with urea has been developed at G.B. Pant University of Agriculture and Technology, Pantnagar (India) which has become popular among the farmers of Tarai region of U.P./Uttaranchal due to its easy and economical treatment procedure. Both chaffed as well as unchaffed roughages could be treated using this technique. The straw could be stored by any of the traditional methods viz. a stack in the open, storage in a hut particularly for wheat 'bhusa' or a room. For making a stack of loose straw in the open, a detachable circular enclosure made of welded wire mesh (20 x 20 x 3 mm size) is used and kept on the flat surface. A known amount of straw is spread uniformly inside the circular mesh or hut and compacted after spraying the required amount of urea solution of desired strength. This process is repeated till either the hump attains the height of 2 m or hut does not allow further compaction of straw. The basic principle involved in deciding the quantity and concentration of urea solution is to raise the moisture content of treated crop residues up to 50% in case of chaffed/ground crop residues/bhusa or 40% (w.b.) in case of un-chaffed crop residues in the long form and at the same time add 5 kg urea/100 kg crop residue DM. For calculating the quantity and concentration of urea solution, the following thumb rules may be used:

Volume of urea solution, lit/100 kg chaffed or ground or bhusa (V_1) = 2 ($50 - M_1$)

Volume of urea solution, lit/100 kg long form of crop residue (V_2) = 10/6 x ($40 - M_2$)

Concentration of urea in the solution (%) = 5 (100 – M_1 or M_2) V_1 or V_2

Where, M stands for the moisture content of crop residue at the time of treatment.

The compaction should be done in such a way that as much of the inside air as possible should come out of the stack. The density of compacted straw ranges between 150 and 200 kg/m³. The top of the hump or hut is made to the shape of a dome so that the rainwater runs off easily and does not leach down into the stack. The treated straw is left for 3 to 4 weeks for reactions to take place and then it is fed to the animals.

Industrial scale ammonia (urea) treatment

I) Treatment of loose straw

The industrial scale treatment of straw with urea has been reported by Bergner et al. (1974). Urea is mixed with straw at the rate of 2% and the mixture is then pelleted using straw thick dies which raise the temperature to about 150°C during pelleting. When urea is decomposed at this temperature, it releases ammonia which reacts with straw during pelleting process. If the temperature measured in the pellets is 90°C, the temperature in the ring dye would be around 110°C which is sufficient to release about 1 kg ammonia for upgrading 1 kg straw pellets (Bergner, 1980).

II) Ammonia (urea) treatment of baled straw

The combine harvesting of cereal crops is gaining popularity in most of northern states as well as in South India for harvesting of paddy. Presently there are over 18000 combine harvesters operating on Indian farms which are being indigenously manufactured by over two dozen manufacturers of Punjab and Haryana. These combines, however, have limitations as they leave a large amount of 'header trash' which must often be removed from the field for agronomic or management regions. However, because of their low bulk characteristics and low market value per unit weight, handling and transportation is generally uneconomical. This in turn presents disposal problems to farmers who find open field burning as a perfect solution. The huge amount of straw which is wasted either by burning in the fields or due to poor utilization could contribute to the income of Indian farmers. The conservation of rice and wheat straw as livestock feed requires a series of on-farm and off-farm operations including collection, packaging, handling, transportation, storage and pre-feeding processing. The field balers are new introduction in Indian agriculture and are being projected as an alternative to open field burning for collection and disposal of straw.

Currently, five models of conventional field balers (Fig. 3) viz. three imported models from CLASS, New Holland and Gallignani, and two indigenous models from Punjab have been introduced on Indian farms. These balers produce rectangular bales of 36 x 46 x 110 cm size with bulk density varying from 115-150 kg/m^3. Under Indian conditions, the cost of field baling has been worked out to be Rs. 800 per ha or Rs. 400-500 per tonne and the collection of bales from fields, transportation and storage at a site about 5 km away from fields require about Rs. 250 per tonne (Thakur and Rajan, 2001).

Fig. 3. Field baling of paddy straw after combining

During past several years, a number of technologies for ammonia (urea) treatment of baled rice and wheat straw have been tried at Pantnagar (Tiwari, 1993 and Ramu, 1994) but recently a technology for ammonia (urea) treatment of stacked baled straw using 'Dripping Method' has been successfully developed and is being popularized (Bamaga, 2000; Rajan, 2001 and Uttam Kumar, 2002). The possible methods for ammonia (urea) treatment of baled straw are presented in Table 1. However, only those technologies which are of practical importance and tested are described over here.

Table 1. Possible methods for ammonia (urea) treatment of baled straw investigated at Pantnagar

S.No	Methods of treatment	Merits/Demerits	References
1.	Soaking of bales into urea	• No control over application rate • Imvevem treatment of bales • Problem of mould infestation	
2.	Spraying of urea solution on crop residues spread over the field using field sprayer before baling	• Harmful effects i.e. corrosion etc. of baler components • Excessive loss of urea • Baling and storage of bales after urea spraying	• Ramu (1994)
3.	Spraying of urea solution on crop residues at the gathering unit or bale chamber of the baler during baling	• Harmful effects i.e. corrosion etc. of baler components • Need for quick covering and storage of bales after baling	
4.	Injection of urea solution into bales in the baling chamber itself	• Better control over application rate • Harmful effects on baler components • More complicated mechanism • Practically difficult and expensive method • Need for quick storage of bales	
5.	Injection of urea solution into bales after baling	• Better control over application rate • No harmful effects on baler components • Possibilities of treatment according to availability of time, labour and as per requirements	• Tripathi A. and Shome, S. (1991) • Tiwari, R.C. (1993) • Ramu J. (1994) • Tiwari, et.al., (2001)
6.	Dripping of urea solution into stacked baled straw	• Better control over application rate • No harmful effects on baler components • Suitable for large scale treatment of baled straw in batches • Possibility for commercialization	• Bamaga, O.A. (1996) • Singh J.P., (1999) • Bamaga, et.al., (1999 and 2002) • Promod Rajan, (2001) • Utam Kumar (2002)

(i) Application of urea solution before baling

This method involves spraying of urea solution on straw swath prior to baling or at the windrow pick-up unit of baler. The method is appropriate when the main objective of the treatment is to preserve high moisture straw which could be baled without field drying. Moisture content of paddy straw may range from 40 to 75% when harvested. It must be field dried to a safe limit of atleast 17% for baling without adding any preservative. Ammoniated straw with relatively high moisture content can be baled and stored without being spoiled (Knapp et al., 1974). As urea releases ammonia, it preserves high moisture hay as well as high moisture corn (Ghate and Bilanski, 1979 and Ghate et al., 1980).

Fig. 4. Spraying of urea solution on a simulated straw swath before baling

Ramu (1994) used a field sprayer for spraying urea solution on paddy straw swath formed under simulated condition (Fig. 4). Immediately after spraying of solution, straw was baled. Two modes of application i.e. (i) spraying of total required amount of urea solution only on top surface of the straw swath before baling and (ii) spraying half of the required amount of urea solution on top surface of swath and the remaining half after overturning the swath before baling were compared. It was found that the distribution of moisture in baled straw did not vary much in either of the modes. In-situ DM digestibility (67%) was significantly higher when spraying was done using second mode of application. Also, the spraying of urea solution before baling was found as good as injection of urea into the bales tried earlier by Tiwari (1993).

(ii) Application of urea solution during baling

The basic concept of this option is to incorporate a chemical application device/system with the baler itself so that the chemical could be applied into straw during formation of bales in the bale chamber or after its formation at the extension of bale chamber. Khalilian et al. (1990) developed a high pressure chemical injector and mounted it into a 88 cm extension of bale chamber of a rectangular field baler. The injector did not penetrate fully into the bales of 40% moisture and more. The injection of propionic acid was found effective in preventing heating and molding during storage with no apparent benefit to the quality of alfalfa hay when injected into bales at moisture levels up to 30%. This method is preferred when quick harvesting and storage of high moisture forage is required so as to prevent losses of nutrients due to rains and to maintain the quality of forage. The method was tried later on for application of urea solution into baled straw with positive results.

(iii) Application of urea solution after baling

There are two potential methods of treatment of baled straw with urea solution after baling. These include (i) injection of urea solution into baled straw at treatment posts (Tripathi and Shome, 1991; Tiwari, 1993 and Ramu, 1994) and (ii) application of urea solution by dripping into baled straw stacks (Bamaga, 1996; Singh, 1999; Rajan, 2001 and Uttam Kumar, 2002). These two methods, which were developed at Pantnagar, are described below:

A) Injection of liquid area into baled straw

A treatment set-up for injection of liquid urea @ 5% (wt/wt) into baled straw to raise its moisture content up to 40% (w.b.) has been developed (Tiwari, 1993 and Tiwari et al., 2001). The treatment set-up consisted of two main units (Fig. 5).

a. The urea treatment set-up

b. Examination of wetted area of a treated bale

Fig. 5. Treatment set-up for injection of liquid urea into baled straw

The first unit was equipped with an engine operated piston pump with discharge rate of 600 cc/s at 900 rpm, connecting pipes and hoses, control valves, pressure gauge and a hand held device for mounting the injector probe. The second unit was a bale holder used for holding of single bale while injecting liquid urea and a trough with a strainer for collection of drained liquid at the bottom of the unit. Various parameters such as type of injectors (flat fan type probe, two transverse orifice probe and triple orifice probe), injection pressure (1-5.5 kg/cm^2), modes of injection (central-horizontal and eccentric- horizontal), number of injecting stations (5 and 10 numbers for bale length of about 60 cm) have been evaluated in terms of liquid retaining efficiency and wetted area of bales. The two transverse orifice probe was found best for injecting urea solution at a injection pressure of about 4 kg/cm^2. The injection stations should be located at every 10 cm spacing on two vertical planes spaced at about 5 cm eccentricity to either sides of the central-vertical plane of a bale for acceptable distribution of liquid. While injecting, the bale should be kept vertical and injector should be positioned horizontally. The evaluation of injector for injecting urea solution into baled wheat straw has shown significant increase in the nutritive value of straw. It increases the DMD coefficient by nearly 13% units, crude protein content by about 9% units and ammonical nitrogen content by about 0.65% units which is very well in the range of the values reported for ammonia (urea) treatment of loose straw.

B) Dripping of urea solution into bales

Bamaga (1996) and Bamaga et al. (1999 and 2000) developed and evaluated the dripping method for application of urea solution into baled straw by using drip irrigation emitters (Fig. 6). The results of experiments on wheat straw bales revealed that the bed depth, discharge rate and specific amount of liquid application, all affected significantly the specific amount of liquid retention, liquid retaining efficiency and final moisture content of baled straw bed. It was found that for bed depth of about 1 m (3 bales), the bottom most bale had the minimum moisture content but had achieved the minimum recommend level of moisture content of 40% (w.b.) when total discharge rate was 500 or 1000 cm^3/min. The corresponding values of liquid retaining efficiency were found as 70% and 62% and that of average moisture content of straw bed as 45% and 43.5% (w.b.). Singh (1999) further conducted experiments on baled paddy straw using the 'Dripping Method'. He studied the effect of spacing between emitters of liquid applicator on the moisture migration pattern and found that decreasing the spacing from 10 to 5 cm had no significant effect on the moisture migration. He found that the liquid retaining efficiency of a bed of three bales decreased from about 67% at specific liquid application of 725 cc/kg DM to about 65% at 1125 cc/kg DM for discharge rate of 500 cc/min. The moisture content of three treated bales was found to vary, therefore, for equalization of moisture among treated bales, it was recommended that the bales with higher moisture should be kept at the bottom and those with lower moisture content should be at the top of the stack and covered with plastic sheets during the curing period.

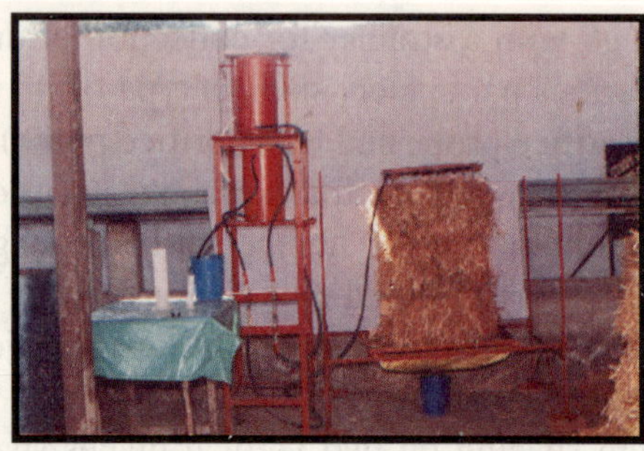

Fig. 6. Dripping method for application of urea solution into a stack of three bales

C) Batch method of ammonia (urea) treatment of stacked baled straw

The methods described above have been found technically feasible but economically unviable at present for treatment of large quantity of baled straw for a dairy herd. Therefore, based on the results of earlier studies and further basic research conducted to establish the treatment regimes, a small-scale treatment set-up as well as a treatment plant were designed, installed and evaluated.

(a) Ammonia (urea) treatment set-up

A treatment set-up for application of urea solution using 'Dripping Method' into a stack of baled straw has been developed (Bamaga, 2000, Rajan, 2001 and Bamaga et al., 2002). It consisted of two different units. The first unit consisted of a 500 litre container for storage of urea solution, monoblock centrifugal pump, pipes and connections, control valves and liquid applicator equipped with drip irrigation emitters. The second unit consisted of a frame for stacking of bales and a liquid applicator (Fig.7a). While treating straw, the water pump withdraws urea solution from the container and pumps it through the main pipe and control valve to liquid applicator. The pressure head in the main line is maintained at a desired level by regulating the back flow through a by-pass valve and pipe. The liquid applicator consisted of polythene pipes of 10 mm diameter each, onto which drip irrigation emitters are mounted. Another polythene pipe of 10 mm diameter is used as a main liquid supply line and provided with elbows and T-joints for mounting of lateral pipes. The distance between emitters on each lateral as well as between laterals has been kept 100 mm. On this set-up the bales are stacked up to a height of 2 m in six layers. A total of 24 bales (4 bales in each layer) with a total weight of about 0.5 tonne could be treated at a time. The bales are treated with urea solution up to a specified period to raise their moisture content up to 40% (w.b.). After treatment, the bales are unloaded and stacked separately until a stack of 2-3 tonnes of treated bales is obtained. Thereafter, the bales are sealed with plastic sheets for a period of 3-4 weeks for curing (Fig.7b). The cost of this set-up has been estimated as Rs.18,000/- and the treatment cost varies from Rs.300 – 350 per tonne of straw depending upon the size of cattle herd.

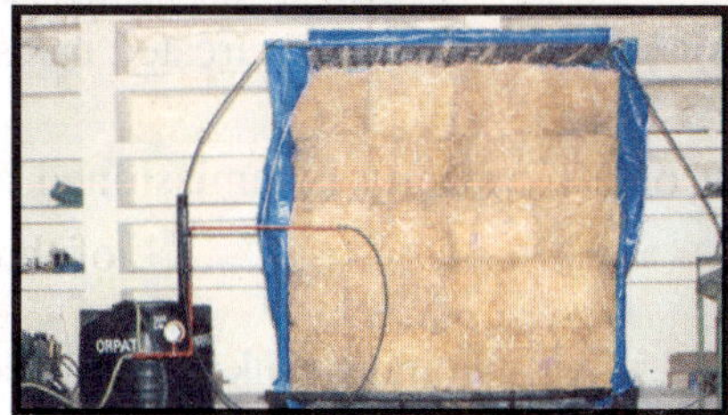

a. The urea treatment set-up (0.5 t capacity)

b. Curing of treated bales in air tight plastic sheet enclosure

Fig. 7. Batch method of ammonia (urea) treatment of stacked baled straw

(b) Ammonia (urea) treatment plant

An ammonia (urea) treatment plant of about 5 tonnes capacity has been designed by scaling up the treatment setup by over 10 times which can accommodate 216 bales of 36 x 46 x 92 cm size and 120 kg/m^3 density (Fig. 8a). Thirty six bales are, therefore, placed in each layer in a predetermined subsequent layers are superimposed one over the other in such a way that there are no direct vertical gaps between different layers for runoff from top layers of stack (Fig. 8b). The plant consists of mainly a platform, main frame and a liquid applicator incorporating dripping system. The platform on which the bales are to be stacked for treatment is kept at a height of about 50 cm above the floor area so that the runoff of urea solution could be collected for reuse. To overcome the excess load acting on the base area, 3 inner brick walls of single brick width have been provided perpendicular to the longest wall of 4.30 m length. The iron rods have been placed above the steel plates across 3.85 m long walls. The floor of the base area as well as the top of the walls have been provided a slope of about 5% so that urea solution obtained as runoff from treated bales could be collected through single outlet. For proper movement, handling and covering of the treated bales with plastic sheets on the plant itself (Fig. 8c), the width of the outer walls is kept larger by approximately 60 cm (Uttam Kumar et al., 2002).

a. Urea treatment plant

b. Treatment of bales

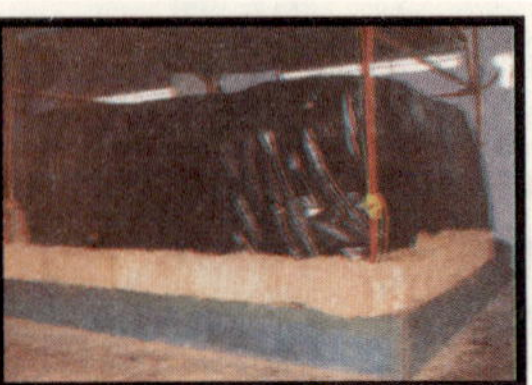

c. Curing of treated bales

Fig. 8. Ammonia (urea) treatment plant of 5 t capacity commissioned at Pantnagar

The excess runoff solution is collected in an adjoining tank constructed close to the plant. Runoff solution is recycled into the main storage tank through a pump. The frame consists of two sections i.e. liquid applicator frame and the main frame. The liquid applicator frame consisted of laterals, main pipes and emitters. The main frame holds the liquid applicator frame with the help of ropes, pulleys and holders. The height of the liquid applicator frame is adjustable with the help of ropes and pulleys. The liquid applicator frame can be held at a particular position with the help of holders which are welded at all the four poles. The centrifugal mono-block pump sucks the urea solution from the storage tank and delivers the solution into the main pipe. The impurity present in the urea solution is removed with the help of screen filter. The discharge in the main pipe is controlled with the help of control valves and the excess amount of solution returns back to storage tank through a by-pass line. A pressure gauge is used to indicate the pressure required to deliver the solution into the main pipe. The lateral pipes are connected to the main pipe and the emitters are fitted on the lateral lines.

For calculating the required quantity and strength of urea solution, the following formulae may be used:

Volume of urea solution required, lit./100 kg baled straw (V_3) $= 10/6 \times (40 - M_3)$

Concentration of urea solution (%) $= 5\ (100 - M_3)\ /\ V_3$

Retaining efficiency (R_e, %) = 80-85

Actual volume of urea solution required, lit. /100 kg baled straw (V_4) = $V_3 \times 100/\ R_e$

Where, M_3 is the moisture content of baled straw at the time of treatment.

As an example, for a batch of 3392 kg DM, the amount of liquid to be added into the stack to raise the moisture content upto 40% (w.b.) will be 2025 kg. The amount of urea assuming a ratio of urea: straw DM as 5% (wt/wt) recommended for ammonia (urea) treatment, will be 83 g/kg water. In view of the fact that not all the applied liquid will be retained in the stack and some amount will flow out of stack as runoff, the actual amount of urea solution applied into the stack will be about 2411 kg assuming a retaining efficiency of 84%. The average actual time required for completion of wetting of baled straw upto the recommended moisture content of 40 % (w.b.) with urea solution, an inflow rate of 35.76 g/min /emitter works out to be about 42 min for wheat straw bales. Although significant variation of moisture content may be observed among different layers of ammonia (urea) treated wheat straw bales, yet the variation of moisture from the top to the bottom layers is found to be within the safe range of treatment. Therefore, stacks of 6 layers have been recommended for ammonia (urea) treatment on the plant of about 5 tonnes capacity costing about Rs. 60,000. The retaining efficiency of wheat straw has been found to be about 87 %. The cost of treatment works out to be Rs. 445, 379, 336 and 324 per tonne for wheat and Rs. 419, 342, 318 and 303 per tonne for paddy straw bales for herd sizes of (75 t) 25, (150 t) 50, (225 t) 75 and (300 t) 100 animals, respectively.

A still better approach to minimize the cost of treatment would be to increase the number of days for which the plant is operated in a year. For this purpose, the baled straw needs to be stacked in sufficient quantity near the treatment plant so as to operate the plant for a minimum of 3 to 4 months at the end of each harvest season. To enable this, the number of balers used for the baling operation will have to be increased to match the requirements of the treatment plant. Assuming that about two batches of 5 tonnes each could be treated every day, the total quantity of baled straw required would be 500 tonnes in each season or about 1000 tonnes/annum.

D) Urea treatment of bruised wheat straw (Bhusa)

In the mid 80's local manufacturers of Punjab developed a machine which harvests wheat straw left in the field by a combine harvester. This machine collects the straw left in swath, cuts the stubbles, process the straw into finely bruised feed i.e. 'bhusa' and then collects it in a trailer attached behind the machine (Fig. 9). The straw combine essentially consists of four units viz. stubble cutting unit, feeding unit, straw bruising unit and straw blowing unit. An enclosed type trailer is also attached to collect the material. The average field capacity of straw combine varies between 0.24 and 0.40 ha/h. The height of cut varies between 5 and 7 cm, and straw output ranges from 1.5 to 2.5 t/h. The combine recovers about 1.8 t/ha of bruised straw as compared to 3.5-4.0 t/ha obtained with manual harvesting and mechanical power thresher system (Singh et al., 1998). The average custom rate for straw combines comes to about Rs.1000 / ha. Also, in some specific crop conditions due to lodging of crop and shattering of ear heads, substantial loss of grain occurs in the fields when using combine harvesters. It has also been reported that the straw combines recover about 0.1 t/ha of wheat grain thereby further reducing its cost of operation. Therefore, a net saving of Rs 1000 /ha by the use of straw combine is obtained.

Fig. 9. Straw (Bhusa) combine in operation after combining of wheat

The economics of various straw recovery systems prevalent on Indian farms have been reported by Bamaga and Thakur (2003). The straw combine based recovery system has been found to have least total cost among all the systems studied for farm size of up to 100 ha. The cost was found as Rs 681 and 604 per tonnes for farm sizes of 50 and 100 ha, respectively. For utilization over an area of 100 ha, the cost of combine alone constitutes to about 55% of the total cost. This indicates that just in the field, the operation of bruised straw making would cost about Rs 334 per tonnes. Transportation up to 5 km distance and unloading would require an additional cost of Rs 128 and Rs 142 per tonnes, respectively.

The bruised straw obtained from a straw combine is found deficient in protein, minerals and vitamins and poor in digestibility. The quality of bhusa is also coarse in comparison to bhusa obtained from a wheat thresher and generally not preferred by animals. The manual method of spraying urea solution for ammoniation of wheat bhusa has been found effective in increasing the digestibility and intake of straw (Agrawal, 1994). The manual spraying method for urea application, however, could not be practiced by the farmers because of high labour requirement and cumbersome process of treatment. A new method has, therefore, been devised for large scale treatment of wheat bhusa by using the basic components of ammonia (urea) treatment plant for baled straw installed at Pantnagar (Verma, 2004).

The set up consisted of two different units (Fig. 10). The first unit consisted of a 300 litre container for storage of urea solution, mono-block centrifugal pump, pipes and connections, control valves and liquid applicator equipped with a herbicide spraying nozzle. The second unit consisted of a cylindrical wire mesh for holding wheat bhusa layers and liquid applicator rod with which the nozzle is attached. While conducting experiment the water pump (1) withdraws water from the container (2) and pumps it through main line pipe (3) and control valve (4) to the liquid applicator (5). The pressure head in the main line which could be noted from a pressure gauge (6) is maintained at desired level by regulating the back flow through the by-pass valve (7) and pipe (8). The liquid applicator consisted of main line pipe of 20 mm inner diameter on which a spraying nozzle is mounted. The height of liquid applicator could be adjusted according to the height of layers of wheat straw (bhusa) by sliding the upper part of liquid applicator nozzle upward or downward and then fixing it at a desired position with the help of stop rods (9). The lower base consisted of an elevated cylindrical wire mesh base frame (10). The diameter of the holes of base frame is 5 mm. A bucket is provided under the base frame to collect runoff liquid and direct it into a runoff collector (11).

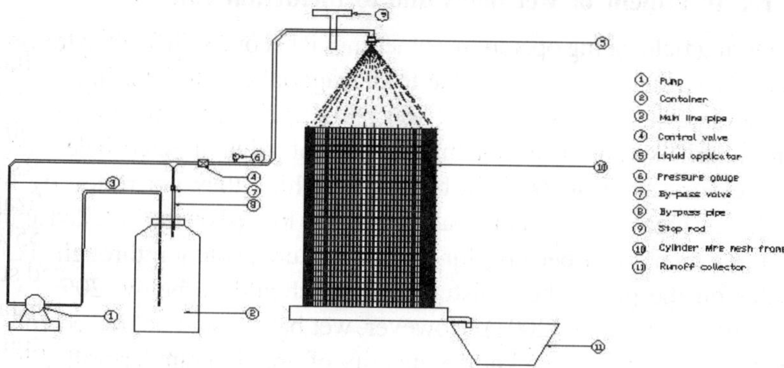

Fig. 10. The ammonia (urea) treatment set-up for bruised straw (bhusa)

Ten layers of 150 mm thickness are needed to treat a column of 1.5 m height. For a constant inflow rate of 1250 g/min, the amount of liquid application in ten layers starting from 1st (bottom) to 10th (top) layers is kept as 7882, 7782, 7806, 7811, 7797, 7836, 7723, 8050, 7102 and 10000 g, respectively for obtaining the final moisture constant of 50% (w.b.).

After urea treatment, the straw columns are covered with plastic sheets (Fig. 11) for a period of 3-4 weeks for curing and thereafter, it is opened and fed to cattle. The evaluation of urea treated wheat straw (bhusa) has indicated significant increase in its nutritional values over untreated straw (control). On an average NBDMD increased from 43.47% (untreated) to 59.14% (treated). The CP content of treated straw was found as 11.89% which was 8.33 percentage units greater than untreated straw (3.56%). Similarly, the AN content was found as 0.838% which was 0.793 percentage units greater than untreated straw (0.045%).

a. Straw *(bhusa)* column during treatment

b. Curing of urea treated straw *(bhusa)* column

Fig. 11. Treatment of wheat bhusa by using the basic facility of treatment plant

E) Urea treatment of wet bales due to incidental rains

During field baling operation, sometimes it becomes difficult to transport all the bales to a safer storage site due to managerial constraints. In case there is rain, the bales left in the field get wet particularly while baling paddy straw during kharif season. The moisture content of rains affected bales has been found to vary as 20 to 60% (w.b.) due to which they get rotten by fungus growth. The high moisture straw bales cannot be treated on the installed ammonia (urea) treatment plant because for leaching of urea solution through six layers of bales on the plant, the moisture content should be below 20% (w.b.) as demonstrated by Rajan (2001). However, wet bales could be treated and saved from rotting by applying a known quantity of urea in granular/prilled form. In this direction some research has been conducted by Alhadhrami et al. (1993) to evaluate the effects of uniformly adding urea (0.8, 1.3, 1.5 and 2.6%) to lucerne hay at baling (Experiment 1) or on top surface of bales (1 kg per bale or 2.1% of DM) subjected to water and rain damage (Experiment 2). Addition of 1.3% urea to high-moisture lucerne hay (24 to 26%) effectively lowered bale temperature, non-enzymic browning, visible mould and chitin content from those of high-moisture, untreated hay. Addition of urea to high-moisture lucerne hay resulted in higher nutritional quality through reduced heating, mould development and lower structural carbohydrate content compared with control high-moisture hay. For meeting the specific need of saving wet bales due to rain, a simple hand held urea granule applicator (Fig. 12) was developed and evaluated (Shinde and

Thakur, 2004). The device was evaluated on wheat straw rectangular bales of 36 x 46 x 100 cm size with an average bulk density of 100 kg/m^3 and initial moisture content of 30% (w.b.) which was added artificially to represent wet bales due to rain. The total amount of urea to be applied per bale on 5% (wt/wt) basis was calculated as 700 g which was applied with the applicator at nine equally spaced stations located on two rows spaced at 20 cm in the centre of a bale. The total time taken for application of urea in a bale was found to about 6 min. The bales were stacked and covered with plastic sheets for curing for a period of 21 days. The changes in nutritional values of straw in terms of DMD coefficient, CP and AN contents were found in the similar range as reported by other researchers. This has revealed that wet straw bales due to rains could be saved successfully from rotting and with enhanced nutritional values by application of urea granules. The technique could also be used for treatment of green fodder bales specially high moisture paddy crop directly without adding water.

Nutritional Changes in Ammonia (Urea) Treated Straw

In a comprehensive review, Jaiswal (1990) reported the factors affecting the animal response in terms of digestibility, intake of nutrients and growth while feeding ammonia (urea) treated crop residues. The results have been summarized in terms of average values of responses compiled from literatures till 1990 and are presented in Tables 2 to 4. The results clearly show a marked improvement in responses due to ammonia (urea) treatment of crop residues which is dependent upon type of straw, methods of treatment, species of animals, type and amount of supplement. The maximum response to straw treatment was observed in case of buffaloes and crossbred heifers. The bull calves/steers showed the minimum differences between treated and untreated straw diet (Table 4). The nutritional changes in ammonia (urea) treated baled rice and wheat straw stacks cured in an airtight polythene sheet enclosure have been found to be in the same order (Bamaga, 1996; Bamaga, et al., 2000 and Rajan, 2001) as has been reported in case of loose straw treatment (Table 5).

Table 2: Effect of ammonia (urea) treatment of crop residue on digestibility and intake of dry matter

Crop residues	DM-digestibility, %		Change %	DM intake, (g/kgw$^{0.75}$)		Change %
	US	TS		US	TS	
Wheat straw	48.7	54.6	5.8	75.2	90.1	14.9
Rice straw	45.7	53.7	8.0	75.2	91.6	16.4
Oat straw	55.6	66.0	10.4	66.1	73.4	7.4
Barley straw	52.6	58.9	6.1	52.7	58.8	6.1

US: Untreated straw TS: Treated straw
Source: Jaiswal, 1990

Table 3: Effect of different supplements to treated (TS) and untreated (US) crop residues based diet on DMD, intake and growth of livestock

S.No.	Supplements	DM digestibility %		DM intake (g/kgw0.75)		Growth (g/day)	
		US	TS	U. S. Nagar	TS	US	TS
1.	No supplement	43.7	53.5	61.1	73.3	-67.7	71.8
2.	Straw + green fodder	43.2	52.1	76.1	96.8	94	234.5
3.	Protein supplement to balance the CP requirement	47.4	49.6	84.7	102.4	175.8	303.3
4.	Concentrates to balance the energy & protein requirement	49.9	58.1	75.4	99.3	97.0	290.5

US: Untreated straw TS: Treated straw
Source: Jaiswal, 1990

Table 4: Effect of feeding treated (TS) and untreated (US) crop residues on DMD intake and growth in different species of livestock

S.No.	Supplements	DM digestibility %		DM intake (g/kgw0.75)		Growth (g/day)	
		US	TS	U. S. Nagar	TS	US	TS
1.	Crossbred heifers	44.2	47.8	77.5	106.6	141.2	336.4
2.	Buffalo calves	50.5	57.9	80.5	93.6	-92.6	194.4
3.	Sheep and goat	46.4	54.6	42.6	52.1	-	-
4.	Crossbred bull/ male calves/ steers	48.8	58.2	81.2	96.0	134.3	270.8

US: Untreated straw TS: Treated straw Source: Jaiswal, 1990

Table 5: Effect of ammonia (urea) treatment of baled straw on nutritional changes

Baled straw	DMD coefficient, %			CP content, %			AN content, %		
	US	TS	% age change	US	TS	% age change	US	TS	% age change
Wheat[1]	50.20	56.00	5.80	3.73	13.42	9.69	0.035	0.573	0.538
Rice[2]	41.35	55.90	14.55	4.54	13.51	8.97	0.052	0.687	0.635

US: Untreated straw TS: Treated straw Source: [1]Bamaga, 1996 and [2]Rajan, 2001

Livestock Feeding Trials with Ammonia (Urea) Treated Straw

The treated crop residue can be fed in the similar way as the untreated crop residue except that the quantity of supplements required decreases considerably. In case of unchaffed treated crop residue, the chaffing is recommended to avoid wastage of improved quality fodder. In view of the improved quality of treated crop residue, the need for its supplementation with all deficient nutrients becomes absolutely essential. The voluntary intake of treated crop residue increases considerably after a week of adaptation during which the introduction of treated crop residue should be gradual.

Experiments have shown that in case of growing cattle a growth level of 400-500 g/day can be obtained by feeding treated crop residue ad lib. with 4-5 kg of green leguminous fodder available in the season, mineral supplements (both macro and micro) and about 250 g of by-pass quality oil seed cake per head/day (Jaiswal, et al., 1988). Similarly in lactating cattle 5-6 kg of milk can be obtained by feeding them with treated crop residue mixed with 10% leguminous fodder ad lib. without any concentrate mixture as evident from Tables 6 and 7 (Agrawal et al, 1988). The major beneficiaries of feeding treated crop residues are the growing and non-producing categories of livestock (Table 8) due to which the age at first calving and service period is reduced (Verma, 1988).

A saving of 1.0-1.5 kg concentrate is possible by feeding treated crop residues to growing animals while in case of lactating animals the saving could be about 2-3 kg/head/day depending upon the level of production. In practice the quality of treated cop residues becomes equal to most of the non leguminous green fodders and as such the farmers can either save land to be used for raising non-leguminous fodders and use it for planting other cash or commercial crop or use it for raising leguminous fodder crops which will help in further improving the utilization of treated crop residue (Table 8). The net result of feeding treated crop residues is that it not only improves livestock production but makes it cheaper.

A feeding experiment conducted with growing crossbred heifers has shown that the digestibility of DM of treated baled wheat straw increased from 43.45 to 50.98 %. Similarly, the voluntary DM intake of roughage increased from 1.6 to 2.15 kg/ 100 kg body weight, showing an increase of 57.70% in digestible DM intake (Pathak, 2002) As a result of this long term growth trial of over six months, it was concluded that by feeding ammonia (urea) treated baled wheat straw alongwith 2 kg green fodder, 30 g mineral mixture and 30 g salt, a saving of about 25% in terms of concentrate mixture is possible for obtaining the normal growth rate. Further, there was no significant difference found in blood glucose and urea concentration in milk of animals fed on untreated or ammonia (urea) treated baled straw even after two years of feeding ammonia (urea) treated straw (Bhatt, 2003). In a livestock feeding trial conducted in villages around IGFRI, Jhansi a saving of upto 35% of concentrate mixture has been reported by feeding ammonia (urea) treated straw for the same level of milk production (Mazumdar et al., 2003).

Generally, because of increased consumption of water by livestock fed on treated crop residues the consistency of the dung decreases and therefore, in places where cow dung is used for making dung cakes for fuel the farmers feels the difficulty in collecting it and using the same for fuel purposes. In such cases the use of a mixture of treated and untreated crop residues in the ratio of 2:1 is recommended to improve the consistency of the dung. In other places where the dung is used in biogas, the consistency of the dung is in fact useful because less water will have to be added from outside. Moreover, because of higher nitrogen content, the dung from ammonia (urea) treated straw fed animal's helps in enhanced gas production.

Table 6: Effect of substituting two levels of ammonia (urea) treated rice straw in berseem diet on milk production and feed cost

Treatment	Ratio of straw : berseem : conc.	Dry matter intake			Milk yield		Feed Cost		Body wt. change kg
		kg/100 kg LW	g/kg $LW^{0.75}$	Conc., kg/d	Total, kg/d	FCM kg/d	Milk	FCM	
Berseem ad lib + 3 kg straw	19:67:14	3.94^{a}	169^{a}	1.92	11.17^{a}	12.24^{a}	1.29	1.18	+15
Berseem + US, 50:50 mixture ad lib	42.5:42.5:15	3.43^{ab}	1.49^{ab}	1.82	10.69^{ab}	11.63^{ab}	0.97	0.89	+10
Berseem + TS, 50:50 mixture ad lib	44:44:12	3.87^{a}	1.68^{a}	1.73	10.75^{ab}	12.18^{a}	1.15	1.01	+03
TS ad lib. + 4 kg berseem	80:8:12	3.15^{b}	136^{b}	1.31	9.91^{b}	10.91^{b}	0.85	0.77	+04

The values bearing different superscripts in the same column differ significantly ($P<0.05$)

Table 7: Effect of different rate of concentrate feeding with ammonia (urea) treated wheat straw on digestibility of DN, intake of DM and DDM, milk yield and fat corrected milk yield and live weight changes in cows

Treatment	DDM %		DM intake, g/kg $w^{0.75}$			Conc. kg/d	Milk Yield kg/d		LW change kg
	Diet	Roughage	Total	Roughage	DDM		Total	FCM	
U.S. ad lib. + conc. To balance DCP and TDN	68.4^{a}	65.2	133	1041	91	2.84	8.58^{b}	9.03^{b}	-12.5
T.S. ad.lib + conc. @ 400 g/kg total milk (T)	64.5^{b}	60.4	153	122	99	3.25	9.28^{a}	9.60^{ab}	+5.2
T.S. ad.lib. + conc. @ 400 g/kg for T-2	64.4^{b}	63.0	138	126	89	1.88	7.67^{c}	8.27^{c}	-27.3
T.S. ad.lib. + conc. @ 400 g/kg for T-4	62.6^{b}	61.8	139	133	87	1.56	8.69^{b}	9.42^{ab}	-8.4
T.S. ad.lib. + conc. @ 400 g/kg for T-6	65.6ab	65.1	135	131	88	0.94	8.81ab	9.79a	-12.6

Values bearing different superscripts in the same column differ significantly ($P<0.05$)

Table 8: Effect of feeding ammonia (urea) treated straw on area under forage crops, milk yield and age at first calving

Period	Livestock population		Quantity of T.S. used		Area under forage crops		Av. milk yied, (kg/d)	Age at first calving (months)
	Adult units (AU)	Adult cows	Tonnes/ annum	kg/head /d	Total	Per AU		
Sept.1997 to Aug.1981 (Pre-trial period)	57.6	29.3	-	-	9.28	0.16	11.0	36.0 (29)
Sept.1981 to August 1984 (Phase I)	50.8	28.0	21.70	1.2	7.44	0.14	10.9	32.7 (40)
Sept.1984 to August 1987 (Phase II)	56.8	34.8	79.20	3.8	6.40	0.11	13.2	34.0 (26)

Figures in parentheses denote number of calvings

Rice and wheat are considered as security food and fodder crops for India. These two crops contribute over 45% of the total crop residues produced in the country. With the expansion of combining of cereals in Indo-Gangetic Plains, less and less straw is available for livestock as most of the farmers burn the straw in the fields leading to not only economic loss but also creates environmental pollution. This necessitates the introduction of appropriate straw recovery systems for collection and utilization of straw as livestock feed which is already in deficit in many states of the country. The technology for ammonia (urea) treatment of loose straw for enhancing its nutritive value as livestock feed is well established. Systematic research has been carried out at Pantnagar for ammonia (urea) treatment of baled straw for last several years. The nutritional changes in treated bales have been found in the same order as that obtained with the treatment of loose straw. A new method known as 'Dripping Method' is being used successfully for ammoniation of 5 tonnes of stacked baled straw at a time on an ammonia (urea) treatment plant installed at Pantnagar. The cost of treatment works out to be Rs 300-350 per tonne. Long term feeding trials for a period of over 2 years have revealed a saving of about 25% concentrate mixture by feeding ammonia (urea) treated wheat straw for the same growth rate of the heifers fed on other diets. Also, there was no adverse effect detected on the reproductive behaviour of heifers and quality of milk.

There is a distinct alternative available to feed mill owners to expand and diversify their existing mills by adding ammonia (urea) treated straw with grain, minerals and molasses with an ultimate aim to produce a complete feed for ruminants in the form of pellets which could be economically transported to long distances or even exported to Middle East countries. Technologies for straw based complete feed are available in European countries and could be imported as per need. It also requires government policy decisions for establishment of various straw based industries such as animal feed, pulp and paper, board making, electricity generation, chemicals etc. on regional basis depending upon the availability of straw. In near past, the U.P. government announced large scale demonstrations of ammonia (urea) treatment technology to the farmers in its agricultural policy documents but it never materialized. There is an urgent need to involve the stakeholders and state government line departments for faster dissemination of technology.

Acknowledgements

The funds provided by the ICAR, New Delhi through NATP on "Baling, Densification and Storage of Grasses/ Fodders/ Crop Residues (PSR–39)," through which most of research studies have been carried out, is thankfully acknowledged. The association of Dr. M. L. Verma, Professor and Head, Animal Nutrition (Retd.), Department of Animal Science, Pantnagar in the projects on ammoniation of straw is highly acknowledged.

Bibliography

Agarwal, I. S.; M. L. Verma, A. K. Singh and Y.C. Pandey (1989). Feeding value of ammonia (urea) treated straw in terms of milk production. Indian J.Anim.Nut., 6(2): 89-96.

Agrawal, U. S. (1994). Development and performance evaluation of treatment structures for ammonification of wheat straw. Ph. D. (PHP&FEngg.) Thesis. GBPUA&T, Pantnagar.

Alhadharmi, G. I. (1993). Effect of preservation and maturity on the nutritional value of alfalfa hay for lactating dairy cows. Dissertation Abstracts, International, B-Sciences and Engineering, 52 (2): 586 B, Abstract of Thesis, University of Arizona, USA, 141pp.

Bamaga, O.A. and T.C. Thakur. (2003). A dynamic programming model for cost analysis and selection of straw recovery systems under Indian conditions. International J. of Agril. Engg. AIT, Bangkok (Accepted).

Bamaga, O.A. (1996). Investigations on batch method of ammonia (urea) treatment of baled wheat straw using dripping technique. M. Tech. (Agril. Engg.) Thesis, GBPUA&T, Pantnagar.

Bamaga, O.A. (2000). Evaluation of mechanized systems for collection, densification and ammoniation of combine harvested straw for livestock feeding. Ph. D. Thesis, Department of Farm Machinery and Power Engineering, GBPUA&T, Pantnagar.

Bamaga, O.A., T.C. Thakur, and M.L. Verma. (1999). An experimental investigation into batch method of ammonia (urea) treatment of baled wheat straw using dripping technique. J. Agric. Engg. 36 (1): 1-17.

Bamaga, O.A., T.C. Thakur, and M.L. Verma. (2000). Effects of ammonia (urea) treatment of baled wheat straw on its nutritive value. Indian J. Animal Nutrition. , 17 (1) 8-12.

Bamaga, O.A., T.C. Thakur, and M.L. Verma. (2002). Laboratory investigations into the wetting of baled wheat straw stacks with urea solution for ammoniation. Transactions of the ASAE, 45(1): 55-64.

Bergner, H. (1980). Chemische Grundlagendes Strohaufschlusses in der Pelletier-presse. Arch. Tierernahr, 30: 239-256.

Bergner, H.; J. Muller; J. Marienbeng,;, K. Adom and H. J. Zimmer (1974). Undersunger zur charak teristerung vonund. 15 N- Mar-Kierten Aufschluss mitten In undihre (cited by Jackson, 1978).

Bhatt Chetna (2003). Effect of feeding ammonia (urea) treated baled wheat straw with different levels of concentrate on nutrient utilization, blood and reproductive parameters in crossbred heifers. M.Sc. (Ag) thesis, Deptt. of Animal Sciences, College of Agriculture, GBPUA&T, Pantnagar.

Dolberg, F. (1992). Progress in the utilization of urea-ammonia treated crop residue: biological and socio-economic aspects of animal production and application of the technology on small farms. Livestock Research for Rural Development, 4(2).

Doyle, P.I.; S.Wales; W.J. Chanpongsang and G.R. Pearce (1987). Variation in the nutritive value of wheat and rice straws. In: R.M. Dixon (Ed.): Ruminant Feeding Systems Utilizing Fibrous Agricultural Residues. IDP of Australian Universities and Colleges, Canberra, Australia.

Gadre, K.R. (1979). A study of urea as a source of ammonia for treating straw to increase its digestibility. M. Sc. (Ag.) Thesis, G.B.P.U.A.& T., Pantnagar.

Gadre, K.R. and M.G. Jackson. (1980). A study of urea as a source of ammonia for treating straw to increase its digestibility. In: Maximum livestock from minimum land, Ed. Haque et al; Bangladesh Agricultural University; Mymensingh, Bangladesh, 66 p.

Ghate, S. R.; W. K. Bilanski and J.B. Robinson, (1980). Urea and chemstor as preservatives for high-moisture Corn. Transactions of the ASAE, 23(6):1569-1572.

Ghate, S. R. and W. K. Bilanski (1979). Treating high moisture alfalfa with urea. Transactions of the ASAE, 22(3): 504-506.

Hamid, M. A.; M. Haque and M. Saadullah (1983). Water hyacinth as a supplement for ammoniated rice straw. In: Fourth seminar on Maximum Livestock Production from Minimum Land, 2-4 May, 1983. Proceedings Ed. by Davis et al., Bangladesh Agril. Univ. Mymensingh, Bangladesh, 32p.

Jackson, M.G. (1978). Treating straw for animal feeding. Animal production and Health, Paper No. 10, FAO, Rome.

Jaiswal, R.S.; M. L. Verma and I. S. Agarwal (1988). Effect of various protein supplements added to ammonia (urea) treated rice straw diet on nutrient intake, digestibility and growth in crossbred heifer. Ind. J. Anim. Nutr., 5(3): 188-194.

Jaiswal, R.S. (1983). A study of optimum conditions for treatment of rice straw with urea as a source of ammonia and development of economic feeding system for growing crossbred heifers based on treated straw. M. Sc. (Ag.) Thesis, GBPUA&T, Pantnagar.

Jaiswal, R.S. (1990). Development and evaluation of optimum conditions for ammonia (urea) treatment of unchaffed rice straw in bulk and a study of methods of conserving / recycling of nitrogen during treatment. Ph. D.(Ani.Sc.) thesis, GBPUA&T, Pantnagar.

Khalilian, A.; M.A. Worrell and D.L. Cross (1990). A device to inject propionic acid into baled forages. Transactions of the ASAE, 33(1):36-40.

Knapp, W. R.; D. A. Holt and V. L. Lechtenberg (1974). Anhydrous ammonia and propionic acid as hay preservatives. Agron. J.,66: 823-834.

Lucht, H. W. (1990). Non traditional feed mill conditions straw with ammonia. World Grain, January/February/March, 1990.

Patil, B. R.; Pradhan, P. K.; Rangnekar, D.V. and Udo, H. 1994. Supplementation of milking cows in Gujarat. In: Kiran Singh and J.B. Schiere, (Ed.). On-farm Research for Testing of Appropriate Technologies in Crop-livestock Production Systems, ICAR, New Delhi and Wageningen Agric. Unv. The Netherlands.

Perdok, H. B ; M. Thamotharam; J. J. Blom; Born, Vanden and C. Valuw Van. (1983). Practical experiences with urea ensiled in Sri Lanka. In: T.R. Preston, C.H. Davies, F. Dolberg, M. Haque and Saadullah (Ed.): Maximum Livestock Production from Minimum Land, Joyedebpur, Bangladesh, P.123-124.

Rajan, P. (2001). Studies on performance evaluation of a field baler and ammonia (urea) treatment of large stacks of baled paddy straw. M. Tech. (Agril.Engg.) Thesis. GBPUA&T, Pantnagar.

Ramu. J. (1994). An investigation into the effects of ammonia (urea) treatment of paddy straw before and after baling. M. Tech (Agril.Engg.). Thesis, GBPUA&T, Pantnagar.

Rashiq, M. N. (1980). Urea treatment of wheat straw; 'In-vitro'and 'In-vivo' studies of the effect of urea treatment on the nutritional value of wheat straw. Development and Cooperation Bureau, Royal Veterinary and Agricultural University, Copenhagen, Denmark.

Saadullah, M.; M. Haque, and F. Dolberg (1981). Effectiveness of ammonification through urea in improving the feeding value of rice straw in ruminants. Tropical Animal Production, 6(1): 30-36.

Shinde, A. T. and T. C. Thakur (2004). Design, development and performance evaluation of a urea granule applicator for ammonia (urea) treatment of wet baled straw. Special problem report, Department of Farm Machinery and Power Engineering, GBPUA&T, Pantnagar, Pp. 1-46.

Singh, J. P. (1999). Investigations on ammonia (urea) treatment and manner of stacking of treated baled paddy straw as animal feed. M. Tech. (Agril.Engg.) Thesis. GBPUA&T, Pantnagar.

Singh, M.; V. K. Sharma and S. S. Ahuja (1998). Selected design and operational parameters of serrated tooth type bruising mechanism of a straw combine. AMA, 29 (3): 33-38.

Sundstøl, F.; E. M. Coxworth and D. N. Mowat (1978). Improving the nutritive value of straw and other low quality roughages by treatment with ammonia. World Anim. Rev. (FAO), 26: 13-21.

Tiwari, R. C; T. C. Thakur and Verma, M. L. (2001). Design, development and performance evaluation of a probe type injector for injecting liquid urea into rectangular wheat straw bales. The Inst. of Engrs (India).Vol. 82 pp 22-26.

Tiwari. R.C. (1993). Investigation on injection of liquid urea into baled wheat straw. M.Tech. (Agril. Engg.) Thesis, G.B.P.U.A&T, Pantnagar.

Verma, M.L. (1988). Availability and intensive utilization of cereal straws and fibrous agricultural residues in South Asia. In: C. Devendra (Ed.): Non-Conventional Feed Resources and Fibrous Agricultural Residues-Strategies for Expanded Utilization. Proceedings of a consultation held in Hisar (India), March 21-29, pp 21-32.

Tripathi, A. and S. Shome (1991). Exploratory studies on possibilities of injecting liquid urea into baled paddy straw. B. Tech. (Agril. Engg.) Dissertation, GBPUA&T, Pantnagar.

Verma, S. K. (2004). Studies on performance evaluation of wheat straw (bhusa) combine and ammonia (urea) treatment of retrieve straw for livestock feeding. M. Tech. (Agril. Engg.) thesis, Department of Farm Machinery and Power Engineering, GBPUA&T, Pantnagar, Pp. 1-143.

Thakur, T.C. and P. Rajan (2001). Availability of wheat straw in combine harvested fields for baling. Agril. Engg. Today. 25(5-6):1-14.

Uttam Kumar (2002). Performance evaluation of a conventional field baler and design, development of an ammonia (urea) treatment plant for stacked baled straw as livestock feed. M. Tech. (Agril. Engg.) Thesis, Department of Farm Machinery and Power Engineering, GBPUA&T, Pantnagar.

Uttam Kumar; T. C. Thakur and R. S. Jaiswal (2002). Design, development and evaluation of an ammonia (urea) treatment plant for stacked baled straw as livestock feed. Proceedings of the XXXVII Annual Convention of ISAE held at CTAE, M.P. Uni. of Agri. & Tech., Udaipur, Jan. 29-31.

CHAPTER - 6
MECHANICAL FORAGE HARVESTING

S.S. Ahuja, S. Khar and A. Sharda

India ranks first in milk production and also has the largest livestock population in the world. Present population of the animal wealth in the state of Punjab is 9.68 million cattle including buffaloes (8.4 million being adults animals). Buffaloes population in the country is about 57% of the total world population whereas other milch cattle population is only 15% of the world population. Consequently fodder production is an important component of this system. In terms of animal productivity our country is almost at the lowest edge. This is mainly due to insufficient supply of good quality fodder and feed available to the animals. The area under fodder crop in Punjab is approximately .78 million ha with an annual production of fodder around 55.5 million tones (Annon. 2003). Compared to this, the national figure presents a very dismal picture. The major fodder crops of Punjab and country are barseem, oats, maize, sorghum, bajra, Lucerne, guinea grass etc.

Harvesting of fodder is highly labour intensive job. Introduction of fodder harvesting machinery would encourage dairy farming and assist in crop diversification efforts. Also timely harvesting of green fodder would not only help efficient forage harvesters are required which could harvest the crop economically. to retain nutritive value but would also save time and cost of forage harvesting. Improved sickles have been used for harvesting fodder since ages and are still being used without much modifications. Harvesting operation requires about 160 man-hrs/ha involving a cost of more than Rs.1200/ ha. In Western countries, most of these operations have been fully mechanized. However, those machines are not of much relevance to our conditions because we feed freshly harvested fodder to the milch animals.

In addition our farm holdings are very small and our socio-economic factors are not very favourable for adoption of such costly machinery system. Fodder harvesting include operation of cutting the standing crop itself, conditioning transport and chopping. Therefore, in India small but

Principle of Cutting:

i) Shear cutting:

To cut the material two forces must be applied at the cutting stem so that the materials fails in shear. During this process small amount of deformation and bending of stem also takes place which causes wastage of energy. A common way of applying cutting forces is to press the material between two shearing elements which move against each other with or without very small amount of clearance. This helps in avoidance of bending process.

ii) Impact Cutting:

This is essentially a single element shearing and is possible only when the surface holds the plant erect from the stem to provide sufficient resistance against the cutting element. Here again cutting must be done close to the ground. Ground itself serves as one cutting element. A high speed cutting blade design provide much less time to stem to bend or deform.

Conventional cutter bar mowers:

Based on shear cutting principle few types of forage harvesters namely tractor driven side mounted mower, front mounted vertical conveyor forage harvester for tall crops and self propelled (engine operated) cutter bar forage harvester have been designed and evaluated (Table 1 & Fig 1). However so far side mounted tractor driven forage harvester has got commercialized on limited scale (for specifications see Appendix I).It has a reciprocating cutter bar, which is used to harvest fodder as well as selected cereals. After harvesting, it spreads the crop on the ground. Tractor mounted and engine operated reciprocating cutter bar are available these days. These are generally rear mounted or front mounted. Front mounted cutter bar have windrowing attachment with the help of star wheels guiding the linear movement of the crop to the side. However, while handling tall crops in between entanglement of the stalk causes choking problem, whereas tender crop like barseem does not move smoothly for side windrowing. But front mounted machines provide better visibility and have easy control, but require additional monitoring and drives to operate. Rear mounted cutter bar without windrowing facilities are more common and have been commercialized. Effective width of cutter bars varies from 1.52 to 2.75 meters. The offset drag force on the outer end of the cutter bar limits its width. Two shoes of the cutter bar help to maintain uniform height of cut. During transport cutter bar can be folded to vertical position.

Table 1: Machine performance of different types of reciprocating forage harvester

Type of Machine		Side Rear Mounted	Front Mounted	Self Propelled Modified	Self Propelled Engine Operated
Cutting width	mm	1720	2040	750	1400
Speed of operation	kmph	3-4	2-2.5	3-4	2-2.5
Height of cut	mm	40-50	80-120	50-70	40-50
Windrowing performance		Smooth cutting	Satisfactory upto 150mm	satisfactory	Scattering and choking problem
Crops tested		Berseem,oats	Oats,bajra, sorghum	Berseem	Berseem
Labour saving w.r.t. manual	%	83	80	73	67
Field capacity	ha/hr	.28	.20	.15	.10
Year of study	-	1987-89	1995-99	1991-93	1999-2001
Price of machine	Rs	18,000	25,000	40,000	50,000
Labour required for gathering	Man-hr/ha	25	30	40	50
Fuel consumption	lpha	10-12	12-15	10-12	12-15

However, tractor driven machines work better under dry field conditions and have more field capacity but cannot perform efficiently during rainy season. In order to meet this situation department has also developed a low weight self-propelled walk behind type of forage harvester. This machine has been operated with a 5 hp diesel engine having a cutter bar width of 1.50 m. Two side guards have also been provided to control and guide the width of cut. Power from the engine to the cutting mechanism has been provided through belts and pulleys using a small gear box.

The machine worked well on barseem. Effective width of cut is 1.45 m and can be operated at a speed of 1.5 to 2.5 kmph. The height of cut is 3-4 cms. In order to windrow the crop, two crop gathering drums having flappers rotating center wards have been provided slightly behind the cutter bar.

Power to the drum has been provided through chains and pulleys. The power to the wheel has been provided through extensions with the help of chains and sprockets. Track width is 104 cms. A castor wheel has also now been provided to improve the stability and maneuverability. The machine is still under final testing.

Fig 1: Different types of reciprocating forage harvesters

Registration and Alignment

A knife is said to be in proper register, if it attains its maximum speed when exactly in between two guards/ledger plates for smooth cutting and proper finish. For the adjustment entire cutter bar can be moved in or out with respect to pitman shaft. Alignment of cutter bar is quite important and it should be kept exactly at right angle to the direction of movement during field operation. Outer end of the cutter bar is normally kept 20 to 50 mm ahead per meter width of cutter bar depending upon forward speed and crop density to compensate for the backward bending of the outer edge of cutter bar.

Operating speed and adjustments

Pitman drives can be operated at 850 to 1000 rpm but this speed can be further increased to 1200 rpm with proper balancing. A typical speed of 1000 rpm at 76.2 mm stroke length gives maximum loop speed of 4 m/sec comparably very low for cutting stems by impact in rotary cutters with a speed of 50 m/sec (linear). Thus close clearance between knife section and ledger plate need to be maintained. Adjusting knife clips can reduce the clearance, whereas large clearance results in clogging of knives consequently resulting in increased power requirement due to stalk deflection. The forward speed ratio to the cutter bar rpm are adjustable in the machines.

Cutter bar can be operated at a slow forward speed in dense crop and at a relatively high forward speed if crop density is lower. But deflection of stalk before cut and bunching of stalk before a cut limit the top forward speed while in most cases these are below 6-7 kmph. Higher forward speed may cause frequent clogging of machine. If crop is not adequately erect, forward speed may have further to be reduced. Sharp and well maintained knives with close adjustment help to maintain good forward speed.

Energy & Power requirements

Energy & power requirements of reciprocating cutter bar are quire low in comparison to rotary cutters especially shredders. However, power requirement and energy vary widely depending type and condition of crop, condition of cutter bar and forward speed, On an average 0.90 to 1.0 kw per meter cutter bar width is required out of PTO at a forward speed of 7 kmph and 100 rpm crankshaft. Peak value may be around 2.4-2.5 kw/m cutter bar width. Out of this only 33% is the power required for crop cutting while the rest is for frictional requirements.

Impact Rotary Cutter

These are of two types; one rotating in vertical plane and other rotating in horizontal plane. Rotary cutter blades rotating in horizontal plane are installed on the periphery of disc or drum rotating on their vertical axis. These also have counter rotating pairs. There can be two or four drums. Effective width of cut is between 1-2 meters. Peripheral speed varies from 50-80 m/second. The counter rotating drum is able to make windrow in between. Blades are hinged through vertical pins on periphery of disc to provide back swing. Drum can be powered from P.T.O and belts.

These types of cutters differ from conventional cutter bars in many ways. Unlike cutter bar these can be operated at high forward speeds (upto 14 kmph). There is hardly any chance of choking if proper rotational speed is maintained. These need comparatively low maintenance, but have high initial price and high power requirements in comparison to simple cutter bar.

Rotary Disk-type forage cutters

This machine is tractor rear/ front mounted type. It has six swinging blade on each of the two rotary disks doing the cutting. Conical drums have been attached on each disk rotating in opposite directions form a windrow and prevent the crop from wrapping around the shaft (Fig 2). Power from pto to the rotating disc is transmitted through V-belts chains, sprocket & gear system. Height of cut can be lowered/ raised by adjusting the position of skidding shoe.

A field capacity of about 0.35 ha/ hour on berseem have been obtained with this machine at a forward speed of 5.4 kmph. Plant damage varies from 0 to 3%and uncut plant population varies from 3 to 5% (Table-2). Plant damage and uncut plant in general decreased with increase in forward speed. Windrow quality is satisfactory at all forward speeds. Blade width of 50 mm is suggested for the highest speed of 5.4 kmph for berseem. On guinea grass a field capacity of 0.27 ha/h was observed at a forward speed of 3.6 kmph. Plant damage and uncut plants were negligible. The machine was also used on bajra, menthe & oats. In bajra crop choking was observed due to creeper type of plants in the field. On oats cutting & conveying was satisfactory.

Flail Shredders/ Mowers

Flail type or shredders have large number of blades staggered in 3-4 lines on a rotating drum on a horizontal axis width of blade varies from 50 to 150 mm and has various shapes (Fig 2). These blades are attached to the drum or rotator shaft

Fig 2: Tractor operated disc & flail type forage harvester cum chopper

through hinges or small ring like to avoid damage if any hard object is encountered. The peripheral speed varies from 40 to 60 m/sec. Even though the field losses are more than the conventional mowers by 5-7 percent, but under lodged condition these have far better performance. Power requirement are quite high of about 25 kw/meter cutter bar width. It also depends upon many other factors like crop density, forward speed etc.

Two types of impact rotating blade cutters have been designed and field evaluated. These included, tractor front mounted horizontal rotating blade type, rear mounted vertically rotating blade type and rear mounted vertically rotating blade with chopping arrangement type flail cutters. (For specification see Appendix-I).

Table 2: Field performance of tractor operated disk & Flail type forage harvester cum chopper

Sr.No.	Parameter		Disk type rear & front	Flail type harvester cum chopper
1.	Speed	Kmph	3-5	1.8-2.2
2.	Effective Width,	mm	1300	1500
3.	Field capacity	Ha/h	.27-.35	0.13 to 0.16
4.	Size of chopped fodder,	Mm		40 to 90
5.	Labour requirement	Man-h/ha	40	15-18
6.	Labour saving (In comparison Sickle harvesting)	%	73	70-90
7.	Height of cut,	mm	50-100	24-38
8.	Fuel consumption	L/ha	12-15	15.8-22.5
9.	Tractor size	HP	25-35	35-45
10.	Crop height	m	Upto 3inch	3-5
11.	Year of investigation	1989-95	1989-95	1999-2001
12.	Tentative price (2005)	Rs	35,000	75,000
13.	Crops tested		Berseem,oats, maize,bajra	Maize, bajra, oats, sorghum
14.	Harvesting and windrow performance	mm	Unsatisfactory for front mounted	

Flail type forage harvester

A tractor operated forage harvester cum chopper is capable of harvesting, chopping and loading the chopped fodder like maize, bajra & oats etc. in an attached trolley. Working width of the machine is 1.50 m. This machine can harvest from 100 cm to 280 cm crop harvest with a stalk density varying from 20 to 80 stalks/ m^2 at a forward speed of 2 kmph. Cut size varies from 4 to 9 cm. (Table-2). Machine can be operated with any tractor of 35 hp or above.

However, none of forage harvester have been picked up for large scale commercialization due to one or other shortcomings. Therefore, it is suggested that further R&D efforts are needed to refine and reform these machines so that these meet the aspiration and requirement of the dairy farmers. Further more, extensive field evaluation and testing is also required to improve the acceptability of these machines and to popularize these technologies.

SUMMARY AND CONCLUSIONS

This paper discusses the need and relevance of forage harvesting in India. Both types of principle used in harvesting have been described. The performance of different types of machine both tractor as well as self propelled have boon discussed alongwith their limitations. Further research, development and field investigations are needed for adoption of these technologies on large scale.

Bibliography

Biennial reports of the Department of Farm Power & Machinery 1985-87, 1987-89, 1989-91, 1991-93, 1993-95, 1995-97, 1997-99, 1999-2001, 2001-03 (9 volumes).

Summer School on "Advances in Seeding and Harvesting Machinery" May 21- June 19, 2001. Dr. Surendra Singh.

APPENDIX - Ia

Specification of different types of reciprocating forage harvester (1985-2005)

Type of Machine		Side Rear Mounted	Front Mounted	Self Propelled Mounted	Self propelled Engine Operated
Type of prime mover	HP	Tractor 25-35	Tractor 25-35	Engine 7.5	Engine 5 hp
Dimension(size)	mm	Na	2120#2600#2870	1500#825#900	na
Prime mover speed	rpm	540	540	3600	na
Blade speed	rpm	540	500-600	600-800	na
Cutting width	mm	1800	2100	800	1460
Gathering mechanism		Na	Star wheel and conveyor	Na	flapper

APPENDIX - Ib

Specifications of rotary cutter forage harvester(1987-95, 99-05)

Machine Type/Parameter	Unit	Horizontal Disk Type Front and Rear Mounted	Vertical Flail Type Rear Mounted Forage Harvester Cum Chopper
Type of prime mover	HP	Tractor 25-35	Tractor 35-45
Speed of prime mover	Rpm	540	540
Cutting width	Mm	800	1300
Flail speed	Rpm(m-sec)	1100 (40)	1118 (42)
Dimensions(size)	Mm	1480#700#545	2120#2600#2870
Length of flail	Mm	75	Na
Bevel angle	Degree	25	Na
No of blades /unit	-	3	10 or 11
No of units	-	2	3
Width of flail	Mm	-	5
Shape of flail	-	Straight	c-type
Weight of machine	Kgs	175-180	350-400
Price/unit	Rs	35000	75000

CHAPTER - 7
SIZE REDUCTION AND PROCESSING EQUIPMENT

D.S. Rajput

The term size reduction is applied to all the ways in which particles of solids are cut or broken into smaller pieces. Forages and concentrates may be reduced in size in many ways, but only four of them are commonly used in size reduction machines/equipment. They are (i) compression, (ii) impaction, (iii) attrition or rubbing, and (iv) cutting. In general, compression is used for coarse reduction of hard solids to give relatively few fines; impact gives coarse, medium, or fine products; attrition yields very fine products from soft, nonabrasive materials. Cutting gives a definite particle size and sometimes a definite shape, with few or no fines.

According to Henderson and Perry (1976), agricultural produce is reduced in size by cutting, crushing, and shearing either singly or in combination. Cutting is separation or reduction which is produced by pushing or forcing a thin, sharp knife through the material to be reduced. Minimum deformation and rupture of the reduced particles results. The new surfaces that are produced by the sharp edge of the knife are relatively undamaged. Cutting is especially well adapted for forages. Juicy part is retained in the material and is therefore, useful for the milch animals. The most satisfactory cutting device is a knife of extreme sharpness and as thin as structurally possible. The motion of the knife should be such that the edge has a sawing component in moving through the material. This provides a smoother cut, probably with less energy.

Crushing is reduction by applying a force to the unit to be reduced in excess of its strength. Failure results by rupture of the material in many directions. The resulting particles are irregular in shape and size. The characteristics of new surfaces and particles are dependent upon the type of material and the method of force application. Limestone and other chemical fertilizers, ground feed for livestock, flour and meal, and fruit and vegetable purees are produced, in part or whole by crushing. Crushing is used to extract juice from sugar cane and to break the structure of forage crops to speed drying. The force used in crushing can be applied statically as is done when cracking a walnut in a vise or dynamically as with a hammer. Crushing by means of a rigid roll or bed such as the sorghum mill is an example of static force application. The hammer mill exemplifies dynamic force application.

Shearing is a combination of cutting and crushing. If the shearing edge is thin and sharp, performance approaches that of cutting. A thick, dull shearing edge performs more as a crusher. Shearing is usually used for reducing materials of tough fibrous nature where some crushing may be advantageous and the resulting units are of large uniform size. Cutting ensilage is an example. The shearing units consist of a sharp knife and a bar. The knife is usually thick to withstand the shock that results when it hits the material. For best performance, the clearance between the bar and the knife should be as small as possible and the knife as sharp and thin as practicable.

Chaff cutters and forage choppers for chopping and feed grinders and mills for grinding are most common processing equipment as used for size reduction of forages and feeds respectively. As per scope of this paper, size reduction and processing equipment for forages and feed concentrates have been covered and described in the following pages.

Size Characteristics

Size of the ground feed may be evaluated by screening the material through a set of sieves. Standard sieves are used for conducting the screening tests of the ground material. The techniques used in screening a sample have been standardized and should be followed if significant results are to be expected. Screen analysis can be used for classifying any granular material.

Fineness modulus indicates the uniformity of grind in the resultant product and is defined as the sum of weight fractions (in percent) retained above each sieve divided by 100. Any standard procedure for determining the fineness modulus may be adopted.

Accepted farm practice show that coarse grinding is advisable for feeding. Finely ground grain feeding to farm animals is not advisable under any condition except for chicks. Chopping of hay and stover is advisable to minimize the amount of material thrown out of the feed bunks by the animals. Chopping of forage is also justified in order to mix less palatable but nutritious forage with other feed.

The performance of a machine for reducing the size of material is characterized by the capacity, the power required per unit of material reduced, the size and shape of the product before and after reduction, and the range in size and shape of the resultant product. Therefore, in order to study performance, a method or methods must be available for determining the size characteristics of any material. Sieves may be used to classify the size of ground material and determine the performance of feed grinders.

Power and Energy Requirement

The power required to operate a forage chopper is utilized in the following ways: (a) to gather, convey and compress the material to be cut, (b) to shear the material, (c) to move the air pumped by the cutterhead and impeller-blower, (d) to accelerate the cut material to approximately the peripheral speed of the impeller, (e) to overcome losses due to friction of the cut material within the housing, and (f) to overcome mechanical losses in the machine.

Distribution of power required in a flywheel-type field chopper as a funtion of feed rate has been reported by Bainer et al (1955). Similarly total energy requirement for forage chopping has been reported by Bainer et al (1955).

Energy necessary to reduce a specific mass of particles from one size to another is

$$E = -C \int_{2}^{1} (dL/L^n) \quad \text{———} (1)$$

Kick's law and Rittinger's law were developed assuming n = 1 and 2, respectively. These equations were developed from studies of materials common to the chemical and mechanical engineering such as coal, limestone, etc. Such materials are different from agricultural materials such as forage, small grains, etc.

Studies on energy requirement for grinding and chopping of agricultural materials have been conducted in the past. Hansen and Henderson (1972) and Hansen and Stewart (1965) reported that the following equation representing grinding energy for those ground materials that yield a straight line on a log-log plot.

$$E = A\,K^{-a} \quad \dots\dots (2)$$

Where E = energy input per unit weight

K = product size modulus (representative sieve size which will theoretically permit

100% of the ground material to pass through)

a = distribution modulus (from equation)

A = a constant

Size Reduction Equipment

Agricultural processing includes those operations which maintain or raise the quality of raw material, change its form, or prepare it for market. Its importance is growing and will continue to increase as farming becomes more commercialized, specialized and mechanized in India. Some of the advantages are:

1. Produce, which is now being wasted because of inadequate transport facilities and excess supply can be preserved and used later.
2. The processing may serve as a small-scale industry, providing additional employment facilities for idle labour.
3. Perishable products can be changed to a stable commodity, e.g. canning of fruits, drying of fruits and fodder, making of silage, etc.
4. Products can be prevented from being affected by insects and pests during storage, rendering them completely unusable.

The reduction in size is brought about by mechanical means without change in chemical properties of the material and uniformity in size and shape of the individual grains or units of the end product is usually desired but seldom achieved.

Forage Chopping Equipment

It is a well known fact that animals are equipped by nature to grind and chew most fodder crops, but still they are fed chopped fodder. This is done: 1. to save storage; 2. to aid in curing; 3. to make it more palatable and 4. to keep the crop from spoiling, while in storage. The machine used for chopping fodder is called chaff cutter or ensilage cutter or silo filler.

Chaff Cutters

Choppers (cutters) are most commonly used for chopping of forages. These include (i) manually operated chaff cutters (ii) animal operated chaff cutters (iii) stationary power operated choppers (iv) field forage harvester cum chopper.

According to the power used for operating these machines they are divided into three different types:

1. Hand chaff cutters.
2. Animal operated chaff cutters.
3. Power chaff cutters or silage cutters.

These are also classified as:

1) cylinder type cutting head
2) fly wheel type cutting head machine

The chaff cutters driven by manual, animal or mechanical power have been basically blades, which chop the stalk with shearing action pressing it with the face of shearing plate.

In any of the chaff cutter the stalk is fed either with the help of a conveyor or manually to the rollers which have teeth to grip the stalk. These rollers feed a compressed layer of stalk to the shearing plate. The cutting blade in a rotating action comes in contact with the layer of stalk and either by impact force or impact and shear force chops it. Thus chopped fodder is blown (in some cases) aside.

A forage harvester has a number of blades hinged in a rotating axis, which while in rotation due to centripetal action becomes as straight and powerful to cut the standing crop in the field as desired. The special characteristic of the harvester is that a single group of cutting- blade performs three tasks at a time.

1) Cutting the standing crop
2) Chopping it successively and
3) Lifting as well as throwing the chopped fodder in a trolley.

The common trouble as listed below may be remedied accordingly.

1) The blades become blunt
2) Worn bearing
3) Due to improper lubrication machine found shricking
4) Loose nuts and bolts requiring more power than the recommended.

The cylinder type heads are mostly found on power chaff cutters. They are also mounted on field forage harvesters, which perform simultaneously three operations. Namely, harvesting chopping and blowing into wagons these machines are provided with four spiral-shaped knives on a revolving cylinder, which looks something like the rotary lawn mower. The knives can be sharpened without being removed from the cylinder. While sharpening the blades with a grinder, care should be taken not to spoil the bevel of the edge and also not to get it overheated by excessive grinding. The length of the cut is changed by changing the speed of the feed rolls provided adjacent to the shear plate. The chopped material falls into a housing from where the blower fan blows the fodder out either into a wagon or into a silo pit. Cylinder head type machines are not in common use in India. The cutting head is not as simple in construction as the other type, hence it tends to be more costly.

The flywheel type of cutting head consists of a cast iron flywheel, blower fan and radially mounted knives. Straight knives are used on power chaff cutters whereas curved knives are mounted on hand chaff cutters. The number of knives varies from two to six on power chaff cutters, but hand chaff cutters are provided with two knives only. The blower blades are mounted along the circumference of the flywheel and there are eight on power chaff cutters. Since these blowers are used for silo filling, they are not provided on chaff cutters, which require two men to operate and a third man to feed the crop. The average working speed of the hand chaff cutter is 50 rpm with one knife in use and 35 rpm when using two knives. The power-operated machines are operated in the range of 600 to 1000 rpm.

To obtain a clean cut, knives must operate close to the shear plate without striking. They can be adjusted for clearance by set screws. There are about six set screws provided on each knife which is fastened to the flywheel by means of two to three counter sunk bolts. The length of cut can be changed by:

1. Changing the speed of the feed rolls. (Power operated machines are provided with a two or three speed gear box. Hand operated machines are equipped with a two speed worm for this purpose.)
2. Reducing the number of knives (on hand operated machines, sometimes one of the two knives, is removed.)
3. Increasing the speed of cutter head (it is possible on power operated machines).

The most common length of cut for silage ranges from 2.5 to 5 cm. The length of cut of green fodder may be more than 2.5 cm.

The feeding mechanism of a power chaff cutter consists of an apron (endless chain) and two feed rolls. Aprons are generally provided with metal or wooden slats to give a positive feed. The length of the table on which the apron moves is about 2m so as to accommodate tall grasses or fodder-crops. Spring-loaded feed rolls are provided to compress and feed the unchaffed material to the cutting head. Generally, corrugated or toothed rolls are preferred. Hand operated machines are not provided with aprons. A feed table or merely a feed box is considered to be enough for this machine and the feeding is done by hand. Feed rolls of the hand-operated machine have variable spacing and they are, operated by the worm mounted on the flywheel shaft.

An animal operated chaff cutter is very much similar to the hand operated one. The only difference is that it is provided with a smaller diameter flywheel on which two or three knives are mounted. By means of a set of gears and universal joints, the slow speed of bullocks is utilized to operate the machine at about 250rpm and thus its capacity becomes 4 to 6 times that of the hand-operated machine.

The capacity the power operated chaff cutters is expressed in metric units by Duffee's formula:

$$C = W\ H\ L\ N\ R\ K \qquad (3)$$

Where C = capacity of machine in metric tonnes per hour

W = width of throat in cm

H = height of throat in cm

L = length of cut in cm

N = number of knives on the fly wheel

R = speed in rpm

K = constant equal to 2×10^{-3}

The constant K includes the value of effective density of maize at 70 per cent moisture content. Here the value of density is 332 kg per cubic metre. The capacity may increase proportionately to the density.

Feed Grinding Equipment

All concentrate feed (grain and oil cakes) should be ground for a dairy herd in order to obtain the maximum feeding value. Otherwise as much as one third to half of the grain may pass through the cows undigested. Grinding is also profitable for many crops with other kinds of livestock. The feed value of grain sorghum to all livestock is greatly increased by grinding. In some countries even the roughage not containing grain is also ground. It is believed that grinding of roughage does not increase its digestibility, but may be desirable to improve palatability, avoid digestive troubles and lessen waste.

Concentrate feed is mostly fed to the livestock in the form of either powder or broken pieces. In order to reduce the size of feed, some type of feed grinder is used. Common practice of reducing the size is to use the hand operated small grain pounding stone (chakki). This small equipment is unable to meet the requirement of large size herd, therefore, power grinders are generally preferred. The two most common types of feed grinders are

1. Burr grinder or Burr mill or Flour mill.
2. Hammer grinder or hammer mill.

(1) Burr Mill

Most of the burr feed grinders are provided with two circular stones or cast iron plates or rollers which have their rough surfaces parallel to each other either in horizontal or vertical planes. One of the plates or rollers is fixed to the drive shaft of the power unit and the other remains stationary, The feed is fed into the centre of the plates. It is ground there to the required fineness before it comes out along the periphery. The fineness of the final product depends upon the roughness of the grinding surfaces and the gap between the two. The gap can be adjusted by means of a hand screw and is locked in the desired position. For increasing the roughness of the grinding surfaces, the plates are taken out and the roughness is increased, by chipping the surfaces of the plates. The cast iron plates wear out at much slower rates and their surfaces are made rough by means of special cutting tools. Generally, the stone plates are worked about 50 hours before their surfaces are chipped. If the desired ground product has not to be in the powdered form, the burr plates are worked for a considerable period. The capacity of the burr grinders depends directly upon the diameter of the plates, fineness of the grinding and the speed. The diameters of the power mills vary from 20 to 45 cm and they are generally operated at 1000 to 1200 rpm. Hand operated machines are also in use but they have a very low capacity.

Grinding feed is fed through a hopper by means of gravity and ground material is received in bags at the bottom. Sometimes an elevator type bagging attachment is also used in large mills. Burr mills are commonly used for making flour in India. They are also used in making rough feed from grain. Generally, they are not suitable for crushing oil cakes which constitute a major part of the live stock feed in India. This machine is also likely to be damaged by presence of undesirable materials like stones, nuts and bolts in the grinding feed.

(2) Hammer Mill

The hammer grinder consists of a rotor, rigid or swinging hammers and a screen. The hammers are fastened on the rotor, which rotates at the speed of about 2000 to 3500 rpm inside the housing. A semi-circular screen is placed below these hammers. The size of the screen determines the fineness of the final product desired. In the process of grnding, the grinding material is fed into the housing where revolving hammers at high speed strike the material to explode into pieces. Due to the constant hammering action the material gets disintegrated till it becomes finer than the screen to pass through holes. The ground feed is either removed from beneath the mill or it is blown into the larger hopper of the bagging attachment. If a machine is operated at the recommended speed, its capacity depends upon (i) rate of feeding, (ii) hardness of the grinding feed and (iii) size of the opening in screen. The hammer mill is favoured because of the following reasons:

1. Foreign material in the feed will not damage the machine parts.
2. Hammers do not require frequent attention such as sharpening, chipping etc.
3. Any kind of feed from course tó very fine can be ground.
4. Efficiency of the machine is not affected by the length of use it is put to.
5. The fineness of the ground material is positively controlled by the size of the screen. It does not need to be adjusted during operation.
6. More than one kind of feed may be ground in any desired proportion at a time.

Due to the obvious advantages, the machine is very commonly used for grinding on large dairy farms. However, it is not available in smaller sizes as a hand operated machine because of the high-speed requirement. Grinding screens are available in the following range of many sizes

Size reduction of forages and feed or concentrates is practiced for their better utilization by the animals. Forages are chopped to reduce their wastages. The feeds are ground to increase their digestibility by the animals. Accordingly, forage and feed processing equipment such as choppers (chaff cutters) and grinders (mills) have been developed in our country and being used to achieve the desired effect. Field forage harvesters which cuts the standing crop, chops into small pieces, elevates it and finally loads into an accompanying trailer, which are also available in our country basically to economies labour requirement and promote large dairy enterprises in India.

Bibliography

McCabe, W.L.; Smith, J.L. and Harriot, P (1985) Unit Operations of Chemical Engineering, 4th Edition, McGraw Hill Book Company, New York, Chapter 27: Size Reduction, PP 767-796.

Henderson, S.M. and Perry, R.L. (1976) Agricultural Process Engineering, 3rd Edition. The AVI Publishing Company, INC, Westport, Connecticut. Chapter 6: Size Reduction,pp 130-159.

Bainer, R; Kepner, R.A. and Barger, E.L. (1955). Principles of Farm Machinery, John Wiley and Sons, INC., New York, Chapter 15: Forage Chopping and Handling , pp 346-369.

Hansen, R.C. and Henderson, S.M. (1972) Development of size distribution, measurement, and representation for ground cereal grains. Trans. ASAE 15, 510-514.

Hansen, R.C. and Stewart, R.E. (1965) Energy-size reduction relations in agricultural grain comminution. Trans. ASAE 8, 231- 237.

Smith, H.P. (1965) Farm Machinery and Equipment, Fifth Edition, Tata McGraw-Hill Publishing Company Ltd. Chapter 17: Forage Harvesting Equipment, pp 317- 328.

Micahel, A.M. and Ojha, T.P. (1966) Principles of Agricultural Engineering: Volume I, Chapter 9 Agricultural Processing and Plant Protection Machinery, pp 269-298.

Srivastava, A.C. (1990) Elements of Farm Machinery, Oxford & IBH Publishing Co. Pvt. Ltd., New Delhi. pp 182-183.

CHAPTER - 8
PRINCIPLE AND DESIGN OF FODDER HARVESTERS

Chandra Shekhar Sahay

Fodder and forage crops are the requirement of dairy farmers. In case of the bigger dairy farms, requirement of fodder may be huge and vary from season to season in the form of fodder, silage fodder or fresh green forage. In the year 2003, total availability of green forage crops in the country was 387.7 m tonnes and the same for dry fodder crops was 437.3 m tonnes (Pathak and Kumar, 2004). This reveals the quantity of forage and fodder crop sowing, harvesting, transporting and handling them before feeding to animals. For the large dairy farms, there is an inevitable requirement of mechanical support to perform the operation. Main operations in forage and fodder handling are harvesting the crop at appropriate time, transporting, chaffing, feeding or making silage as the case may be. In case of small dairy farmers, the labor and energy requirement for performing the operations is fulfilled manually be their family members or hired laborers. For manual operations, suitable implements are required to perform the operation in an efficient way. Harvesting of the fodder crop is one such operation, which involves intensive energy input to the fields. This paper describes main harvesting principles and different machinery available for performing harvesting of straw and fodder.

Principles of Harvesting

Mechanical harvesting of a plant involves separating or cutting the plant or residue from the root material using a mechanical means. Cutting takes place when a system of forces acts upon the plant in such a way that the upper portion of the plant is sheared off with rest of the plant. This shearing may be provided either by sharp, opposite moving blades or by imparting the energy to the plant material in such a way that the plant stem gets crushed and breaks into pieces. These two modes of harvesting are discussed as following.

Shear plates

In this mechanism there are two sharpened plates moving in relatively opposite direction and the plant material comes in between these two blades and gets cut. The support for cutting plants is provided by the other edge as in case of a common scissor. Here the considerations of design importance are shear strength of the plant material, stiffness of the plant against bending, operating speed of the blades, clearance in between two relatively opposite moving blades and conveyance of material after cutting. This principle is applied in the entire conventional mower, which are having reciprocating cutter bar. The power transmission chain of machine consists of carrying the power from the engine to reciprocating cutter bar. In case of these cutter bars, after cutting of material, it is to be disposed in such a way that there is easy flow of the uncut crop on the reciprocating cutter bar.

Impact Cutters

In these cutters, energy for shearing the plant is imparted by giving an impact to the plant material. This principle is applied in rotary cutters and flail mowers where the impact energy is provided by the kinetic energy of the moving blade. When energy is provided in the impact form, there is an essential action of compressing or crushing of plants before cutting. Therefore the plant material cut here is more crushed as compared to shear cutting. In this case, the stiffness for cutting the plant is provided by the standing strength of the plant itself.

Straw and Fodder Harvesters

Different straw and fodder harvesting machines which are used for harvesting, cutting, chopping are following.

Tractor operated straw chopper cum spreader

The common practice after harvesting of rice in rice-wheat growing belt is to burn the straw stubbles left after harvesting of the field using combined harvester. For burning the straw, it is first cut and spread in the field. The machines like harrow and rotavator are unable to operate in the harvested field with stubbles. Therefore for straw management and incorporation of the straw in to the field, a tractor-operated straw chopping machine (Fig.1) has been developed to chop the straw and spread it in the field. This machine harvests the straw left after the operation of combined harvester and chops it into pieces for spreading in the field in a single operation. The chopped and spread stubbles are buried easily in the soil by the operation of disc harrows in two passes. After burying of straw other crop operations like sowing of wheat could be taken as usual.

Fig. 1 Flail straw harvester cum chopper

The tractor drawn straw chopper consists of a rotary shaft mounted with flail type blades to harvest the straw and uses knives for chopping the harvested straw. The straw after cutting by flail passes on to the chopping mechanism. Harvesting unit has 38 flails mounted in three rows. Chopping mechanism has 660 mm diameter cylinder with 6 rows of serrated knives and 4 counter rows each having 22 knives at the bottom. The working width of the machine is 2280 mm. The capacity of the machine is 4 ha/day and a 30 kW tractor operates it. The cost of machine is approximately Rs.30,000. Performance results of straw chopper cum harvester are shown in Table 1.

Table 1: Performance results of straw chopper cum spreader

Parameter	Observation
Combined harvested crop residue of crop	Rice
Height of stubbles(mm)	400-500
Moisture content of the straw (WB), %	50-70
No. of hills/ m^2	30-35
Parameter	**Observation**
Straw quantity, kg/m^2 (loose and standing)	0.87-1.25
Speed of operation, km/h	2.5 to 3.0
Fuel consumption, l/h	5.0-5.5
Chopped straw size (mm)	70-100
Flail speed (rpm)	1000
Chopper speed (rpm)	2000
Labour requirement, man-hr/ha	1.25

After operation of this chopper, there are some problems of choking when zero till drill or strip till drill is used. This problem is overcome by one pass of rotavator and applying irrigation before using zero or strip till drill. However, Traditional seed drill operates well after operation of this flail type chopper harvester and it does not further need the rotavator operation and irrigation.

Since the incorporation of combined harvested rice straw checks environmental pollution providing rich organic manure, there is great acceptability of this machine in the rice wheat growing area and this tractor drawn chopper spreader is gaining popularity.

Tractor Mounted Flail Type Forage Harvester Cum Chopper

Tractor mounted flail type forage harvester cum chopper (Fig. 2) is a machine developed for in-situ harvesting, chopping and loading of chopped fodder in the tractor trolley in a single pass of operation. This machine consists of a rotary shaft mounted with blades to harvest the crop, auger for conveying the cut crop, cutters cum blowers for chopping and conveying chopped fodder through outlet on to the trailer. The blades on the rotary shaft are staggered in 3 rows of 13 blades on each row on a horizontal axis perpendicular to the direction of motion. After the blades cut the crop, it comes to the auger, which conveys it to the chopper unit. The chopping mechanism having 3 blades cuts the crop in to pieces and the chopped material is blown out for filling in the trailer.

Fig. 2 Flail type fodder harvester cum chopper

Using this machine, the crops like maize and sorghum could be chopped to a size of 80 mm. The field capacity of this machine is 0.44-0.52 ha/hr and labour requirement is 3.84 to 4.54 man-hr/ha. Fuel consumption of this machine is 4-4.5 l/hr. It costs Rs.70000 and cost of operation in the field is Rs.1113 /ha. Performance results of this machine are summarized in Table 2.

Table 2: Field performance of flail type mower cum chopper

Parameter	Observation
Speed of operation km/h	1.99
Working width, m	1.12 to 1.26
Field capacity, ha/h	0.20
Size of chopped fodder, mm	80-90
Labour requirement	One person in addition to tractor driver
Labour saving as compared to manual harvesting and chopping , per ha	90
Cost of operation, Rs./ha	1113
Height of cut from the ground level, mm	135
Time required to harvest trolley load of fodder, min	22
Stalk density of the crop, plants/m^2	72.2
Crop height, mm	2235

Self Propelled Fodder Harvester

Self propelled fodder harvester (Fig. 3) is a machine used for cutting soft fodder crops like berseem. This machine consists of a 1450 mm reciprocating cutter bar which is operated by a 3.75 kW diesel engine mounted on the chassis of the machine. Power from the same engine is used through chain and sprocket to drive the ground wheels so that it becomes self-propelled machine. It is like a self-propelled vertical conveyor reaper with the difference of collecting the cut crop. There are two crop gathering drums having vertical sheet flaps, which rotate inward. The vertical side covers around the transmission system are provided to avoid blockage of the fodder and extended vertical guides were provided for proper windrowing. The ground clearance of the machine is 480 mm in order to facilitate easy passage to the fodder in such a way that the cut crop is gathered and windrowed in the center of the cutter bar. The track width of this machine is 1040 mm and a caster wheel is provided at the rear to improve stability and maneuverability.

Fig. 3 Self Propelled Fodder harvester

Performance of self propelled fodder harvester in berseem crop produced field capacity of 0.18 ha/hr at 2.62 km/hr forward speed. It cuts the crop at 30-40 mm height. The working width of machine is 900 mm in a field where 640-672 plants are there in one m^2.

Tractor Drawn Rotary Disc fodder harvester

Tractor Drawn rotary disc fodder harvester (Fig. 4) has been developed at IGFRI with specific objective to harvest soft, green forage crops like berseem grown in the leveled field without cross bunds. It is hitched on the three-point linkage of tractor and the whole unit is side mounted. While cutting operation, tractor runs in the field where crop is already cut and rotary disc cutter cuts the crop in the side of the tractor.

The cutting unit of this machine consists of two rotary discs. On the periphery of these rotating discs, horizontal flat, sharpened flails are attached which cut the crop. While operation, these two discs rotate inward and the cut crop is windrowed in between two discs when the whole unit moves forward with the tractor. The power in this machine is fed through tractor PTO wherefrom a belt pulley mechanism drives a chain and sprocket transmitting power to rotary discs. The field capacity of the machine was 0.3 ha/hr with a field efficiency of 74%.

After the operation of this machine, the crop is cut and windrowed in the field. While operation of the tractor, care is to be taken that the tractor does not move on the cut crop otherwise there is some crushing of the soft forage crop, which increases unpalatability of the crop to the animals. The field where operation is being done should be free from any kind of bunds, stones, pebbles etc. In case of bunds, soil also comes out with the cut crop however small stones and pebbled have got the tendency to be thrown outside which may prove to be hazardous.

Fig. 4 Tractor operated rotary forage harvester

Engine operated grass harvester

The engine operated push type grass cutting machine (Fig. 5) has been developed at the ICAR Research Complex for North Eastern Hill Region, Barapani, Meghalaya. This machine uses a petrol start kerosene run engine of 3400 rpm with 1 kW power. Engine power is transmitted through a V-belt drive from engine to vertical cutter blade holding shaft assembly.

A square angle iron chassis supported on three wheels was made to hold the engine, cutter blade assembly and handle. Major portion of the weight of machine was supported on a two wheels on which machine runs. A third wheel was provided which is supported by a spring for facilitating variation of height of cut up to 10 cm. On a plane surface, height of cut of grass is 5 cm. Since the rotational speed of the blade is high (1200-1400 rpm), a safety skirt was made of 15 numbers of 0.5"Æ m.s. rods bent in the form of arcs covering the blade. The average field capacity and fuel consumption of the machine is 0.0732 ha/h and 1.04 l/h respectively. The cutting efficiency of this machine is 100% because no weed was found uncut after the operation and the machine was able to cut perfectly on the planes of slopes up to 30°. The cost of operation using this machine is Rs.278 /ha. It is able to cut grass/bush up to thickness 2 cm and height 1.5 m.

Fig. 5 Engine operated grass cutter

Harvesting of straw and fodder is an energy consuming operation, which in bulk requires a suitable machine for operation. The harvesting of crop is done on the two basic principles viz. shear cutting or impact cutting. Different types of machines are used to perform the operation in specific situations. By use of these machines, cost of operation is reduced as well as there are many other in-built advantages like timeliness of operation, reduced labor requirement, reduced dependency on labor and reduction in drudgery of operation.

Bibliography

Pathak, P. S. and Sunil kumar, 2004; Forage and Grazing Resources in Different Agro-climatic Regions of India; book chapter in 'Buffalo Production Under Different Climatic Regions', Indian Grassland and Fodder Research Institute, Jhansi(UP), India pp 1-42.

Kepner, R. A., Roy Bainer and E. L. Barger; Principles of Farm Machinery; CBS Publishers and Distributors, 485, Jain Bhawan, Bhola Nath Nagar, Shahdara, Delhi 110032, India, pp:313-339.

Pandey, M. M.; Research Highlights-2004, Technical Bulletin No.CIAE/2004/105; All India Coordinated Research Project on Farm Implements and Machinery, Central Institute of Agricultural Engineering, Nabi Bag, Berasia Road, Bhopal 462038, India, pp: 1-115.

Sahay, C. S., Agrawal, K. N. and Satapathy, K. K. "Development and Evaluation of Grass/ Bush cutter for Hilly terrain" Indian Journal of Hill Farming, Vol. 10, No. 1&2, June & December' 1997 pp67-72.

CHAPTER - 9
CROP RESIDUE MANAGEMENT

P.D. Gupta

Wheat crop is grown in more than 24.73 million ha in India, out of which Uttar Pradesh (34.38 %), Punjab (21.05 %), Haryana (12.76 %), Madha Pradesh (11.19 %) and Rajsthan (8.91 %) contribute maximum to the total production in the country . Similarly paddy is grown in about 44.6 million ha out of which major states growing paddy include West Bengal (15.59 %), Uttar Pradesh (14.43%), Andhra Pradesh (11.72 %), Punjab (9.75 %), Bihar (8.65 %) and Tamil Nadu (8.08 %) in 1999-2000. India produces about 21.5 per cent of global paddy and ranks second in world after china (34.5 %) (Singh and Chandra, 2002). The total production of agricultural crop residues is more than 200 million tonnes in India which could be used as animal feed, power generation, briquetting and for paper or board making industries.

Increased mechanization has made the straw recovery and it utilization from paddy and wheat field difficult for the farmers. The use of grain combine harvester in paddy- wheat region of country is increasing day by day at a faster speed. Off-field utilization of paddy and wheat straws continues to get attention due to enviroment impact from open field burning of the crop residues for its disposal. The baling/densification of crop residue will help in reducing the environmental pollution as burning of straw results in severe environmental pollution resulting in low visibility during dusk and dawn hours especially in winter. This phenomena has also been witnessed in developed countries like USA and Britain, particularly in California. In our country paddy straw being used as cattle feed in some states like Bihar, Madhya Pradesh, Eastern U.P. and some parts of Andhra Pradesh and Orissa. Beside these states about 5% paddy straw is used as cattle feed, 2% for making farm structures , 5% as raw material for making paper and card board, 7 % as packaging material for horticultural crops and other industrial goods like china ware , etc. and rest 80% is burnt in the field causes environment hazards and unhealthy atmosphere (Singh et al ., 2001). The field balers is recent introduction in India for recovery of left over combine harvested straw for its as utilization .The knowledge of straw availability and losses are of paramount importance for determination of economics of the system. Study conducted by Thakur and Rajan (2003) on three wheat varieties revealed that at the prevailing height of cut of 35-90 cm by combine , the straw availability was in the range of 1400-2000 kg DM /ha out of potential straw availability of 4900-6400 kg DM/ha . Also the average linear density of swath was found to be about 0.98 kg DM / m as against required capacity of 1.72 kg DM/m for full loading of baler, thereby, suggesting the introduction of field rakes for windrowing or lowering the height of cut of combine.

Crop residues wheat and paddy left over in the field after combine operation can be harvested and collected by sickles but this is time consuming and uneconomical. Labour shortage at the time of harvesting is also acute. To overcome this problem scientist has worked on this aspect and developed technology of harvesting, collection and baling/ densification of these residues for economical transport and storage.

Harvesting and Collection of Wheat stubble

IGFRI, Jhansi developed technology for harvesting and collection of wheat crop residues left over after combine operation in totality i. e. collect the chaffed thrown stubble(about 35%) by combine as well as standing one (about 65%) . An assessment of left over stubble after combine harvesting of the wheat crop was conducted which indicates that 36.5 q stubble (bio-mass) remains stand in the field after combine harvesting, which could be utilized for feeding the cattle. The initial length of standing stubble as obtained after combine harvesting of crop was 57 cm. This stubble if not harvested timely causes tremendous losses of available biomass. If these stubble harvested manually the average length of cut stubbles was 54.68 cm weighing 0.35kg/m^2 (35q/ha) indicating 2.32 cm leftover remains in the field causing a loss of .015kg/m^2 (1.5q/ha). A thumb rule in harvesting wheat stubbles was that about 300m^2 area per day could be harvested by one manual labour requiring 33 mandays/ha @Rs. 86.81/manday. In the study the working capacity of one manual labour for harvesting wheat stubble was found as 100kg/day. Thus, about Rs. 3038.35/ per hectare is required for harvesting wheat stubbles by manual labour. A Tractor operated IGFRI flail mower was used to harvest and collect the combine harvested wheat stubble for saving time and money. The tractor operated flail-type forage harvesters suitable for harvesting all types of forage crops and grasses was used to collect left over wheat crop stubble after combine operation. After cutting the crop at the base by impact of the rotating flails, it chops and lacerates the crop for easy intake by the animals and places the chopped forage in the accompanying trailer. The flail-type machine is simple in design and produces chopped lacerated fodder, which is suitable for fresh feeding and silage making.

The capacity of is flail mower is 1.5 ha/day or 40q/day. The machine left an average of 15.24 cm stubble, weighing 0.098kg/m^2, which means a loss of

9.8q/ha occurred by using the flail mower. It also shows that the machine was able to cut 41.76 cm of wheat stubbles with an additional advantage of chopping and laceration of the crop 3 times as the cut length obtained was 12.71cm, which may be useful for direct feeding to the animals. The economics of flail mower indicates that Rs.32/q is required in harvesting wheat stubble. Harvesting a hectare field of wheat stubble, therefore required Rs. 853/- by flail mower and Rs.3038.35 by manual labours (Table1). A saving of Rs. 2185.35 per hectare was thus obtained by using flail mower in comparison of manual harvesting. If 8.3 q more loss of wheat bhusa @Rs80/q by flail mower is considered in comparison to manual labour the loss in term of money will be Rs. 664 which also shows a saving of Rs. 1521.35 by mechanical harvesting of wheat stubbles. However, it can not be said as loss because now a days in case of zero till seeding the leftover crop residues in the field is beneficial as it is incorporated in the field.

Table 1: Comparison of harvesting wheat stubbles by manual and flail mower (Jhansi)

S. No.	Content	Units	Flail Mower	Manual
1.	Capacity	q/h	5.000	0.125
2.	Capacity	ha/h	0.187	0.00357
3.	Machine or man hours/ha	-	5.34 h	35 man days of 8 h
4.	Cost	Rs./q	32.00	86.81
5.	Cost	Rs./ha	853.00	3038.35
6.	Cost	Rs./h	159.70	10.85

The technical details of tractor operated IGFRI Flail forage harvester is given below:

IGFRI Flail forage harvester

Prime mower	:	45hp tractor
Type of drive	:	PTO (540 r. p. m.)
Power transmission system	:	PTO shaft and V-belt
No. of flails	:	16 in rows
Width of flail, cm	:	6.3
Length of flail	:	16
Rotor speed r. p. m	:	1550
Tip speed of the flails, m/s	:	46
Flail sharpening angle, deg	:	45
Stubble height, cm	:	5 to 15 (adjustable)
Effective cutting width, cm	:	100
Effective field capacity, ha/h	:	0.2
Discharge height of cut crop, cm	:	265
Cost of machine	:	Approx. Rs. 60000/-

Collection of wheat crop residues by field baler and straw reaper

CIAE Bhopal conducted field experiments on collection of wheat straw and evaluated baler with crop gatherer for baling of wheat straw in the field. Working of DASMESH-841 straw reaper for harvesting of left over wheat stubble after combining the crop was tested at Farm Machinery Training and testing Institute, Hissar. Quality of work, rate of work, ease of operation, handling and adjustment, and design details of the machines was recorded.

After combine harvesting of wheat crop (variety HI-8498), the condition of left over stubbles/combined straw in the field (size and weight with related moisture content) were measured (Table 2). Moisture content of straw and stubble as collected was 8.54 and 8.56% w. b., respectively. The distribution pattern of straw by weight in three size grades are (<50, 50-150 and >150 mm long).

Performance of CLAAS MARKANT-55 Baler was tested under field conditions with average height of stubble left after combining 349.3 mm, average weight of stubble left after combining 152.8 g/m^2, weight of grain 27.7g/m^2 and average moisture content of straw 7.34% (wb). Forward speed of baler was 2.80km/h with collecting width of 1.23m and numbers of bales produced per min 2.46 having average density of bale101.05 kg/m^3. Diesel consumption in operation was 3.33 l/h with 2 labours required. Collection and gathering losses in combine harvested wheat field was ranged from 38.46 to 43.32%. (Table 3).

Operation of straw reaper/combine at CIAE Bhopal. The straw reaper comprising of cutter bar, feeding unit, straw bruising unit and blower with specified manufacturer specifications was hitched to 45 hp PTO driven tractor with two-wheel trailer covered with net cloth for collection of straw. The machine has 2134 mm length of cutter bar with minimum height of cut, as 25-50 mm. Rated field capacity was 0.4-0.8 ha/h with straw output as 0.75-1.5 t/h and unit price of Rs.75000. The performance of the machine was recorded at an average forward speed of 2.64km/hr, effective width of operation as 1910mm, average height of cut of stubbles as 48mm and turning radius with or without trailer as 5510 and 3250mm respectively and details are given in Table3. It was noted that only 52% of the total stubble available in the field could be harvested and the rest was beyond the reach of the cutter bar. It was seen that 20% and 6% respectively were broken and suppressed under the wheels of the grain combine/ tractor and 20% of the loose straw (<150 mm long) were dropped on the field surface during grain combining and 2% of straw were retained on the field in the form of cut stubbles. The total straw/ stubble available was estimated to be 5.81 t/ha after combining the crop out of which 3020 kg/ha straws was recovered by straw reaper. Scattered/shattered spike lets left in the field after grain combining was also collected by straw reaper amounting to 22.0 kg of grain/ha. Considering the prevailing sale price as Rs.0.75/kg for wheat straw and Rs.6.50/ kg for grain during the season, the net income from straw harvesting was estimated to be Rs. 1690/ha with cost of operation as Rs.718 per ha. The tractor energy requirement was calculated from the fuel consumption converted to equivalents direct energy use as 591 MJ/ha. Thus harvesting of straw after grain combining is considered to be a better proposition for income generation and also making available the straw as cattle feed.

Table 2: Details of left over wheat stubbles/straw in the field after grain combining (Bhopal)

S. No.	Parameter	Values
1.	Size of loose straw fallen (Length x Diameter), mm	211.0 x 3.5
2.	Stubbles (Length x Diameter), mm	320.0 x 3.8
3.	Weight of loose straw, g/m^2	431.0
4.	Weight of stubbles, g/m^2	150.0
5.	Weight of straw having less than 50 mm length, % of total straw	6
6.	Weight of straw of 50 -150 mm length, % of total straw	36
7.	Weight of straw having more than 150 mm length, % of total straw	58

Table 3: Field performance of straw baler and straw reaper after grain combining in wheat fields (Bhopal)

S.No.	Performance parameter	Straw baler	Straw reaper
1.	Average height of cut stubbles, mm	195.0	16
2.	Effective field capacity, ha/h	0.95	0.34
3.	Straw output, kg or bales/h	No. of twine tied bales (size: 61X36X47)=87 per h	1027
4.	Field efficiency, %	70.5	67.4
5.	Fuel consumption, l/h	3.33	3.5
6.	Average cost of operation (including cost of operation of tractor and trailer), Rs/h	360	244

Harvesting and collectiopn of paddy stubble

PAU Ludhiana reported that during the year 2002 about 13.91 lakh tonnes of paddy straw was available in the Punjab state from 2.61 m-ha area under paddy cultivation and about 80 percent of paddy is combine harvested. Potential of paddy straw after combine harvesting and before operating baler in the field was therefore studied. A stubble shaver was operated to reduce the stubble height and to maximize the straw collection that cuts the standing stubbles to a height of 7-9 cm. Amount of recoverable (loose) paddy straw using field baler at a capacity of 0.26 ha/hr was 80-86% (Table 4). The non- recoverable paddy straw includes stubbles + left over by field baler amounting to 14 to 20 percent which can be incorporated in the soil to increase the soil organic matter. Combine operation saves tremendous time in harvesting the paddy in comparison with manual method.

Table 4: Status of paddy straw recovery in Punjab (Ludhiana)

S. No.	Parameter	Average
1	No. of plants / m^2	19
2	Plant height (cm)	98
3	Height of cut with combine harvester (cm)	35
4	Height of cut with stubble shaver (cm)	7-9
5	Weight of loose straw collected after combine harvesting (gm) / m^2	772.1
6	Weight of stubbles without roots after combine harvesting of paddy (gm) / m^2	619.83
7	Weight of stubbles without roots after operation of stubble shaver (gm) / m^2	207.16
8.	Recoverable paddy straw by field baler, %	80-86
9.	Non – recoverable paddy straw, %	14-20
10.	Per hectare man- hr requirement by	
	- Combine harvesting	4.5
	- Manual method	250

GBPUA&T, Pantnagar investigated effect of forward speed on performance parameters of field baler in paddy crop field immediately after combining. The straw from the headland and a plot containing 6 swaths of 26 m width and a length of 120 m was demarcated by collecting the loose straw with the baler. A swath after every plot was left so as to collect data on straw availability. Likewise 9 plots of 6 swaths each were prepared to study the effect of forward speed on baler performance such as fuel consumption, field efficiency, field capacity, wheel slippage and baling losses. The results indicated that the average field capacity, field efficiency ,fuel consumption, wheel slippage and baling losses of conventional field baler were 1.09 ha/h, 70.97% , 3.255 l/h, 10.737% , 118.20 kg DM/ha respectively for the forward speed of about 3.41 km/h (Table 5). This speed could be taken as the recommended speed of operation of field baler in combine harvested paddy fields under Indian conditions.

Table 5: Effect of forward speed on performance parameters of field baler (Pantnagar)

Forward speed Gear No.	Km/	Fuel Consumption l/h	Field efficiency %	Field capacity, ha/h	Wheel slippage %	Baling losses kg DM/ha
1st	2.89	3.196	69.456	0.922	9.889	113.316
2nd	3.41	3.255	70.97	1.090	10.737	118.20
3rd	4.50	3.226	66.19	1.341	11.913	131.86

Baling machines

Baling/densification machines are useful for baling and densifying grasses, crop residues and fodders. Different types of machines are available depending upon their mobility and use like stationary, portable type and field balers and shape of bales to be made like rectangular or round. Few machines developed in country are given below:

Stationary type

IGFRI Jhansi, used existing densifying machine, which produced rectangular bales of size 330x300x500 mm with capacity ranging from 300-600 kg/h for different materials. The machine is equipped with crank-piston drive mechanisms and rectangular piston cross-section. The machine is operated by a 15 hp electric motor with V-belt and pinion gear power transmission system. Density of bales was achieved 180-400 kg/m^{3}with average weight ranging from 12-20 kg per bale depending upon material type fed and can be adjusted with the help of spring tension mechanism provided at the outlet point of the machine. A force fed mechanism has been provided for pre-compression of the loose herbage in hopper, which in turn increases the hay input inside the compression chamber and was pressed by reciprocating motion of piston. The tying mechanism of bales was manual by using 3 mm steel winding wire. However, during experimentation with wheat bhusa, it was observed that the bales thus tied were loosen. A 15 cm widen tape was therefore used to tie the blocks longitudinally from top to bottom at two places. The technical details of machine are given below:

IGFRI Densifying Machine

1. Overall length : 6130 mm
2. Overall width : 1260 mm
3. Main frame : MS angle 75x75x8 mm
4. Top and bottom chassis : MS channel 200x80x8 mm
5. Ram: Heavy duty ram fabricated out of channel : 250x75x10 mm plate 10 mm
6. Ram speed : 35 strokes/minute
7. Connecting rod : 1800 mm long sturdy connecting rod steel (I-beam) with brass bushes at each end and I-beam 170x85x6 mm
8. Cross sectional areas of compression chamber: 305x330 mm
9. Length of stroke: 680 mm
10. Length of compression chamber: 2650 mm
11. Gear wheels : 2 numbers. steel gears of 80 teeth of 125 mm length and 48 mm pitch with gun
12. Main drive gear: 1 numbers steel gear of 12 teeth of 125 mm length and 48 mm pitch with gun metal bushes .
13. Main driven gear: 1 numbers steel gear of 25 teeth of 125 mm length and 45 mm pitch.
14. Pinion: 2 numbers steel pinion gears of 19 teeth of 125 mm length and 45 mm pitch
15. Power requirement: 15 HP, 1440 RPM
16. Size of bales: 310x310x500 mm (length adjustable)
17. Average weight of bales: 10 kg.
18. Average density of bales: 200 kg/m^3
19. Tying system: steel wire rope tying arrangement
20. Feeding mechanism: Manually in feeding chamber
21. Tying : Semi-automatic
22. Material: Wheat stubble
23. Bale Density adjustment: Through spring tensioning

CIAE Baler

CIAE Bhopal improved unloading mechanism of the bale with the help of supporting guards for easy unloading and tying mechanism of existing machine developed in the institute and evaluated for crop residues. For increasing the density of straw and easy unloading of bales, some adjustment in length of main shafts and cut off device were made. Due to increase in length of shaft, the compression part could compress the straw around 3-4 cm from both sides and the bale could be taken out easily from the machine door. It was so because of reduction in length of bale. The overall dimension of machine was 4000 mm x 800 mm x 200 mm with power requirement of 3 hp, single phase, 1440 rpm electric motor and gives compression ratio of 3:1 (by volume). It produces bales of size 520 mm x 450 mm x 450 mm weighing 20 kg at 8% m.c. and having density of 200 kg/m^3. The machine had out put capacity of 70 to 84 kg/hr taking 10-12 minute/bale.

PAU, Ludhiana Rectangular Baler

PAU, Ludhiana used a rectangular baler powered by tractor PTO (25 hp) or by electric motor (10-15 hp) with a V- belt and pinion gear mechanism and made by Sherpur Fabricator,Ludhiana. This baler has a center top mounted pickup mechanism to which straw is fed manually and is delivered to the baling chamber with the help of packer finger. The machine is equipped with the crank piston mechanism. The density of bales was 131 kg/m^3 for wheat and 146 kg/m^3 for paddy straw and can be adjusted with the help of spring tension mechanism, provided at the exit point of compression chamber. Overall dimension of machine was 500 cm X 142 cm having swath feeding platform width of 53cm, size of bale chamber as 95 cm x 53 cm, outlet support (length) of 60 cm and total working stroke of 14 strokes/minute. The length of bales can be varied by introducing wooden planks at different intervals, which has a provision of two grooves on either side to facilitate the movement of GI wires to tie the bales. The machine has capacity of 12 to 15 bales per hour weighing 27 kg per bale for wheat straw. Four persons are required to operate the machine.

Presently the farmers of Punjab state are densifying the wheat straw manually by kneading it. PAU, Ludhiana, therefore compared the bulk density of loose wheat straw with the wheat straw densified with different methods of densification. Their study reveals that the wheat straw can be densified about 2.5 times using portable rectangular baler when compared to loose straw and about 1.9 times when compared with manually densified wheat straw. The wheat straw having loose density of 49.5 kg/m^3 was compacted manually up to 68.4 kg/m^3 and when baled was compacted up to 130.88 kg/m^3. Bales of 45.72 X45.72 cm cross-section and length 96 cm was made having volume 0.2011 m^3 and weight 26.3 kg by portable rectangular type baler. Paddy straw bales were also made by portable rectangular baler with cross sectional dimensions of 52 X 52 cm and length varied from 87 cm to 113 cm with an average of 97 cm. The average density of the formed bale was 150 kg/m^3 while that of conventional tied paddy straw was 70 kg/m^3. Nearly 12-15 bales of 36.55 kg average weight and 0.2691 m^3 of average volume were baled in one hour with an average time of 4-5 minutes for baling single bale. Two persons were engaged in feeding of the straw and two in tying of bales. Wheat straw was baled by field baler, which produces bales of average dimensions of 87X42.5X33 cm and average weight of 19 kg with the average density as 158.3kg/m^3.

Bibliography

Gupta, P.D.; Dwivedi,P.N.;Pathak,P.K. and Deshmukh,G.R.2000-2003.NATP Final Report " Baling, Densification and Storagc of Grasscs, Foddcrs, & Crop Rcsidues ", Indian Grassland and Fodder Research Institute, Jhansi

Singh,J; Kaur,P; Kaur,G. and Bagha,C.S 2000-2003. NATP Final Report " Baling, Densification and Storage of Grasses, Fodders, & Crop Residues ", Punjab Agricultural University, Ludhiana.

Singh, G and H.Chandra(2002). Production and Economic Factors Growth in Indian Agriculture. Technical Bulletin No. CIAE/2002/91 pp.17-19. Published from CIAE, Nabi bagh , Berasia Road, Bhopal, M.P.

Kulkarani, S.D. and Mangaraj , S 2000-2003. NATP Final Report " Baling, Densification and Storage of Grasses, Fodders, & Crop Residues ",CIAE, Nabi bagh , Berasia Road, Bhopal, M.P.

Thakur,T.C.; Verma M.L. and Jaiswal,R.S. 2000-2003. NATP Final Report " Baling, Densification and Storage of Grasses, Fodders, & Crop Residues ",G.B. Pant University of Agriculture and Technology, Pantnagar, Uttaranchal .

Thakur , T.C. and Rajan , P. 2003.Avalability of wheat straw in combine harvested fields for baling. Agri. Engg. Today . Vol. 25 (5-6): 1-14.

Singh, A.K ; Sehgal, V.K. and Kaur,G.2001Densification of Agricultural Residues.Punjab Chapter, Indian Society of Agricultural Engineers, Annual day Souvenir, PAU, Ludhiana :35-39.

CHAPTER - 10
DENSIFICATION MACHINES

Amar Singh, S.K. Jha, J. S. Pawar and A. Pandeya

Mechanical densification of biomaterials is a value adding process. The densification is done to create fuel pellets, briquettes, feed wafers and feed blocks. It offers several advantages viz reduced unit transportation cost, improved safety during transportation and less storage space and cost. While the briquetting of structural materials for the building and road construction industry is very extensive, the compaction of agricultural materials into blocks has been negligible compared to its importance in the country.

Densification is commonly done by baling, palleting, briquetting and block making. A comparative study of different densification processes is given in Table 1. In baling which is commonly done in roughage sources, the volume reduction is comparatively quite less. Degree of densification is very high in palleting but it is an energy intensive process because of chopping and grinding requirement of straws. The briquetting process densifies the residues by 10 to 12 times by the application of pressure and heat. The existing palleting and briquetting machines cannot handle feed mixture having larger size straw and high proportion of the crop residues. Densification through complete feed block preparation is a new concept in India but offers several advantages over other densification processes as indicated in Table 1.

India currently produces 540 million tonnes of crop residues and there is an increasing trend every year with increase in crop production. These residues could be used as feed for our vast livestock resources. However, the handling, transportation and storage of the crop residues are a problem and costly practice because of their low bulk density. The problem can be solved only through a planned scientific management of transportation, storage, and management of crop residues and hays in different regions of the country. There are regions with deficit and excess production of these crop residues in the country. The feed management, with roughage diets, would improve drastically if the deficit areas could be stocked with processed and densified animal feed in the deficit and disaster prone areas.

Table 1 Comparative studies of different densification processes

S. No.	Item	Block making	Briquetting	Pelleting	Baling
1	Machine cost (Rs) Approx.	13,00000.00	22,00000.00	20,00000.00	30,0000.00
2	Capacity (kg/h)	500	700	1000	350
3	Horse power requirement	25	80	50	10
4	Suitability for	Straw + concentrate	Straw + concentrate	Straw + concentrate	Straw + urea + molasses
5	Density of feed (kg/m^3)	400	900	400	150
6	Loading per truck (MT)	10	15	10	4 (6% molass)
7	Chop length required	Below 3 inch	Below 1 inch	Below 2.5 mm	6-12 inch
8	Molasses - maximum dose	Up to 12%	Up to 6%	Up to 12%	Up to 30%
9	Temperature rise of feed	nil	>120°C	70°C	nil
10	Permissible moisture in raw material	Up to 20%	Up to 15%	Up to 12%	Up to 40%
11	Moisture in finished feed	12-15%	5%	12-15%	15-20%
12	Process cost (Rs/kg)	0.8	0.9	0.7	0.65
13	Power consumed (kWh/ton)	35-40	50-55	30-35	12-15
14	Annual handling potential (MT) [10 h per day, 200 days/year]	900	1400	1600	700
15	Possibility of capacity increase by improvements	100%	25%	25%	30%

It is possible to formulate complete animal feed blocks using straws and diet supplements such as molasses, concentrates, minerals and salts, which can be stored over long period of times, transported economically over long distances and could be stored large volumes in small protected spaces. These feed blocks could be made in the customized sizes, shapes and formulations for different size animals according to prescribed diet of the animal.

Densification Machines

A brief description of the various densification machines viz baler, palletter, briquetting machine and feed block formation machine is discussed in following pharagraphs:

Balers

A baler is a machine used to compress hay or straw into bale for easy transport and storage. Balers are classified in several ways like rectangular or round balers based on shape of bale; field or stationary balers based on mobility; tractor mounted or self propelled field balers; and manual, semi-automatic or automatic balers based on feeding, tying, bale length controlling mechanism.

The rectangular balers are essentially plunger type balers. The field balers has a unit to either pick-up windrow and elevate, collect and elevate; a conveyor to move paddy straw to bale-chamber entry; a packer to place paddy straw in the chamber while plunger is on its retracted stroke; a reciprocating plunger to compress paddy straw and move it through the bale chamber; means for applying force to resist movement of paddy straw through bale chamber thereby controlling degree of compaction and bale density; a metering device for controlling bale length; means for separating bales and placing wire or string around the bale; and tying and cutting device. Generally stationary balers are not provided with first two components. Instead of first two components, it has a feeding chute for manual feeding. As plunger moves on its compression stroke, the paddy straw in the bale chamber is compressed until the plunger force becomes large enough to move compressed mass through chamber. On return stroke, the compressed paddy straw is held by spring loaded dogs that project into bale chamber and new paddy straw is taken in. The bale density varies from 80 to 200 kg/m^3 depending on moisture content of paddy straw. Most balers have chambers with cross section ranging from 36 x 46 cm to 43 x 58 cm and bale mass is 20 to 70 kg. Large balers, however, have bale size of 1 .5 x 1.5 x 2.4 m and 800 kg mass. The energy requirements vary widely from 0.57 to 1.05 kWh/tonne. Generally, balers require a tractor of 40 HP or more.

Briquetting machine

Briquetting techniques could be classified into two broad categories based on the operating conditions: hot and high-pressure densification or cold and low-pressure densification. Depending on the type of equipment used, hot and high pressure densification can be categorized into four main types: piston press densification, screw press densification, roll press densification and pelletizing. Products from the first three types of densification are of relatively large size and normally called briquettes.

Palleting machines

There are two types of palleters namely, disc die and ring die.

Disc die palleters

It consists of a screw conveyor, a gearbox, a stirrer and a palleting device. The screw conveyor, whose rotating speed is adjustable, is mainely used to control the charging rate. Inside the disc there are 2 to 4 pressing rollers and a disc with many holes. In operation, the disc rotates at a speed of 210 rpm. After falling on to the disc die, the cooked meal is distributed evenly by a spade and forced through the round holes of the die by the press rollers. Cylindrical pallets from the die are then cut into pieces by a blade under the disc. The disc palleters can be classified into three types: movable disc, movable roller and movable disc and roller type. Hole diameter in the disc can be 4, 6 and 8 mm. The diameter of roller ranges from 160 to 180 mm.

Ring die palleters

The main parts are a screw conveyor, a stirrer, palleting device and a drive mechanism. The mixed meal goes into the palleting device, which consists of a ring disc and pressing rollers. In operation, the ring disc rotates, making the roller turn and press the material through the cylinder holes of the ring die. The cylindrical pallets rotate with the ring die, and are cut by a blade. The larger the hole diameter, the higher the output of the machine and lower the energy consumption. Hole diameter is determined by animal requirements.

Machines for Feed Block Formation

Initially in India, the feed block formation was started using pillar type hydraulic machine and screw press. These machines were useful for laboratory studies but not for mass scale commercial purposes. Some of the earlier machines used for feed block formation are discussed in the following paragraphs:

Pillar type hydraulic machine

A two pillar type manually operated machine with up and down strokes and fitted with 3 hp electric motor was developed for compressing straw based feed filled in a cylindrical die. The die (21.5 cm diameter) made up of mild steel sheet of 5.05 mm thickness was fitted on the platform of the machine. The feed mixture filled in the die was compressed by moving piston downward. After compression the piston was moved upward to release the pressure from block. The die was than opened into two halves on hinges and block removed manually. The die was closed and fitted for reuse. A pressure of 0.68 MPa to 2.70 MPa and dwell time of 0.5 minutes to 5.0 minutes was used for block formation. The disadvantage of machine was its too low capacity as well as its batch operation.

Manually operated screw press

The press consisted of a screw-rotating wheel at one end and a punch on other end. The screw is fitted with the help of a frame. A die of desired shape and size was used for feed block formation. The die filled with the formulated feed was is compressed with a punch attached to the screw. The screw could be manually operated. Long handle was attached to the wheel for applying more pressure on the feeds. After releasing the pressure and opening the die, the block could be removed from the die. This was a slow process machine and was used for making blocks for research experiments only. Dies and punch of different shape and sizes were used to suit the requirement of various animals. The disadvantage of this machine too was its too low capacity, low-pressure generation and its batch operation. These machines were thus not fit for commercial purposes.

Developments in Animal Feed Block Formation Machines

Commercial scale feed block formation machines for moulding animal feed into blocks using a whole range of crop residues, feed supplements and suitable binders were recently developed at Indian Agricultural Research Institute (IARI), New Delhi. It could form compressed blocks in the sizes appropriate and economical for handling, transportation and storage of large volumes of animal feed. A comparative study of different feed block formation machines developed at Indian Agricultural Research Institute, New Delhi is given in Table 2. A brief detail of these machines are given in the following paragraphs:

Animal Feed Block Formation Machine (Semi-automatic)

This machine was developed in the year 1990 at Indian Agricultural research Institute, New Delhi in a project with collaborating partner Indian Veterinary Research Institute, Izzatnagar. The machine (Fig. 1) consisted of a hydraulic cylinder, power pack, compaction chamber and frame. The compaction chamber is made up of 50 mm mild steel plate, the outlet of which is closed with 50 mm thick plate during compaction. The hydraulic cylinder operated with 7.5 HP 3-phase motor with a stroke of 750 mm and formed a block of size (30 x 30 x variable thickness) cm. The block thickness could be adjusted as per requirements. The machine is designed to exert a maximum pressure of 200 kg/cm^2. The major limitation of the machine was its low capacity and the manual process of block removal. Hence, it could not be found suitable for commercial use.

Fig. 1 Animal Feed Block Formation Machine (Semi-automatic)

Table 2. Comparative studies of different feed block formation machines developed at IARI, New Delhi

S. No.	Type of Machine	Power required, HP	Capacity kg/h	Block size, cm x cm x cm	Price Lakhs	Labour required	Operation cost, Rs./ ton
1	Animal Feed Block Formation Machine (Semi-automatic)	7.5	100	30 x 30 x VT	1.5	3	657
2	Low capacity Animal Feed Block Formation Machine	5.0	40	12 x 12 x VT	2.0	2	1275
3	Animal Feed Block Formation Machine (Automatic)	25	250	20 x 20 x VT			
	25 x 25 x VT	6.0	1	420			
4	Animal Feed Block Formation Machine	40/30	500	30 x 30 x VT	11.0	2	350
5	High capacity Animal Feed Block Formation Machine	45	1000	37 x 37 x VT	14.0	6	162
6	Mobile Feed Block Formation Machine	6.5	100	15 x 15 x VT	4.0	1	600

(VT stands for variable thickness)

Animal Feed Block Formation Machine (Automatic)

A prototype of commercial scale Animal Feed Block Formation Machine (Fig. 2) of capacity 250-300 kg/h was developed in the year 2000 at Indian Agricultural Research Institute, New Delhi under National Agricultural Technology Project. The machine was designed for a maximum compaction pressure up to 420 kg/cm^2 with a cross section of 20 cm x 20 cm for feed block. It consisted of 2 hydraulic cylinders, power pack, feed hopper, compression chamber and electric control panel. The movement of hydraulic cylinders could be governed by the electric control panel. The machine had a provision to operate

Fig. 2 Animal Feed Block Formation Machine (Automatic)

both in automatic and manual mode. The machine is powered with 25 hp electric motor to run its hydraulic system. The processing cost of feed block formation using this machine was found to be Rs 420/ton.

Low Capacity Animal Feed Block Formation Machine

Keeping in view the requirement for small animals like rabbit, goat etc. a low capacity animal feed block formation machine (Fig. 3) of capacity 100 kg/h was developed. The machine consisted of hydraulic cylinder for pressing, operating gate, upper hopper plate/lid and feed hopper, power pack for supplying power to hydraulic cylinders, D C valve and mild steel frame. The movement of the cylinders could be governed using a hand operated lever. The machine forms a feed block of (12 x 12 x variable thickness) cm size. The weight and thickness of the feed block could be adjusted as per the requirement varying from 0.5 to 1 kg. The machine runs on 5 HP 3-phase electric motor.

Fig. 3 Low capacity Feed Block Formation Machine

High Capacity Animal Feed Block Formation Machine

To meet the high industrial requirement, high capacity feed block formation machine (Fig. 4) of capacity 1 ton/h (1 TPH) was developed at IARI, New Delhi. The machine consists of hydraulic cylinder for pressing, operating gate, upper hopper plate and feed hopper, power pack for supplying power to various hydraulic cylinders, electric control panel and mild steel frame. The machine forms a feed block of (37 x 37 x variable thickness) cm. The weight and thickness of the feed block could be adjusted as per the requirement. The machine is operated with two 3-phase electric motor (30 HP and 15 HP).

Fig. 4 Animal Feed Block Formation Machine (1 TPH)

Mobile Feed Block Formation Machine

To facilitate easy transport and storage of fodder in hilly region, a mobile animal feed block formation machine (Fig. 5) was developed at IARI, New Delhi. The machine consisted of two hydraulic cylinders, power pack and body. The machine is mounted on a four-wheel trolley of 3 m x 1.5 m size having pneumatic wheels. The machine is operated with 6.5 HP air-cooled diesel engine. This unit could be mounted and used in different locations and situations. The feed block size is 15 x 15 cm in cross section and of varying thickness. The feed block weight varies from 0.5 kg to 2.5 kg. The capacity of the machine ranged between 1.0 to 1.25 q/h. The cost of feed block formation was found to be Rs 600 per ton.

Fig. 5. Mobile Feed Block Formation Machine

Main features of the Animal Feed Block Formation machines

- It is very simple in operation and only one person (unskilled / semiskilled) is required to regulate entire operations of the machine.
- The compaction pressure of the machines could be varied upto 420 kg/ cm^2.
- It can compact all kinds of feed materials to square shape blocks of varying thickness and weight.
- The bulk density of the roughage based feed blocks from these machines was found to be 4 to 42 times more than the original feed material depending upon the crop residue indicating a significant incentive in transportation and storage of crop residues.
- The feed blocks from these machines are stable and do not require tying.

Bibliography

Gupta P D, Goyal R K, Chattopadhyay P S. 1998. Performance evaluation of IGFRI, densifying machine for grasses. Journal of Agricultural Engineering 35(4): 33-38.

Kundu J F, Sharma D D, Gupta PD, Reddy M R, Pachauri V C. 1995. Densification of roughage. In Handbook for Straw Feeding Systems, Eds. Kiran Singh and J.B. Schiere. ICAR, New Delhi.

Kundu,S S, Goyal R K, Sharma D D. 1998: Densification for block making and evaluation of calcium hypo chlorite and alkali treated wheat straw based feed blocks in comparison to urea treated straw feed. Buffalo Journal 3: 291-302.

Ram A K, Mehra U R, Chalta J. 1990. Preparation and evaluation of complete feed mix block for goats. Indian Journal of Dairy Science 43(4): 458-461.

Singh A, Panwar J S, Kumar A, Jha S K, Pandeya A. 2002. Development of a densification machine for crop residue based animal feed blocks. Journal of Agricultural Engineering 39(1): 9-15.

Verma A K, Mehra U R, Dass R S, Singh A. 1996. Nutrient utilization by Murrah buffaloes (Bubalus bubalis) from compressed complete feed blocks. Animal Feed Science and Technology 59: 255.

The
of very
ethnobot
medicinal
is tradition
India is 1
15000 p
(Krishna
with the lo
parts of
of Local
civilizati
entire flu
plants. T
world's
al. 1989)

Intro
on medic
1000 BC
on medici
informatio
1991,199
Schultes (
other docu
investigat
directly w
number of
documente

CHAPTER - 11

MEDICINAL PLANTS FOR ANIMAL PRODUCTION

J.N. Gupta

The presence of multiethnic groups of ancient lineage and the occurrence of very diverse vegetation make India one of the richest repository of ethnobotanical knowledge. Ethnomedico-botany focuses on the knowledge of medicinal plants that rural people have developed over generations. The country is inhabited by over 50 million tribals belonging to about 550 tribal communities. India is one of the World's 12 regions having the largest biodiversity. It has 45000 plant species of which 15000-20000 possess proven medicinal value (Krishna Kumar 1996). The local health traditions originated in close association with the local vegetation before centuries and stabilized through ages in different parts of India over 6000 plant species (Bulletin of Foundation for Revitalisation of Local Health Tradition, 1991) are used as biomedicines.. The vane of human civilization is rooted in the exploitation of vegetation. During ancient times the entire human life support system was based on the judicious use of the native plants. The World Health Organisation estimates that as many as 80% of the world's population depend on plants for their primary healthcare (Farnsworth et al. 1985).

India is also rich in ancient literature which can be tapped for information on medicinal plants. Rigveda and Atharvaveda, which date back to 2000 to 1000 BC are the oldest literature sources.They contain valuable informations on medicinal plants, which were in vogue during that period.Ethnobotanical informations has also been gathered from Puranas (Sensarma 1984,1987,1988, 1991,1992). Particularly herbal veterinary medicines from the Garud Puran. Schultes (1981) writes that "this region possesses in the Vedas and a variety of other documents an ethnobotanical continuum that enables the contemporary investigator to delve in to the distant past and often to link modern folklore directly with that of ancient peoples and cultures" Since last three decades a number of workers have surveyed tribal communities and collected and documented ethnobotanical informations (Binu et al.,1992).

Ethnobotanical study in relation to veterinary medicines has also been made by a number of workers (Chauhan,1991;Gaur et al.,1992; Jain,1981 ; Jain et al.,1984 ; Kumar et al.,1997; Mandal and; Pal,1980,81,92;,Sebastin, 1984; Singh1987; singh1995a,b; Singh et al.2001.) In addition to ethnobotanical study plants used as veterinary medicines are also reviewed by a number of workers (Jaiswal et al, 2004; Sharma and Joshi, 2004, Singh et al.,1998)

Indian livestock keepers raise domestic animals such as for milk, food, agriculture, and markets. There are only a few veterinary hospitals. They are located at district and block levels far from rural and remote areas. Therefore many people treat their sick animals by herbal medicines that are locally available but vary seasonally.Secondly many of the allopathic drugs used for different ailments are costly,have problems of drug resistance,many are toxic in higher concentration,may be harmful to humans handling them and may contaminate the environment as well. As against this scenario we have herbal medicines,which are now gaining significance due to their low costs,ease of use and are generally free from the risk of harming human handling them as well as they hardly contaminate the environment.

In the present paper ethnobotanical informations on medicinal plants used as veterinary medicines have been gathered from literature and presented disease wise for easy grasp by animal scientists and extention workers who will disseminate the knowledge to the users. Plant species used in Latin America as veterinary medicines which are available in India also described.

Botanical and local names, habit and distribution of plant species referred in the text are listed in the appendix.

Wound healing

Tissue healing is an important process which can be enhanced by using several herbs/herbal drugs. There are certain plants ,which have been proved helpful in wound healing.

Azadirachta indica (Neem) is an important medicinal plant found in Indian sub continent, its leaves,bark and fruits are used for wound healing. It has antiseptic,astringent.insecticidal,larvicidal and a little anti inflammatory property (Rawat,1995). Paste of its leaves,oil from its seeds and powder made from its bark is used for this purpose. Decoction of its leaves is also used for washing the wounds.(Singh et al.2001). Paste of the bulbs of Allium sativa is applied on worm infested wounds of cattle. The wound is washed with the wood extract of Shorea robusta. An ointment made from dried leaves of Annona squamosa in white petroleum jelly for healing of wounds and powder of its seeds is used to treat maggoty wounds (Pal, 1981; Sudarsanam et al.1995) Root powder of Lygodium flexuosum applied on septic wounds of cows and buffaloes in eastern India. Seed oil from Madhuca longifolia is applied on wounds between hooves of cows, goats and sheep. Aerial parts of Hypericum hookerianum are used for treating burns and wounds by peoples of Shola forest of Tamilnado (Mukherjee et al., 2001).Tugnaiyat et al,(2000) states that ether extract of seed powder of Helianthus annus can be used in ointment form for better wound healing. Leaf paste of Moringa oleifera after slight warming is applied as poultice on boils of cattle. Bark paste is also applied locally in West Bengal (Mondal and Chauhan, 2000). Ointment made from alcohalic extract of dried roots of Withania somnifera in white petroleum jelly used in wound healing (Singh et al.2001).

Seed oil from Pongamia pinnata with leaf paste of Justicia adhatoda applied as balm on hump of cows and buffaloes. An ointment made from 10% leaf juice of Prosopis juliflora has been found very effective for wound healing in Calves. The paste prepared from the rhizomes of Curcuma domestica using tori (Brassica napus) oil is used for healing fresh wounds and also to check microbial infection.An ointment is also made from C.amada or C.domestica in petroleum jelley to treat abscess and wounds.

The juice extracted from young leaves of Prunus persica or plant of Crotalaria linifolia mixed and pounded with common salt and applied to remove maggots from the wound of cattle. A paste of twigs of Pteris aquilina alongwith few tender leaves of Artimesia vulgaris (Titapati) finds application in any type of cut wounds. It causes immediate coagulation of blood. The pounded root of Eclipta prostrata are applied externally with the oil of Kusum (Carthamus tinctorius) as an antiseptic to cure ulcers and wounds in cattle.(Pal,1981).Jain and De (1966) have reported that the leaf juice of Eclipta prostrata is applied to cure sores on the shoulders caused by carrying heavy loads. In wealth of India it is reported that the root of this plant is antiseptic to ulcer and wound. Oral feeding of finely grinded fresh leaves or powder of dried seeds of Coriander sativum is used in healing of chronic ulcers (Singh et al., 1998). Accordingly to Jain (1971), if a piece of the root of Jatropha gossypifolia is tied on a thread and is put around the neck of cattle, it is believed to cure their wound.

Pounded whole plant of Agrimonia pilosa, crushed leaves of Chenopodiun album,root paste of Cirsium verutum , latex of Euphorbia pilosa, leaf juice of Solena heterophylla, bark paste of Betula alnoides, Buxus wallichiana,Debregeasia salicifolia,Ulmus willichiana,Urgenia indica is applied locally for wound healing in Uttar Pradesh Himalaya (Gaur et al.,1992). Veterinary crude drugs used in Rayalseema,Andhra Pradesh for wound healing is reported by Sudarsanam et al.,1995.According to him leaf juice of Erythrina variegata is squeezed locally for quick healing, Stem bark and fruit of Syzygium cuminii and leaf infusion of Tarenna asiatica is given orally. Root decoction of Smilax zeylanca is given for sores, swelling and abscesses. Decoction of the leaves of Woodfordia fruticosa is applied on sores and on ulcers of the cattle. Paste of leaves of Adhatoda vasica or Gardenia gummifera is applied in wound cattle for healing.

Skin diseases

Skin diseases in animals comprise scabies bacterial and fungal infections,bruises and sprains,inflammation,burns,sore feet,ringworm,eczema,skin eruptionsurticaria and infestation of tick,mite,lice and maggots. Leaves of Abrus precatorius with mustard oil give relief to painful areas.Juice mixed with Plumbago zeylanica roots is applied to the leucodermic spots.It is also applied to the bare skin in alopacia,stiffness of shoulder joint and paralysis. Powdered gum of Acacia nilotica mixed with egg white is applied in burns and scalds. The bulbs and seeds of Allium cepa are locally used in irritation of insect or scorpion bites and also in skin diseases.Mixed with mustard oil it is highly effective inflammatory swellings. Mustard or coconut oil in which garlic has been fried is an excellent antiseptic application for scabies and maggots infesting ulcers, ulcerated surfaces and wounds. Its juice mixed with salt is applied to bruises and sprains and also to relieve neuralgia and earache. Garlic is applied externally for deafness and pain. Garlic juice mixed with 3 or 4 parts of ordinary or distilled water is used as a lotion for washing wounds and foul ulcers. Seed and leaf oil of Azadirachta indica possess insecticidal and antiseptic properties. It is used as a dressing for foul ulcers, eczema and skin diseases like ring worm, scabies and mange in dogs. As insecticide it is used for the destruction of lice. Dry seeds are used for killing pediculi and powdered kernel for washing of hair. Toddy or fermented sap of leaves is also useful valuable in skin diseases. A preparation called Pancha tikta grita made from boiling 800 g each of neem bark, leaves of Momordica dioica , Solanum surattense , Tinospora cordifolia Adhatoda vasica in 5-6 litre of water, till it is reduced to quarter and then adding 3.5 litre of butter and about 3 kg myrobalans is recommended as one tablespoonful with little hot milk internally twice daily in chronic skin diseases. Infusion or decoction of the root of Beta vulgaris when applied to the temples it stops inflammation of eyes and in burns it is used in combination of oil and Alum. Mustard seeds and oil have a cleansing effect on the skin make soft, remove the grease and are useful in general pustular condition. Mustard oil is used as a base for many herbal preparations and in combination with several other medicinal plants is used in skin ailments of various degrees. The physiologically active substance is found in the milky juice of Calotropis gigantea . Flowers are digestive, tonic and have powerful action on the skin. An oily preparation (Arka taila) made by boiling together 8 parts sesamum oil, 16 parts Calotropis juice and one part turmeric is useful in eczema and skin diseases. In scorpion and insect bites it relieves pain and burning. It is also useful in ringworm infection. Equal part of branches, leaves, milky juice and flower are taken to prepare pills and given every morning to treat various kinds of skin diseases. A powder of dried leaves mixed and boiled with sweet oil and turmeric is applied to eczema and other skin ailments with high efficacy. Wood of Cedrus deodara yields an oleoresin known as Kelanka tel. Which is applied to ulcers and other skin diseases especially in manage in horses and sore feet of cattle? In pruritus and eczema camphor (Cinnamomum camphora) ointment is very useful. Its strong odour protects animals from the menance of insects. Lime juice is applied externally for relieving irritation of insect bites. A local application made up of lime juice, potassium carbonate and copper sulphate is useful for warts and other skin eruptions. Lemon oil mixed with glycerine is applied to pruritus of vulva and scrotum. Lemon oil with camphor is also used in various skin diseases.

A paste of turmeric and leaves of Adhatoda vasica with cow urine is rubbed on the skin affected with prurigo and eczema. When mixed with ginger oil and applied to the skin it prevents skin eruptions. A paste of turmeric along with the pulp of neem leaves is used in ringworm infection itching, eczema and other parasitic diseases of the skin. In chronic eczema and itches ointment made of turmeric, hemp leaves, onion and warm mustard oil gives immediate and immense relief. In chronic skin ailment turmeric 640g, clarified butter 500g, milk 1.5 litre, sugar 120g are boiled and black pepper, ginger, cinnamon are mixed to it. The dose includes 10ml every morning in prurigo, boils, urticaria and chronic skin eruptions. The dried leaves, gum exudation from the stem and oil distilled from the fresh leaves of Eucalyptus globulus are used against skin diseases. Externally fresh young leaves are applied as local stimulant in small wounds. The leaf extract suitably diluted is used as disinfectant lotion and antiseptic lotion in certain chronic skin diseases and in erysipelas of the face, leg and scrotum. Seed oil of Jatropha curcas is highly effective for itches, ringworm, scabies, herpes and eczema and it is a cleansing agent for wounds, sores and ulcers. Locally the Peppermint oil is used as a powerful anaesthetic, antiseptic and germicide and is very useful in pruritus.

The juice of whole plant of Momordica charantia is mixed with Cinnamon, Long pepper, Rice and Chaulmogra oil to forms a good ointment for psoriasis, scabies and other skin diseases. Externally it is applied to scalp in the burns and boils. Nicotine salycylate found in the plant of Nicotiana tabacum is the remedy for certain skin ailments. Decoction of tobacco is applied locally to relieve pain, irritation in swellings, syphilitic nodes, other skin diseases and to reduce orchitis.

In orchitis the upper surface of the leaf painted with silarasa is applied to the painful swollen part. A paste made with snuff, lime and bark of Calophyllum inophyllum is also applied in orchitis. The leaves of Ocimum sanctum made into paste are used to cure parasitical diseases of the skin, especially the ringworm. A 12% decoction of the plant acts as a parasiticide and antiseptic and makes the larvae inactive. In the form of paste Piper nigrum is applied externally to the boils, alopecia and other skin diseases. Massage with pepper and oninons makes hair growth on bald patches left after ringworm. Seed oil of Pongamia pinnata is applied to scabies, sores and eczema. A paste prepared with karanj seeds, Cassia tora Linn.seed and the root of Saussurea hypoleuca Spreng. Is mixed with cow urine and applied in various kinds of skin diseases with high success.

Sharma and Joshi (2004) reported a product of IVRI which is an outcome of several years of research and trial on thousands of animals on the skin diseases of in dog, sheep and goats, Ollinall, a new herbal formula contains Vetiveria zizanioides , Annona squamosa , Argemone mexicana , Butea monosperma and Azadirachta indica. In specific proportions. In mange in dogs, fungal infections of skin in sheep, pyoderma of goats, ollinall gave promising results.

Lice on the body

Root powder of Peganum harmala is applied to kill lice. The leaves and fruits of Annona squamosa are made into paste with Nicotiana tobacum leaves and used as insecticide. In wealth of India it is reported that the unripe fruit, seed, leaf and the root of A. squamosa are used for destroy insect and lice. Juice extracted from tender parts of Crotalaria alata is externally applied. The paste prepared from the tuber of Dioscorea deltoidea is applied. Leaf decoction of Nicotiana tobacum finds its application in removing both acute and chronic lice, Leeches and other insect infections. Seeds of Annona squamosa are pounded with the cake of Madhuca longifolia and the pounded mixture is applied to remove lice or other external pasasitic insect from the body. The paste made from the leaves of Lawsonia alba with the leaves of tobacco used as an insecticide. Powder of root of Echinops echinatus with mustard oil is applied to kill lice.

In Latin America the juice of mashed leaves of Cassia tora is given to an animal, the ticks will jump off. This anti-tick folklore seems exaggerated (Duke and Vasquez 1994). In Mexico the gum obtained from the bark of Mammea americana is used to extract chiggers from the skin and kill ticks and other parasites external parasites of domestic animals (Standley 1920-1926)

Digestive disorder

The seed paste of Cassia fistula given orally in Keonijhar (Orissa). Leaf of Bambusa arundinacea with black pepper and common salt to cattle to cure diarrhoea. (Kirtikar and Basu 1935).Leaf paste of Cympogon flexuosus is given to cattle in orissa to core indigestion (Saxena et al. 1981). Leaf of Bambusa spinosa fed to cattle to cure dysentery. (Mishra et al 1996)Bread of mixed flour of Eleusine coracana with seeds of Trachyspermum copticum fed twice a day to goat, cow of buffaloes to cure dysentery. Grains of Hordeum vulgare with chickpea cooked and given as diet to cure bloody dysentery in Darbhanga (Bihar). Baked seeds of Oryza sativa fed as main diet to goats to cure dysentery.Fresh bark extract of Shorea robusta is given in dysentery and when there is no moon in October the bark is removed. The tender parts of Bombax cibaricum are given as fodder. Dried leaves and inflorescence of Cannabis sativa are orally and ministered in indigestion, dysentery, diarrboea and Pneumonia in cattle. The decoction of the aerial parts of Atylosia scarabaeoides and the seeds of Dolichos biflorus are given in dysentery of cattle. The plant of A.scaraeboides as such is given to cattle in diarrhoea. Cattle suffering from dyspepsia are allowed to inhale the smake of the fibrous dry mesocarp of the fruit of Luffa cylindrica. In wealth of India it is reported that in China the ripe fruit is used after burning and pulversing as a carminative and anthelmintic. About 10 to 15 red roots of Smilax zeylanica, 20 to 30 leaves of Paederia foetida and five to seven roots of Celastrus paniculatus are pounded together. The pounded mixture is given orally with water twice daily to use amoebaeoisis in cattle. The decoction of the bark of Soymida febrifuga is mixed with Terminalia bellerica fruits and the mixture is prescribed orally in diarrboea in cattle. Tribals of Eastern India use the mixture of the pounded old seeds of Sal and boiled rice suspended in cold water, twice daily to cure loose motions, especially in goats. If there is no improvement the resin of Sal is prescribed in next dose for definite cure. Seeds of Celosia argentea are given to cattle for diarrboea .Whole plant of Orobanche aegyptica is given to cattle for diarrboea.

Inflamation

Root extract of Adiantum caudatum is given in throat swelling by Hakims in Bangladesh. Seed oil of Moringa oleifera is applied as swelling of legs. The heated leaves of Calotropis gigantean are applied on any swellings of the legs of cattle. Jain and De (1966) have stated that warmed leaves of C.gigantea are applied on swollen checks in Sannipat disease of cattle. Paste prepared from the seeds of Entada pursaetha is applied locally for inflammatory glandular swellings. The cotyledon of the seed is also made into paste with the ashes of Bracket fungus which grows on the bark of living Sal tree. The paste is applied twice daily on swelling of the throat in cattle. The paste is also prescribed orally in serious diarrhoea in cattle. Juice of the leaf of Agave sisalana also applied on the inflammation of the cattle throats.

Insect Bite/Sting

Hakims in Bangladesh use the root paste of Adiantum caudatum in dog bite to bovine group of animals. Juice of the leaves of Marsilea minuta is applied on stings of bees, wasp etc. The leaves and crushed bark of Betida alnoides mixed with cattle feed is used as anti-rabbies. Tender fruits of Capsicum frutesens are rubbed against the site bitten by snakes.

Intestinal Worms

Tuberous root powder of Cyperus rotundus given with fodder. Root of Cynotis tuberosa or Imperata cylindrical given as anthelmintic for 2-3 days (Borthakur and Sarma 1996). Seed oil of Dillenia indica, Shorea robusta and Sterculia foetida (Pal, 1992). Hakims in Bangladesh use the seeds of Cordia dichotoma to the bovine groups of animals to expel roundworms. The decoction of leaves of Ananas camosus is given once or leaves fed regularly once a day for 10 to 15 days. Crushed rhizome of Paris polyphilla mixed with cattle feed is given as anthelmintic. In Latin America Peasants use leaf decoction of Caladium bicolor to get rid of external cattle festers caused by worms (Balbach 1980), Leaves of Chenopodium ambrosioides are used in Brazil to expel worms in animals (Di Stasi et al. 1989). Small pox. Juice of fresh leaves of Melia azadirachta is applied in bovine group of animals. The tuberous roots of Solena heterophylla are made into a paste with the water of habul babul and the paste in applied on sores of small pox in cattle.

Increasing Lactation

About 250 grams of crushed root of Asparagus racemosus mixed with cattle fed to the newly pastorated cow to enhance early lactation. The fresh rhizome of Curcuma caesia is pounded and is prescribed orally by the Lodhas to cattle for more milk. About 25 g roots of Argyreia nervosa are pounded with 5 g seeds of Ajawain (Trachyspermum ammi), 150 g Masur (Lens culinaris) and 150 g of Molasses. The pounded mixture is given orally for lactation of cattle. Powder of bark of Lagerstromia parviflora mixed with forage is given to cattle for increase of lactation. Grassland or cropland weeds like Borreria hispida,B. stricta, Melilotus alba are fed to bufallow for increasing milk.

Bone fracture

The paste of seeds of Annona squamosa and a paste of the Cuscuta reflexa plant with lime 3:1 is applied with roasted leaves of Calotropis gigantea are bandaged on swellings as well as dislocated joints, sprains and fracture of bones. The rhizome of Drynaria quercifolia, three internodes of Cissus quadragularis and 100 g of Zingiber officinale are pounded. The pounded mixture is applied on bone fracture in cattle. It necessary, two three strips of bamboo are placed around the fractured past and dirtied with a piece of cloth for 15 days. Bark paste of Grewia optiva,Pinus wallichiana ,Prunus cerasoides and root paste of Osyris wightiana is applied as plaster for 2-3 weeks to aid the union of fractured bone. Direct application of root paste of Symphytum officinale is applied for bone fracture in U.P. Himalaya (Gaur, et al. 1992). Paste of stem or leaves of Bambusa arundinacea is used orally or topically for healing fracture. Aqueous extract of crushed leaves is also used (Asgar, 1999). The barks of Careya arborea, Madhuca longifolia, one piece of Cissus quadragularis, plant and few aerial root tips of Ficus benghalensis are made into a paste and applied on bone fracture in cattle. It is bandaged with the leaves of Bauhinia vahtii and tied with the bark of the same plant. Jain and De (1966) report that the leaves and stem of Drycible paniculata are boiled in water and the thick decoction is applied on sprains of cattle. The sprained parts are first cleaned with water. According to them juice of Glossogyne bidens is also considered to have healing effect on fractured bones in cattle. Pulp of fruit of Cassia fistula is given a laxative, pulp and seeds of this plant have antibacterial activity. Aqueous extract of the rhizome of Drynaria quercifolia possesses antibacterial properties. In Rayalseema,Andhra Pradesh Leaf paste of Dodonaea viscosa with egg albumen and lime are applied to aid in bone setting.Leaves of Murraya paniculata pounded with egg albumen are applied as plaster (Sudarsanam et al., 1995).

Explusion of placenta

Leaf of Bambusa tulda fed to animals after delivery to hasten removal of placenta by Nepalis in Assam (Borthakur and Sarma, 1996). Leaf of Dendrocalamus strictus fed to animals with the nation that it causes a more rapid. Fresh leaves of Saccharum spontaneum given as fodder alongwith young twigs of Justicia adhatoda/to help timely explusion of placenta. The root of Adhatoda vasica is also tied around the neck of cattle to losten the same. Flauss of curcuma longa is orally fed to smooth the explusion of placenta (Chettri et al 1992). An infusion of the bark of Ceiba pentandra is given to cattle after delivery to help expel the placenta (Duke 1986)

Vitality

Hakims in Bangladesh used the root paste of Paeonia emodi to bovine group of animals for their vilality. In Brazil the pulp of the seeds of Bixa orellana L. is given to bulls before fights to make them more active and ferocious. The plant may contain some excitant which has not yet been investigated (Standley 1920-1926). Powdered tobacco is mixed with aguardiente (liquor) and given to dogs to make them better hunting animals (Duke and Vasquez 1994).

Mastication

The plant of Euphorbia thymifolia is made into a paste with the leaves of Scoparia dulcis, this paste or Juice of Casearia elliptica leaves is given orally to cattle when they stopped mastication. The Lodhas, tribals of eastern India prescribe the mixture of pounded seeds of Centralherum anthelminticum, honey and seeds of Trachyspermum ammi to the cattle when they stop mastication.

Food Poisoning

Boiled fruit juice of Citrus medica is orally given crushed seeds of Raphanus sativus are mixed with animal feed and orally administered indigestion. The cattle is also fed with its fermented root against food poising.

Haematuria

The young plants of Tinospora cordifolia are given as fodder when the animal suffers from the discharge of blood with urine. About 500 gms of plant material of Cuscuta reflexa mixed with cattle fed is given.

Cough and cold

Leaves of Bambusa arundinacea are given to horses. Jain and De reported that the heated fruits of Cassia fistula are applied on the neck of the cattle to cure swelling due to cold. When cattle are suffering from eye disease (watering) the juice of the tuber of Curculigo orchioides is put as an eye drop.

Fever

Pounded tubers of Cynotis tuberosa given with common salt to cure fever in cattle. It is given in prolonged fever (Pal 1981). Juice of the roots of Cessamplos pareira about a cup is given twice daily to quadruped.

Tongue / Mouth sores

Plant paste of Apluda mutica for tongue and mouth sones given for 3-4 days in Hazaribagh (Bihar) (Tarafder and Rai, H.N. Chaudhuri 1981). Roots of

Cocculus hirsutus are boiled in Kusum oil and given in mouth diseases. Root juice of Phoenix paludosa with cow milk is given in mouth diseases.

Other disorders

Jain and De (1966) reported that the paste of Scoparia dulcis is applied in gout pains in cattle. Jain and De (1966) have reported that the plant of Solena hetsophylla is kept hanging in front of herds of cattle to keep away evil spirits and diseases like Malaria. Crushed roots of Woodfordia fruticosa are given to cattle for rheumatism. By tribes of Nilgiri Hot water extract of leaves of Swertia corymbosa is given to buffaloes as a stomach wash. Leaves of Alysicarpus rugosus are used as a drench for calves with scours.

The plant paste of Atylosia scarabaeoides is given orally in Keonjhar (Orrisa) (Pal 1981) in tuberculosis in cattle.

Table 1 : Botanical Name, Habit and Distribution of plant species

Botanical Name	Local Name/ English Name	Family	Habit	Distribution/ grown in India
Abrus precatorius L.	Tatti	Papilionaceae	Climbing shrub	All over India
Acacia nilotica (L.) Will	Babul	Mimosaceae	Small tree	Arid and semi arid region
Allium cepa L.	Piyaj, Onion	Liliaceae	Annual herb	All over India
Allium sativum Linn.	Hindi Lahsun	Liliaceae	Annual herb	All over India
Adhatoda vasica Nees	Arusa	Acanthaceae	Diffuse evergreen hrub	All over India
Agave sisalana Perr.	Agave	Agavaceae	Xerophytic perennial	Tropical region
Annona squamosa L.	Shrifa	Annonaceae	Shrub or small tree	All over India
Argemone mexicana Linn	Kateli	Papavaraceae	Annual herb	All over India
Adiantum caudatum L.	Fern	Adiantaceae	Fern	N.W. Himalaya
Agrimonia pilosa Ledeb.	-	Rosaceae	Perennial herb	Kashmir Himalaya
Apluda mutica L.	Bhanjuri	Graminae	Annual erect herb	Throughout India
Artimesia nilagirica (Clarke).Pamp.	Nagdouna, Titapati	Compositae	Aromatic shrub	W.Him,Sikkim,Khasi hills,W.Ghats
Atylosia scarabaeoides	Bankulthi	Papilionaceae	Perennial climber	Arid and semi arid
Argyreia nervosa (Burm.f.) Boj	Ghabel	Convolvidaceae	Climbing shrub	All over India
Asparagus recemosus Willd.	Satavar	Liliaceae	Straglling spinous shrub	All over India
Ananas camosus(L.)Merr.	Ananas	Bromeliaceae	Biennial herb	TN,Coastal AP,Assam Kerala,WB, Tripura, Orissa
Alysicarpus rugosus DC.	-	Papilionaceae	"	Tropical region
Azadirachta indica A.Juss.	Neem	Meliaceae	Large tree	All over India
Bambusa arundinacea Willd.	Bans	Bambusaceae	Spiny bamboo	Throughout India

Botanical Name	Local Name/ English Name	Family	Habit	Distribution/ grown in India
Betule alnoides Buch.Ham	Bhojpatra	Betulaceae	Tree	Temperate & sub-trop. Himalaya
Bixa orellana (L.)	latkan, Sinduria	Bixaceae	Small tree	Cultivated in South India
Bombax ceiba L.	Semal	Bombaceae	A tall tree	All over India
Beta vulgaris Linn.	Chukander	Chenopodiaceae	Biennial herb	North India (cultivated)
Borreria hispida L.	Gathua	Rubiaceae	Perennial herb	All over India
Borreria stricta L	Chhota gathua	Rubiaceae	Annual erect herb	All over India
Brassica napus L.	Toria	Cruciferae	Annual erect herb	Throughout India
Brassica juncea (Linn.) Czern. & Coss.	Sarson	Cruciferae	Annual erect herb	Throughout India
Bauhinia vahlii Wight & Arn.	Malijhan	Caesalpiniaceae	Woody climber	Assam, M.P., Bihar, Punjab
Buxus Wallichiana Baill	Papri	Buxaceae	Tree	W. Himalayas
Bambusa tulda Roxb.	Peka,Chau	Bambusaceae	Bamboo	WB, Assam
Butea monosperma Lamk	Dhak	Papilianaceae	Small tree	All over India
Cinnamomum camphora (Linn.) Nees & Eberm.	Kapur	Lauraceae	Tree	Ootachmund
Cedrus deodara (Roxb.) Loud.	Deodara	Pinaceae	Large tree	N.W. Himalayas
Chenopodium ambrosioides L.	American worm seed	Chenopodiaceae	Erect aromatic herb	Bengal, Kashmir, Maharashtra
Ceiba pentandra (L.) Gaertn	safed semal	Bombacaceae	Medium-sized tree	All over India
Caladium bicolor Vent.	-	Araceae	Rhizomatous herb	All over India
Curculigo orchioides Gaertn.	Kali musli	Hypoxidaceae	Perennial herb	Sub-tropical Himalaya Khasi Hills
Centratherum anthelminticum (L) O.Kuntz	Somraj	Compositae		Throughout India
Crotalaria alata Buck.Ham.	-	Papilionaceae	Suberect undershrub	All over India
Curcuma domestica Val.	Haldi	Zingiberaceae	Perennial herb	Bihar, Chhota Nagpur
Cassia fistula L.	Amaltas	Caesalpiniaceae	Orna. Tree	All over India
Cannabis sativa L.	Ganja, Charas, Bhang	Cannabinaceae	Erect herb	All over India
Crotalaria linifolia L.	-	Papilionaceae	Annual herb	Tropical tract
Citrus aurantifolia Swingle	Kagdi Neebu	Rutaceae	Shrub	All over India
Citrus limon (Lin.) Burm.f.	Nimbu	Rutaceae	Large shrub	All over India
Cassia tora L.	Pawar	Caesalpiniaceae	Annual erect herb	All over India
Celosia argentea L.	Murgha	Amarantaceae	Annual erect herb	Tropical tract
Carthamus tinctorius L.	Kusum	Compositae	Annual erect herb	All over India
Celastrus paniculatas Willd.	Mal kangini	Celastraceae	Climbing shrub	All over India

Botanical Name	Local Name/ English Name	Family	Habit	Distribution/ grown in India
Cyperus rotundus L.	Motha	Cyperaceae	Perennial herb	All over India
Calotropis gigantea Ait.	Ak	Asclepiadaceae	Erect undershrub	All over India
Capsicum frutescens L.	Dhangacch mirch	Solanaceae	Branched shrub	All over India
Cyanotis tuberosa L.	-	Commelinaceae	Perennial herb	All over India
Cuscuta reflexa Roxb.	Amarbel	Convolvulaceae	Parasitic twinner	All over India
Curcuma caesia Roxb.	Kalihaldi	Zingiberaceae	Perennial herb	Bengal
Cordia dichatoma L.	Indian cherry	Ehretiaceae	Small tree	All over India
Curculigo orchioides Gaertn.	-	Hyposidaceae	Perennial herb	All over India
Curcuma amada Rob.	Ama haldi	Zingiberaceae	Rhizomtes erect	Central, P. India
Chenopodium albumL.	Bathua	Chenopodiaceae	Annual herb	All over India
Cirsium verutum L.	-	Compositae	Annual herb	Temperate Himalaya
Calopjhyllum inophyllum L.	Surpan	Guttiferae	Evergreen tree	Sea sores
Cymbopogon flexuosus Wats	Lemon grass	Araminae	Perennial grass	Travancone, Malber
Careya arborea Roxb.	Kalikatbhi	Burringtoniaceae	Large tree	All over India
Cissus quadragularis L.	Hadjora	Vitaceae	Sacculent twinner	Warmer parts
Casearia elliptica L.	-	Flacourtiaceae	Shrub	Tropical tract
Cocculus hirsutus (L.) Diels	Jamti ki bel	Menispermaceae	Climbing shrub	Tropical and subtropical tract
Dolichos uniflorus Lam.	Kulthi	Papilionaceae	Annual trailing herb	A.P., Karnataka, T.N. Central India
Dillenia indica L.	Chalta	Dilleniaceae	Medium sized tree	Sub-Hima.Tract, Assam WB, MP, South India
Drynaria quercifolia J.Smith	Fern	Polypodiaceae	Fern	All over India
Dioscorea deltoidea Wall. Ex Kunth	-	Dioscoriaceae	Climbing shrub	Punjab, Bangla, Kashmir, N.W. Himalaya
Debregeasia saticifolia L.	-	Urticaceae	Shrub	N.W. Himalayas
Dodonaea viscosa (L.) Jacq.	Vilayati Menhodi	Sapindaceae	Green shrub	All over India
Dendrocalamus strictus (Roxb.) Nees	Bans Kaban	Graminae	Bamboo	N. India, M.P. Gujrat, Maharashtra, S. India
Entada pursaetha DC	-	Mimosaceae	Woody climber	C & E Himalaya
Ectipta prostrata L.	Bhangara	Compositae	Perennial herb	All over India
Eleusine coracana (L.) Gaertn.	Ragi	Graminae	Annual herb	India
Echinops echinatus Roxb.	Utakanta	Compositae	Spinous under shrub	All over India
Erythrina variegata L.	Dadap	Papilionaceae	Tree	All over India
Eucalyptus globulus Labill.	Tasmanium blue gum	Myrtaceae	Tall tree	Hill stations
Euphorbia thymifoli L.	Chhoti dudhi	Euphorbiaceae	Perennial herb	All over India
Ficus benghalensis L.	Bargad	Ficaceae	Large tree	All over tropical region

Botanical Name	Local Name/ English Name	Family	Habit	Distribution/ grown in India
Gardenia gummifera L.f.	Dikamli	Rubiaceae	Shrub	W. peninsula
Glossogyne bidens DC	-	Compositae	Perennial herb	Tropical tract
Grewia optiva Buck-. Ham.	Bhemal	Tliaceae	A large tree	Temp. subtrop. region
Helianthus annus L.	Suryamukhi	Compositae	Annual herb	All over India
Hypericum hookerianum W & A	-	Hypericaceae	Shrub	W. Himalaya
Hordeum vulgare (L.)	Jau	Graminae	Annual herb	All over India
Imperata cylindrica (L.) P.Beauv.	Vaughan	Graminae	Perennial grass	All over India
Jatropha curcas Linn.	Jangli-erandi	Euphorbiaceae	Evergreen shrub	Tropical tract
Jatropha gossypifolia L.	Bherenda	Euphorbiaceae	Branched shrub	All over India
Luffa aegyptiaca Mill.	Ghia torai	Cucurbitaceae	Annual training herb	All over India
Luffa aculangula Roxb.	Jhinga	Cucurbitaceae	Annual training herb	All over India
Lagerstromia parviflora Roxb.	Dhaura	Lythraceae	Large tree	All over India
Lawsonia inermis L.	Henhadi	Lythraceae	Shrub	Arid and semi arid tract
Lens culinaris Medik.	Masur	Papilionaceae	Annual herb	All over India
Melia azedarach L.	Bakain	Meliaceae	Large tree	Tropical tract
Melilotus alba Lamk.	Sainji	Papilionaceae	Annual herb	All over India
Marsilea minuta L.	Common fern	Marsiliaceae	Fern	All over India
Madhuca longifolia (L.) Maeb.	Mahua	Sapotaceae	Large tree	U.P., Bihar.AP, Karnataka, WB, Maharashtra
Momordica dioica Roxb. Ex Willd.	Kakora	Cucurbitaceae	Clmbing shrub	Arid and semi arid tract
Mammea americana L.	Mammey apple	Guttiferae	Small or medium-sized tree	All over India
Mentha piperita Linn. Emend. Huds.	Peppermint	Labiatae	Perennial herb	Maharashtra Kashmir, Punjab
Moringa oleifera Lamk.	Sahjan	Moringaceae	Small tree	All over India
Murraya paniculata (L.) Jack.	Kamini	Rutaceae	Shrub	All over India
Momordica charantia Linn.	Karela	Cucurbitaceae	Trailing herb	All over India
Nicotiana tobacum L.	Tambaku	Solanaceae	Viscid annual herb	AP, Maharashtra, Karnataka, MP, UP, WB
Oryza sativa L.	Dhan	Graminae	Annual herb	All over India
Ocimum sanctum Linn.	Tulsi	Labiatae	Perennial herb	Tropical and sub tropical region
Paris polyphylla L.	-	Liliaceae	Erect herb	Temperate Himalaya tract
Paederia foetida Merr.	Somraji	Rubiacae	C and E Himalaya	
Pongamia pinnata Pierre	Karanj	Papilianaceae	Small tree	Tropical tract
Prunus persica L.Batsch	Aru	Rosaceae		Kashmir, HP, Kumaon
Paeonia emodi., Wall	Udsalap	Paeoniaceae	Perennial herb	W. Himalaya
Peganum harmala L.	Gandhya Harmal	Zygophyllaceae	Per. Herb	All over tropical tract

Botanical Name	Local Name/ English Name	Family	Habit	Distribution/ grown in India
Plumbago zeylanica L.	Chita	Plumbaginacae	Annual Herb	Cultivated All over India
Phoenix paludosa Roxb.	-	Palmae	Tree	WB, Orrisa, Andman
Prosopis juliflora(SW.)DC	Vilayati babul	Mimosaceae	Small tree	Arid and semi arid tropics
Piper nigrum L.	Kali mirch	Piperaceae		Western ghats, Karnataka Maharashtram, Assam, Kerala
Orobanche aegyplica Pers.	Tokra	Orobanchaceae	Herb	All over India
Pinus Wallichiana A.B. Jackson	Chil	Pinaceae	Tree	W.E. Himalayas
Prunus cerasoides D.Don Wild	Himalayan cherry	Rosaceae	Shrub	Temperate Himalayas
Osyris wightiana (Wall.) ex Wight	-	Santalaceae	Evergreen shrub	Sub-Hima., C.India, Hills of S. India
Symphytum officinale	-	Boraginaceae	Perennial herb	All over India
Raphanus sativus L.	Muli	Solanaceae	Annual herb	All over India
Saussurea hypoleuca spreng.	-	Compositae	Biennial herb	Temperate Himalayas
Syzygium cumini L.	Skeels Jamun	Myrtaceae	A small tree	All over India
Saccharum spontaneum L.	Kans	Gramineae	Perennial grass	All over India
Soymida febrifuga Juss.	Rohan	Meliaceae	Large tree	Bundelkhand, Chhota Nagpur
Scoparia dulcis L.	-	Seraphulariaceae	Herb	Tropical tract
Shorea robusta Gaertn.	Sal	Dipterocarpaceae	Large tree	N, E and Central India
Smilax zeylanca L.	Ram dataun	Smilacaceae	Shrub	E. Himalayas, P. India
Solanum surattense Burm.F	Kateli	Salanaceae	Perennial spiny herb	All over India
Sterculia foetida L.	Jangali badam	Steculiaceae	Large tree	S. India
Swertia carymbosa Weight	-	Gentianaceae	Perr.herb	Deccan Peninsula
Tinospora cordifolia (Willd.) Miers ex Hook.f. & Thoms.	Gulancha	Menispceae	Climbing shrub	Tropical tract
Trachyspermum ammi (L.) Sprague	Ajwain	Umbelliferae	Annual herb	Tropical tract
Trachyspermum copticum	-	Umbeliferae	Annual herb	Tropical tract
Terminalia bellerica Roxb.	Bahera	Combetaceae	Small tree	Tropical tract
Ulmus wallichiana Planch	Mored, Pabuna	Ulmaceae	A large tree	NW Himalaya
Urginea indica (Roxb.) Kunth	Jangli piaz	Liliaceae	Perennial herb	W Himlalya, Bihar
Vetiveria zizanioides (Linn.) Nash	Khus grass	Gramineae	Perennial grass	All over India
Woodfordia fruticosa (L.) Kurz	Dawi	Lythraceae	Shrub	All over India
Zingiber officinale Rosc.	Adrak	Zingiberaceae	Rhizomatis herb	All over India

Bibliography

Anonymous. 1948-1976. Wealth of India (Raw Materials). Council of Scientific and Industrial Research, New Delhi, India.

Asgar A ,1999.Studies on osteogenesis under the influence of Bambusa arundinacea and Cissus quadriangularis-Uraria picta-Lepidium sativum on radius-ulna fracture in dogs,MVSC thesis submitted to G B Pant University of Agric. & Tech.,Pantnagar.

Balbach, A. 1980. A flora nacional na medicina domestica (vol.2). Itaquacetuba, Sao Paulo, Brazil

Bhattarai NK, Folk use of plants in veterinary medicine in Central Nepal, Fitoterapia, 1992, 63, 497-506.

Binu,s,T.S.Nayar, and P.Pushpangadan. 1992. An outline of ethenobotanical in India. J. Eco. Tax. Bot. Addi.Ser. 10: 405-428.

Borthakur, S.K. and U.K. Sarma. 1996.Ethnoveterinary medicine with special reference to cattle prevalent among the Nepalis of Assam, India. Pp. 197-199. In: S.K. Jain (ed.) Ethnobotany in Human Welfare. Deep Publ., New Delhi.

Chettri, R., B. Rai and D.B. Khawas. 1992. Certain medicinal plants in the folklore and golklife of Darjeeling and Sikkim hills, India-I Used for treatment of domestic animals. J. Eco. Tax. Bot. Addi.Ser. 10: 393-398.

Chopra, R., N., S. L. Nayar, and I. C. Chopra. 1956. Glossary of Indian Medicinal Plants. Council of Scientific and Industrial Research, New Delhi, India.

Di Stasi, L. C. et al. 1989. Plantas medicinais na Amazonica. UNESP, Sao Paulo, Brazil.

Duke, J. A. 1986. Isthmian Ethnobotanical Dictionary. Scientific Publishers, Jodhpur, India

Duke, J. A. and R. Vasquez. 1994. Amazonian Ehnobotanical Dictionary. CRC Press, Inc. Boca Raton, Florida, USA.

Farnsworth, N. R., O. Akerele, A.S. Bingel, D. D. Soejarta, and Z. Eno. 1985. Medicinal plants in therapy. Bulletin of the World Health Organization 63(6):965-981.

Gaur RD, Bhatt KC, and Tiwari JK, An ethonobotanical study of Uttar Pardesh Himalaya in relation to veterinary medicines, J Indian Bot Soc, 1992, 72, 139-144.

Jain, A.K. 1992. Ethnobotanical studies on Sahariya tribals of Madhya Pradesh with special reference to medicinal plants. J. Eco. Tax. Bot. Addi.Ser. 10: 227-232.

Jain, S. K., B. K. Sinha, and R. C. Gupta. 1991. Notable Plants in Ethnomedicine of India. Deep Publications, New Delhi, India.

Jain,S.K. 1981. Ethnobotanical Research Unfolds new vistas od traditional medicine . pp. 13-36. In: S.K. Jain (ed.) Glimpses of Indian Ethnobotany. Oxford & IBH Pub. Co., New Delhi

Jain,S.K. and C.R. Tarafder. 1970. Medicinal plant lore of Santals. Eco. Bot. 24: 241-278.

Jain,S.K. and J.N. De 1966. Observations on ethnobotany of Purulia, West Bengal. Bull. Bot. Surv. India 8: 237-257.

Jain,S.K.,V.Mudgal,D.K.Banerjee,A.Guha,D.C.Pal and D.Das.1984.Bibliography of Ethnobotany (Botanical Survey of India).

Jain,S.K.1971.Some Magico-Religious beliefs about plants among adivasis of Orissa. Adivasi 12(1-4) :39-44.

Jaiswal,S,S.V.Singh,Bhupendra Singh and H.N.Singh.2004.Plant used for tissue healing of animals. Natural Product Radiance,3 (4) :28-292.

Kirtikar, K.R. and B.D. Basu. 1935. Indian Medicinal Plants. (2nd revised ed. E. Blatter, J.F. Caius and K.S. Masker) Lalit Mahan Basu, Allahabad.

Krishna Kumar, P. R. 1996. Indian medicine industry under the Emerging patent regimes. Ancient Scinece of Life 15(3):161

Kumar D, Tripathi HC, Tandan SK, Lal J and Malik JK, Ethno-veterinary phytomedicines used in India and Nepal in treatment of fracture wound and allied disorders: An update, Indian J Vet Surg, 1997, 18 (2), 65-72.

Mishra, Sunita, V. Jha and S. Jha. 1996. Plants in ethnoveterinary practices in Darbhanga(North Bihar) pp. 189-193, In: S.K. Jain (ed.) Ethnobotany in Human Welfare. Deep Publ., New Delhi.

Mondal MK and Chauhan JPS, A survey of ethno-veterinary medicine practices in West Bengal, Indian J Vet Med, 2000, 20 (1), 90-91.

Mukherjee PK, Saritha GS and Suresh B, Antibacterial spectrum of Hypericum hookerianum, Fitoterapia, 2001, 72 (5), 558-560.

Pal, D.C. 1981. Observation on folklore about plants used in veterinary medicines in Bengal,Orissa and Bihar.Bull.bot.Surv.India. 22(1-4) :96-99.

Pal, D.C. 1981. Observation on folklore about plants used in veterinary medicines in Bengal, Bihar and Orissa -II. J. Eco. Tax. Bot. Addi.Ser. 10: 137-141.

Pal, D.C. 1981. Plants used in treatment of cattle and Birds among tribals of eastern India. Pp. 245-257. In: S.K. Jain (ed.) Glimpses of Indian Ethnobotany. Oxford & IBH Pub. Co New Delhi

Pal, D.C. and S.K. Jain .1989. notes on Lodha medicine in Midnapur District, west Bengal, India. Economic Botany 43: 464-470.

Rajan, S. 1992. Medicinal plants of Ootacamund, Nilgiris, Tamilnadu. J. Eco. Tax. Bot. Addi.Ser. 10: 429-460.

Rao, R.R. 1981. Ethnobotanical studies on the flora of Meghalaya- Some interesting reports of herbal medicines. Pp. 124-136. . In: S.K. Jain (ed.) Glimpses of Indian Ethnobotany. Oxford & IBH Pub. Co., New Delhi.

Rawat GS, Neem (Azadirachta indica) Natures drug store, Indian for, 1995, 25, 145-152.

Saxena, H.O, m. Brahmam and P. K. Datta. !981. Ethnobotanical studies in Orissa. Pp. 232-244. In: S.K. Jain (ed.) Glimpses of Indian Ethnobotany. Oxford & IBH Pub. Co., New Delhi.

Schultes, 1981. Foreword to the Glimpses of Indian Ethnobotany, Ed. By S.K. Jain. Oxford & IBH Publishing Co. New Delhi.

Sebastian MK and Bhandari MM, Some plnats used as veterinary medicines by Bhils, Int J Trop Agric, 1984, 2 (4), 307-310.

Sensharma, P. 1984. Ethnobotanical Investigation in the Indian Puranas: V. The Kurma Purana. J. Econ Tax. B t. 5 (3): 634-644.

Sensharma, P. 1987. Ethnobotanical Investigation in the Indian Puranas: V. The Vamana Purana. J. Econ Tax. B t. 9 (2): 287-299.

Sensharma, P. 1988. Ethnobotanical Investigation in the Indian Puranas: The Vayu Purana. Man in India 68 (2&3): 278-291.

Sensharma, P. 1991. Herbal Veterinary Medicines in an ancient Snaskrit work- The Garuda Purana. Ethnobotany 3: 49-53.

Sensharma, P. 1992. An ethnobotanical Investigation into the Linga- Purana. J. Econ Tax. Bot. Addl.Ser.10: 271-283.

Sharma M.C. and Chinmay, Joshi. 2004. Plant used in skin diseases of animals. Natural Product Radiance,3 (4) :293-299.

Sharma M.C. and Dwivedi SK, Efficacy of a herbal preparation against dermatomycosis in cattle and dogs, Indian Vet J, 1990,67 (3) 269.

Singh G, Khate K, Bujarbaruah KM, Mondal SK, Pal DT and Kumar S, 2001a, Phytomedicines used for animal treatment by tribals in remote area Nagaland, Indian Vet Med J, 2001, 25 (3), 291-292.

Singh H Jadon MS, Sharma VK, Singh SP, Gupta RS and Kumar S, 1998a, Research on some aspects of Bone and Joint Affections in animals' Published by Directorate of experiment station, G.B. Pant Univ. of Agri. & Tech., Pantnagar- 263145, 1998.

Singh H, Herbal medicines in the management of fracture in small animals, Intas Poliver, 2002, 3 (2), 152-155.

Singh H, Sharma VK and Chauhan RS, 2001b, Herbal preparation and Bio- stimulators in Tissue repair, Research bulletin published by Directorate of Experiment Station, G.B. Pant University of Agriculture & Technology, Pantnagar- 263145, 2001.

Singh MP, Tribal medicinal plant used an animal diseases of Chotanagpur, Indian For, 1987, 113, 758-759.

Singh V, 1995b, Ethno-veterinary medicinal plants used in Jammu- Kashmir, Ladakh and Morni Hills (Haryana), India Fitoterapia, 1995, 66, 356-359.

Standley, P. C. 1920-1926. Trees and Shrubs of Mexico. Smithsonian Press, Washington, D.C., USA.

Sudarsanam G, Reddy MB and Nagaraju N, Veterinary crude drugs in Rayalaseema, Andhra pardesh, India, Int J Pharmacogn, 1995, 33, 52-60.

Tarafder, C.R. and Rai H.N. Chaudhuri. 1981. Less known medicinal uses of plants among the tribals of Hazaribagh district of Bihar. Pp. 208-217. In: S.K. Jain (ed.) Glimpses of Indian Ethnobotany. Oxford & IBH Pub. Co., New Delhi.

Tugnaiyat AK, Bhargawa MK, Chandrapuria VP, Pandey SK and Quadiri MA, Efficacy of Curcuma longa and Helianthus annus on tissue repair in calves: Clinical and biochemical studies, Indian J Vet Srug, 2000, 21 (2), 76-78.

CHAPTER - 12

IMPROVEMENT OF POOR QUALITY ROUGHAGES

S.S. Kundu, A.B. Mojumdar, K.K. Singh and M.M. Das

Fibrous crop residues are an important component of the feed base for ruminants in India. Although animal production is growing rapidly but the pace of productivity has to match the needs of increasing human population. The present trend is shifting towards foods of animal origin and cereal requirement per se is decreasing in the cities. The deficit in supply of green, and dry forage along with concentrate has been reported. Crop residue like straws and stovers being low in nitrogen content and high in complex cell wall contents possesses low digestibility. A specific composition of carbohydrate polymers, cross linkage with lignin and the physical arrangement of the cellular and vascular tissues restrict the utilization of available feed resource (Table 1).

The utilization of the straw can be increased by the supplementation of nitrogen (Singh and Samanta, 1998, Kundu and Sharma, 2003), minerals and vitamins (Pal et al., 1980), but to harvest cell wall energy further it has to be exposed to physical, chemical or biological treatments (Jackson, 1977, Sundstol et al., 1978). Due to inherent limitations, exclusive feeding of straw even do not support the maintenance requirements (Kundu, 1989). Feed intake of wheat straw is improved in cattle and buffalo, after soaking with water (Chaturvedi et al., 1973).

Ammoniation or use of sodium hydroxide exhibited promising results on straw treatment but these methodologies have not been adopted by farmers or feed industry. Chlorine dioxide is possibly the most efficient oxidant of lignin and has been reported to reduce 75% lignin from wheat straw (Sullivan and Hershberger, 1959). Most prominent oxidizing chemicals like sodium hypochlorite, sodium sulphite, bleaching powder have been used to delignify crop residue (Oven and Kategile, 1984). However, systematic studies on the use of these chemicals on wheat straw are scanty.

The density of straw/ stovers as such is very low (55 to 83 kg/m^3). Enhancement in rumen digestibility of straw has been reported when nitrogen and energy sources are added through concentrate ingredients (Yadav et al.,

1990). Feed intake as well as feed conversion efficiency tends to increase further when treated straw is fed in the form of complete diet (Venkanna et al., 1997). The large capacity of the rumen and the presence of large microbial population subsisting on plants materials are especially evolved for the digestion of cellulosic feed. This versatile symbiotic relationship has enabled the ruminants to play a major role in the economic development of agricultural sectors of the world.

The *ad libitum* feeding system should be designed based on the complete diet principle. Complete diet provides constant environment in the rumen throughout the day. But it is difficult to transport straw to a long distance from surplus to scarcity regions due to escalation of transport cost. However, the compounded complete feed having major component of treated straw need hardly 1/3rd space after compaction.

Complete feeding system ensure the consumption of roughage and concentrate in a proper proportion and results in efficient rumen fermentation. The adoption of such feeding system not only reduces the cost of feeding but also enhances voluntary feed intake and productivity in ruminant (Pirie and Greenhalgh, 1978, Sihag et al., 1991, Kundu et al, 1998).

Table: 1 Availability of major crop residues (2000-2001)

Crop	Grain yield (million tonnes)	Factor	Crop residue / fibrous by products (million tonnes)
Wheat	69.68	1.3	90.58
Rice	93.20	1.3	121.16
Barley	1.43	1.3	1.86
Maize	13.16	3	39.48
Millets	11.23	4	44.95
Sorghum	7.55	4	30.22
Total	**196.27**	-	**328.26**

Methods of improvement

Ammonia treatment

The upgrading effects of ammonia on low quality roughages have been known for a long time. The chemical was not used in practice to a great extent till 1970. Since then a number of methods have been developed (Table 2)

Anhydrous ammonia

Anhydrous ammonia is hazardous to handle because of low boiling point (-33.4 °C) at atmospheric pressure. It has to be kept pressurized. At concentration between 15 and 28% in air, it is explosive upon ignition.

Table: 2 Methods for treatment of poor quality roughages with ammonia

Form of ammonia	Procedure for treatment	Conditions (optimum)
Anhydrous and aqueous ammonia	1.Stacks of rectangular or round bales covered with polyethylene and injected with ammonia	3-3.5% NH_3 on DM basis, 15-20% moisture in the material, 1-8 weeks of treatment depending on temperature
	2.Round bales packed in polyethylene bags and treated individually with ammonia	As above
Anhydrous	1.Treatment in ovens with added heat	Treated with heat (90°C) and ammoniated (3-5% NH_3) air for 17 h, standstill 5 h and fresh air circulation
Urea	1.Loose material ensiled in earthen pits, trench silos, bamboo baskets or stacks	3% urea in a water solution, mixed with material at a 30:70% ratio ± at >20°C, three weeks or more
	2. Chopped or ground material added urea before pelleting in an industrial plant	2-3% urea, at least 133°C, 15-20% moisture
	3. Wheat straw in stock of 100 kg applied with urea and kept air tight	3% urea, with 30 to 35 liters of water for 3 weeks or more.

Anhydrous ammonia is commonly used for treatment of straw in Norway, Denmark and some other European countries (Sundstol et al., 1978, Kernan et al., 1979). The stacks of straw as such or in the form of bales are wrapped with sheets of polyethylene and sealed. Thereafter, a perforated injection pipe is forced through the plastic into the middle of the stack where the anhydrous ammonia is administered at about 30 kg per tonne. The liquid ammonia evaporates gradually and the ammonia gas penetrates the stack. In Northern Europe, a treatment time of up to eight weeks is often recommended. In warmer climates, which prevails in the tropics, the treatment time may be considerably shorter and only one to two weeks may be sufficient. Ammonia treatment can also be carried out in airtight boxes, rooms or houses (Solaiman et al., 1978, Kay, 1982). Ammonia is an efficient preservative of high moisture straw, stover etc. In humid climate a check on mould growth is one of the greatest advantage of this method. However, a too low moisture content (below 10%) may, on the other hand be a disadvantage. At low temperature, the ammonia is effectively bound to the moisten material and difficult to aerate. Under cold environment high moisture ammoniated straw when brought in to a warm barn and causes environment pollution.

Aqueous ammonia

Aqueous ammonia is used commercially for straw treatment and it is easier to handle ammonia in aqueous form. If the straw is extremely dry, the added water may also have positive influence on the treatment effect. At normal pressure, one kg ammonia solution can hold 285g at 30 °C and 342g at 20 °C. Solution containing 25-35% ammonia are commonly available. The solution, sufficient to give 3.0-3.5% ammonia in the straw, is normally injected into the stacks through a tube mounted at a metal pipe. The treatment time of about one week may be sufficient.

Urea

Urea is more convenient chemical to handle than ammonia. It is also readily available in all parts of the country. Straw treatment with urea may be carried out on an industrial as well as on farm scale.

Industrial scale urea treatment has been developed in the Germany where urea is added to chopped or ground straw before entering the pellet press. The heat developed in the pressure causes the decomposition of urea into carbon dioxide and ammonia. At 133 °C the decomposition of urea is almost complete. The liberated ammonia will then act upon the roughage.

Farm scale urea treatment has been tried in several places in the world but the most practical approach was developed in Bangladesh (Saadullah et al., 1981, Dolberg et al., 1981). Five percent urea is dissolved in water and the solution is mixed with equal amount of rice straw (weight basis) and the treated material is put into the silo pit of which the bottom and walls may be lined with banana leaves or the like material to prevent contamination with soil. Banana leaves, mats or a piece of plastic may be placed on top of the straw before it is covered with soil. Initially a treatment period of three weeks is recommended, but more recent results indicated that treatment time may be reduced. Other type of containers like bamboo baskets or simply stacks of loose straw may be used. However, silo pits are superior than bags or baskets (Ibrahim et al., 1983, Mojumdar et al., 2003).

Treatment with sodium hydroxide

An overview of some methods for NaOH treatment of low quality roughages are presented in Table 3.

Wet treatment

The modified Beckmann procedure for treatment of straw has been used by Norwegian farmers for more than 40 years. However, the problem related to the disposal of the rinsing water forced a majority of the user to stop this treatment. For high yielding dairy cows as ammonia treated straw did not seem to be a good substitute for the Beckmann –treated straw.

TABLE 3: Methods for treatment of poor quality roughages with sodium hydroxide (Sundstol, 1988)

Moisture level	Procedure for treatment	Conditions (optimum)
Wet treatment	1. Immersion of material in a NaOH solution followed by rinsing (Backmann procedure)	1.5-2.5 NaOH solution, time of soaking about 12h, rinsing to neutrally.
	2. Immersion of material in a NaOH solution without rinsing (dip treatment method), but with a period of storage	1.5% NaOH solution, time of soaking 0.5h, time of 'ripening' 3-6 days
Semi wet treatment	1. Ensiling of the material with a NaOH solution in a container (silo) or stack.	40-70% moisture, 3-5% NaOH on DM basis, air tight, minimum one week of treatment
	2. Ensiling of straw with NaOH in a stack	40-50% niustyrem 1% NaOH and treatment period of 3 to 4 weeks
Dry treatment	1. Loose material sprinkled with a NaOH solution from a garden can or the like mixed.	Solution of 2% NaOH per 100 kg material, treatment time approximately 24 h.
	2. Treatment of material during collection or baling	

Moisture level	Procedure for treatment	Conditions (optimum)
	3. Spraying of material with a NaOH solution in the bales.	0.8-1.0 kg of a 50% NaOH solution per bale of 10 kg, 1 week storage
	4. Chopped and sliced material mixed with a NaOH solution and blown into a heap where the temperature rises upto 80-100°C (Straw treatment machines)	150-180 kg of a 27% NaOH solution per ton DM, minimum 3 days of treatment
	5. Chopped or ground material is mixed with a NaOH solution and pressed into cubes or pellets (Industrial plants).	Conditions vary considerably. Solution containing from 27-47% NaOH is added to give 4-5% NaOH on DM basis. Pressed at more than 100 atmosphere temperature 70-90°C, then cooled.

For large scale treatment the solution may be pumped into another vessel to leave the straw in the first vessel for drainage before storage. The time needed for 'ripening' is dependent on the temperature, and a temperature above 10 °C may be desirable during storage. The amount of sodium hydroxide needed for replenishment of the lye solution is normally 6-6.5 kg NaOH per 100 kg straw, but may vary with the type of material treated. It is therefore advisable to titrate the solution time to time to check that the strength is about 15g NaOH per liter (Greenhalgh, 1976).

Semi – wet treatment

Ensiling of straw and other low quality roughages with NaOH is used when the material has a moisture content between 30 – 60 % (Wilkinson, 1984b). This method may be of particular interest when weaker chemicals are applied and a long treatment time is needed, or the cost of NaOH is prohibitive.

Dry treatment

Industrial procedures for dry treatment of straw with NaOH have been developed in many countries (Rexen and Bach-Kundsen, 1984). The general procedure is that the material is chopped or milled before a solution containing 200-300 g NaOH per litre is sprayed on the material. After a thorough mixing, the material is allowed to pass through a pellet press. The high temperature developed in the press (ring die) causes an almost immediate reaction between the material and the alkali. After leaving the press, 1 –5 % of the moisture evaporates off. The material may be fed as it is or mix with concentrates either before or after pelleting to make 'complete feed'(Venkanna et al., 1997). The optimum concentration of NaOH seems to be 4 – 6% of straw dry matter.

'On farm' procedures are available for dry treatment of straw and other low quality roughages with sodium hydroxide. The simplest form of treatment is that a sodium hydroxide solution is sprinkled over long form of straw, which is thereafter mixed thoroughly by hand using a fork (Jackson, 1978). A lye solution may be sprayed over chopped straw in a mixer wagon ensuring a sufficient mixing of the material before feeding (Kundu and Sharma, 1995). Straw may also be treated during the process of baling which can be done by placing a sprayer boom over the pick-up organ of an ordinary baler. Standing hay, whole crop cereals and even straws may be treated by means of forage harvester (Wilkinson, 1984a). In this case, nozzles may be placed at the inner wall of the harvester for spraying the lye over the material as it is blown into the wagon.

Biological treatment

Enhancement in the digestibility of fermented straw has been observed with white rot fungi (Zadrazil, 1977, Levenon et al., 1983). Some fungi grow on the straw but reduce the digestibility of fermented product (Ulmer et al., 1981). *Agrocybe aegerita* is unable to break the lingo-cellulose complex thereby resulting in decreased substrate digestibility (Zedrazil, 1980). Such organisms, which are generally brown rot and sweet rot fungi, utilize cellulose, hemicelluloses and other soluble carbohydrate fractions for their growth leaving behind residues of low digestibility (Kundu and Mudgal, 1985, Kundu et al., 1988).

Eleven white rot fungi were inoculated on barley straw, pea straw, sugarcane bagasse and sunflower hulls. The results revealed that fungi need to be selected specifically for particular by-product and conditions of incubation may need to be manipulated in order to achieve substantial increase in digestibility (Ibrahim and Pearce, 1980). Wheat straw digestibility increased from 16 to 34% at the expense of the lose in lignin and cellulose by culturing *Sporotrichum pulverulentum* (Nikhat et al., 1983) and even a higher digestibility of 40-50% could be obtained (Zadrazil and Brunnert, 1982). With the removal of 52% lignin from Aspen wood through fermentation with Merulius tremellosus, cellulose digestibility increased from 18-53% (Reid, 1985). Wheat straw incubation with edible mushroom, *Pleurotus ostreatus* increased digestibility by 6 percentage units, through an incubation period of 28 days. However, at 50% DM with doubled incubation period with Erwinia carotovora, the increase was from 32.7 to 47.7% (Table 4). In a feeding trial digestibility of *Pleurotus* harvested wheat straw was not improved but buffaloes consumed it between 10 to 11 kg per day. Water washed spent straw left after Agralcus blsporus cultivation however, had

57.5 and 6.3% total digestible nutrients and digestible crude protein, respectively. In most of the studies, dry matter losses varied widely from 6 to 40% depending on the organism used, duration of fermentation, type of substrate and environmental conditions (Reade and McQueen, 1983, Agosin and Odier, 1985). The losses should be minimum during fermentation process, because removal of digestible fractions reduces the nutritional quality and quantity of residual substrate.

Since biodegraded straws are rich in cellulose, hemicellulose, and fungal mass (protein), which may be a suitable feed for ruminants because of their fermentative digestion in the rumen, whereas separated fungal mass would serve as protein feed for monogastric animals. Some of the fungal proteins are less digestible, while others are highly digestible. Well studied white rot fungi, P. chrysosprorium contained about 23% lysine rich protein which gave apparent digestibility of 64% and protein efficiency ratio as 1.94%, revealing high nutritive value for ruminants than non-ruminants. Direct studies on effect of biodegraded straw on rumen and intestinal microbes are lacking. However, indirect observations suggested that some of the liberated phenolic acids may hamper growth of cellulose utilizing bacteria. But the quantum of liberated phenolic acids may not be enough and rumen anaerobic environment may not allow the growth of inoculated fungi, still direct studies on this aspect may reveal true picture. Thus, it is evident that lignin degradation, dry matter yield and protein enrichment are the factors responsible for increasing nutritive value of lignocellulosic materials.

Growth Studies on Animals Fed Treated Straw

Growth in animals is governed by genetic and environmental factors. Among environmental factors, nutrition plays an important role. Straw treatment with caustic soda improves digestibility and intake by 24 and 40%, while straw treatment with urea (NH_3) enhances digestibility (Table 5 and 6) between 20 and 27%, respectively (Jackson, 1978, Mojumdar et al., 1997). Thus, urea treated straw would be able to support maintenance requirement and alkali treated wheat straw, supports an additional gain of about 100 – 200 g/ day. These figures indicate that to achieve more production in the form of live weight gain or milk production, the supplementation of nutrients is inevitable (Wanapat et al., 1985, 1986). Wheat straw and rice straw treated with 4% NaOH (Dry method) when compared with Lucerne ad libitum. DMI (g/kg $w^{0.75}$) was 111, 118 and 110 g/day and daily gain was 320, 320 and 410 g, respectively (Singh and Jackson, 1975). The diet in straw group had 15 % GN cake and 10% molasses.

Table 4 Digestibility of fungi colonized straws

Organism	Substrate	Digestibility control	Change in digestibility (% unit) due to fungal growth	Remark
Polyporus tulipiferal	Oat straw	38.9	+14.1	At 30^0 C
Polyporus spp.	Oat straw	33.8	+ 9.0	30/35^{0}C
Phanerochaete chryosporium	Oat straw	34.1	+ 9.3	35^{0}C
Agrocybe ageerita	Wheat straw	-	- 20.2	60 day*
Lentinus edodes	Wheat straw	-	+ 27.5	60 day*
Lentinus edodes	Wheat straw	-	+ 19.4	160 day*
Pleurotus eryngii	Wheat straw	-	+22.3	60 day*
Pleurotus eryngii	Wheat straw	-	+19.7	90 day*
P. salmonso florida	Wheat straw	-	+11.0	60 day*
P. salmonso altranineus	Wheat straw	-	+9.6	60 day*
Stropharia	Wheat straw	-	+13.6	60 day*
Rugosoannulata	Wheat straw	-	+12.4	60 day*
Agaricus bisporus	Wheat straw	46.2	-12.2	Calculated from animal trail data
Sporotrichum pulveruletum	Wheat straw	37.8	+16.3	
S. pulveruletum (CBS 841.73)	Wheat straw	45.0	+5.0	20 day*
S. pulveruletum (CBS 841.73)	Wheat straw	23.0	+22.0	10 day*
-do-	Wheat straw	35.0	-5.0	60 day*
Diciomitus squalens	Wheat straw	60.0	+20.0	30 day*
Diciomitus squalens	Wheat straw	28.0	-12.0	10 day*

* Time span after culturing and estimation of digestibility.

Performance of calves on untreated and treated (3.3 kg NaOH/100 kg) straw was compared in calves receiving 800 g and 1000 g GN cake in respective groups. DM intake in treated group increased by 32% and gain were 0.25 and 0.42 kg/day, respectively. In more difficult situations where little or no supplementary forage or concentrate were fed, live weight gain of 180 g in treated straw group against 100 g/day in untreated group was reported (Kehar, 1954). This experiment was conducted in Indian village environment and straw treatment was done by Backmann method. Buffalo heifer fed treated (5%NaOH) wheat straw prepared by dry method improved daily weight gain (70 g/day). Thus, in animals on treated straw supplemented with 2% urea, the weight gain of 150 g/ day against 130 g on wheat straw was observed (Khurana and Naik, 1978). Each animal in this experiment was also given a green maize or sorghum forage supplement containing 0.5 kg DM/100kg body weight/ day.

The report on Hereford cross steers (14 month age, 360 kg weight) fed straw from 4 barley varieties and one wheat variety untreated but 2% urea supplemented and all these five substrates treated with 30g anhydrous ammonia/ kg, indicated that gain in body weight were 110, 130, 400, 200, 270 g in untreated and 360, 330, 610, 600, 520 g/day in treated groups, respectively. In addition to straw diets, all the steers were given 1.5 kg concentrate consisting of 750 g sugar beet pulp, 200 g fishmeal and 50 g minerals (Reid et al., 1988). Twelve crossbred calves between 3 to 5 months of age were divided into two groups, where group I calves were offered wheat straw and one 1 kg concentrate whereas the calves in group II were offered treated (3.2 %urea) wheat straw along with similar quantity of concentrate. Average daily gain was 196 g/day in group I and 256 g/day in group II. There are reports suggesting higher weight gains on straw based diets, but in these cases quantity of concentrate was very high and probably straw utilization in such cases would be expected to be lower (Sundstol et al., 1979, Wilkinson, 1984a).

Table: 5 Digestibility, TDN/ME and DMI of untreated wheat straw and treated/ supplemented, as reported by various workers

Treatment/feed	Straw DMI/100 kg BW	Straw DMI g / $w^{0.75}$ kg BW	Straw DM Digestibility %	Straw TDN Kg
Wheat straw (6 Sahiwal x HF)	1.27	48.77	49.54	44.92
Wheat straw + UMMB *(0.445 kg/day) (6 Sahiwal x HF)	1.95	75.55	48.65	
Wheat straw (3 Karan Fries cattle)	1.12	44.0		40.12
Anhydrous ammonia (3%w/w) treated straw (3 Karan Fries cattle)	1.56	62.5		56.04
Wheat straw (4 Bos Indicus)	1.10	49.4		46.85
Urea treated wheat straw 1.60 (4% urea) (4 Bos Indicus)	72.20		54.77	
Wheat straw (4% urea treated) 1.8 (4 Haryana calves one year)	65.3	48.15	47.98	
Wheat straw (+ 0.4 kg DM/100 kg Body weight) (Buffalo heifers)		53		46
Wheat straw (2% urea)		65		48
Wheat straw (NaOH 5%)		51		55
Wheat straw (NaOH 5%+2%urea)		81		57

* 15% urea and 10% cotton seed cake resulting 9.41 % N (urea molasses mineral block)

Table: 6 DMI, digestibility and average daily gain (ADG) in animals fed urea treated straw rations

	Digestibility			**DMI ($g/kg\ W^{0.75}$)**			**ADG (g)**
	UT	**T**	**Increase %**	**UT**	**T**	**Increase %**	**Average daily gain(g)**
(1)	**(2)**	**(3)**	**(4)**	**(5)**	**(6)**	**(7)**	**(8)**
1) Rice straw Buffalo calves. 4 week treated, 4%urea (40% concentrate in ration)	31	62	100	69	95	38	
2) Rice straw 5% urea	41	52	27			42	
3) Rice straw (3%) urea treated	41	51	24			33	
4) Rice straw (4%) urea treated		52[2]			40		
5) Rice straw (5%) urea treated	45[2]	56	24	46	61	33	
6) Rice straw (3%) urea treated	45	54	20	46	52	13	
7) Rice straw (5%) urea treated	48[2]	55	15				
8) Rice straw	42[1]		26	65		46	
9) Rice straw	53[1]		11			34	
10) Rice straw 5%urea (Buffaloes)	50[1]	52	4	2.11[3]	2.34	11	182 - 79
11) Rice straw 5%urea (Cattle)	44[1]	52	18	2.68[3]	2.70		34 – 7
12) Rice straw 3% urea (Buffaloes)	43[1]	53	23	2.03[3]	2.17	7	130 – 50
13) Rice straw 6% urea treated (Buffaloes)	43[1]	55	28	2.03[3]	2.52	24	130 – 210
14) Rice straw 4% urea treatment (Cross-bred cattle)	41[1]	48	17			44	
15) Rice straw treated with 6% urea + 1.5 kg concentrate						21	79 - 431
16) Rice straw 5% urea treated (6 Buffalo bull)				79.6	86.7	9	90 - 97
17) Wheat straw (Treated ammonia in commercial oven)	45[2]	56	24				
18) Wheat straw treated 3.2% urea (Buffaloes)	41.2[1]	49.2	19.4	1.31	1.69	29	

1-DM digestibility, 2- OM digestibility

3- DMI/100 kg body weight

UT=Un-treated; T= Treated

Mineral required for increased fibre utilization

There are fifteen mineral elements which are essential in the animal system, however only sulphur and phosphorus have been shown to play an important role in the fibre utilization. Sulphur is an essential macro mineral required to ensure optimal growth of microbes. In fact microbial biomass contains as much as 8g sulphur/ kg DM (Orskov, 1982). Since sulphur containing amino acids are present in a constant proportion of microbial amino acids. The accepted ratio of N to S is 10 to 14: 1 (ARC, 1980), and the microbial requirement is generally met from the feeds and forages under natural feeding condition (Singh et al., 2002). There is little evidence to suggest that microbial growth is limitcd by thc supply of phosphorus under normal conditions. Phosphorus forms an essential part of RNA and DNA structure. In addition, it also forms a part of ATP and GTP, which are essential in all energy transaction within the cell. The phosphorus content of microbial matter is 2-6% on dry matter basis (Hungate, 1966) and the optimum N : P ratio is 8 : 1. This means that the requirement of phosphorus is even higher than the requirement of sulphur. It is estimated that the dietary requirement of phosphorus is about 4g/ kg digestible organic matter fermented in the rumen. Like all living organisms, the rumen microbes have a specific requirement for macro and trace mineral to allow for normal cell function and metabolism. Total ash content of rumen bacteria and protozoa has been reported 18 and 14 %, respectively, indicating the higher mineral needs for microorganisms.

Many studies have been carried out during last three decades on ways and means for upgrading low quality roughages as feed for farm animals. Physical treatment, such as chopping/ chaffing has been used in practice for a long time and the main effect is the increased feed intake. Steam treatment has shown the positive effect on the nutritive value of fibrous materials but the feasibility of this method is still questionable.

Soaking with water is common in many regions that lead to higher DM intake. Sodium hydroxide was the first chemical used in practical treatment of straw. Soaking of material in NaOH solution undoubtedly results in the improvement in the digestibility of straw, the corrosiveness of the chemical and the pollution due to final product are some of the disadvantages that make this method less attractive. Dry processes of NaOH treatment may be more convenient but the improvement is generally lower. The cost of NaOH is also prohibitive in India. Currently ammonia or chemicals, from whch ammonia can be liberated, seems to be of great interest. In temperate and industrialized countries the stack method for treatment of straw with anhydrous or aqueous ammonia is predominant, whereas ammonia added to straw in the form of urea seems to be more appropriate for tropical and developing countries. Even if ammonia treatment does not result in the improvement of the digestibility. It has a number of other advantages such as: (a) combined treatment and NPN supplementation, (b) combined storage and preservation (high moisture straw), (c) treatment is simple (d) final product is easy to handle (bales) (e) excess ammonia passed to urine, and may be used as manure. Besides these, other chemicals, such as $Ca(OH)_2$, KOH, SO_2, H_2O_2, H_2SO_4, Na_2CO_3, O_3, ClO_2, have been applied with variable sources. The cost of chemicals makes the processes less adoptable. Microbiological treatment has not been used at commercial scale so far and an appreciable amount of research remains to be done before a practical method is designed. No doubt, this approach has a potential, and high priority may therefore, be given to research on this aspect. The process is likely become industrially viable in future.

The catalytic supplementation by green forage as subabool leaf meals or green legume such as berseem remains of practical value to many of the resource poor farmers. The stack method has been standardized in the country and ammonia may be used in industrial process. The enriched straws thus prepared could be compressed in the form of blocks after using little molasses and concentrates. Market situation will ultimately lead to final application and selection of the technology.

Bibliography

Agosin, F. and Odier, E. 1985. Solid state fermentation, lignin degradation and resulting digestibility of wheat straw fermented by selected white rot fungi. Appl. Microbiol. Biotechnol., 21: 397-403.

ARC, 1980. Agricultural Research council, U.K.

Chaturvedi, M.L., Singh, U.B. and Ranjhan, S.K. 1973. Effect of feeding water soaked and dry wheat straw on feed intake, digestibility of nutrients and VFA production in growing zebu and buffalo calves. J. Agric. Sci. (Camb.) 80: 393-397.

Dolberg, F., Saadullah, M., Haque, M. and Ahmed, R, 1981. Storage of urea-treated straw using indigenous material. World Anim. Rev., 38: 37-41.

Greenhalgh, J.F.D., 1976. Improving the nutritive value of straw by alkali treatment. ARC Research Review, 2 (3): 67-70.

Hungate, R.E. 1966. Rumen and its Microbes, Academic Press, London, U.K.

Ibrahim, M.N.M. and Pearce, C.R. 1980. Effect of white rot fungi on the composition and in vitro digestibility of crop residues. Agric. Waste 2: 199-205.

Jackson, M.G., 1977. The alkali treatment of straws. Anim. Feed ScL Technol., 2: 105-130.

Jackson, M.G., 1978. Treating straw for animal feeding - an assessment of its technical and economic feasibility. World Anim. Rev. 28: 38-43.

Kay, M., 1982. Chemical treatment of straw for animal feed. 6th Straw Utilization Conference, Oxford 24-25 Nov. 1982, pp. 13-18.

Kehar, N.D. 1954. The effect of feeding alkali treated straws on the growth of young cattle. Indian J. Vet. Sci., 24: 189-216.

Kernan, J.A., Crowle, W.L., Spurr, D.T. and Coxworth, E.C., 1979. Straw quality of cereal cultivars before and after treatment with anhydrous ammonia. Can. J. Anim. Sci., 59: 511-517.

Khurana, R.P. and Naik, D.C. 1978. Edited by Jackson M.G. (1978) Treating straw for animal feeding FAO Anim. Prod. And Health 10: 51-52.

Kundu, S.S. 1989. Improving the nutritive value of wheat straw for buffaloes by urea treatment. Tropical Agriculture 66: 321-325.

Kundu, S.S. and Mudgal, V.D. 1985. Chemical changes and digestibility of chemically treated wheat straw. Indian j. Anim. Nutr. 2: 157-161.

Kundu, S.S., Kundu, S., Prasad, D.N. and Jakhmola, R.C. 1988. Lignin biodegradation to improve digestibility of straws. Indian J. Anim. Sci., 3: 1-16.

Kundu, S.S. and Sharma, D.D. 1995. Influence of bleaching powder and other chemical treatments of wheat straw on in sacco degradability and nutrient utilization in buffaloes. Buffalo J. 1: 1-13.

Kundu, S.S., Das, M.M. and Sharma, D.D. 1998. Densification for block making and evaluation of calcium hypochlorite and alkali treated wheat straw based feed blocks in comparison to urea treated straw feed. Buffalo J. 3: 291-302.

Kundu, S.S. and Sharma, D.D. 2003. Composition, in sacco degradability, feed intake and nutrient utilization in reverine buffaloes. Buffalo J. 19: 127-137.

Levenon, M.E., Bone, D.M. and Oeugulie, A.J. 1883. Solid state fermentation and fractionation of oat straw by Beaidiomycetes. Eur. J. Appl. Biotechnol. 18: 120-123.

Mojumdar, A.B., Das, M.M., Maity, S.b. and Anil Kumar. 2003. Nutrient utilization and performance of crossbred cows fed urea treated straw based diets. Amin. Nutr. Feed Technol. 3: 173-178.

Mojumdar, A.B., Singh, A.P., Mishra, A.K. and Radotra, S. 1997. Studies on utilization of ammoniated dry grass in crossbred heifers. Proc. National on feeding strategies for ecofriendly animal production. IVRI, Izatnagar pp. 41-42.

Nikhat, P., Kehlon, S.S., Sethi, R.P. and Chopra, A.K. 1983. Solid substrate fermentation of wheatstraw into animal feed. Indian J. Anim. Sci., 53: 1191-1194.

Orskov, E.R. 1982. Protein Nutrition in Ruminants. Academic Press, London.

Owen, E and Kategile, J.A. 1984. Straw etc. in the practical rations for sheep and goats. In: Sundstol, F and Owen, E. (editors). Straw and Other Fibrous By-products as Feed. Elsevier, Amsterdam, pp. 454-486.

Pal, V., Sachdeva, K.K. and Sengar, O.P.S. 1980. Effect of mineral supplementation on the feed digestion in sheep. Nutr. Abst. Rev. 52: 850.

Pirie, R. and Greenhalgh, J.F.D., 1978. Alkali treatment of straw for ruminants. 1. Utilization of complete diets containing straw by beef cattle. Anim. Feed Sci. Technol., 3: 148-154.

Reade, A.E. and Mc Queen, R.E. 1983. Investigation of white rot fungi for the conversion of popler into a potential feedstuffs for ruminants. Canadian J. Microbiol. 29: 457-463.

Reid, G.W., Orskov, E.R. and Kay, M. 1988. A note on the effect of variet, type of straw and ammonia treatment on digestibilityand on growth rate in steers. J. Anim. Prod. 47: 157-160.

Reid, I.D. 1985. Biological delignification of aspen wood by solid state fermentationwith the white rot fungi. Appl. Environ. Microbiol. 50: 133-139.

Rexen, F.P. and Bach-Knudsen, K.E., 1984. Industrial-scale dry treatment with sodium hydroxide. In: F. Sundstol and E. Owen (Editors), Straw and Other Fibrous By-Products as Feed. Elsevier, Amsterdam.

Saadullah, M., Haque, M. and Dolberg, F., 1981. Effectiveness of ammonification through urea in improving the feeding value of rice straw in ruminants. Trop. Anim. Prod., 6: 30-36.

Sihag, Z.S, Rathee, C.S. and Lohan, O.P. 1991. Factors affecting complete feed block formulation of straw based rations. Indian Journal of Animal Sciences 61: 1218-1221.

Singh, K.K. and Samanta, A.K. 1998. Effect of sources and levels of nitrogen supplementation on the utilization of C. ciliaris. Indian J. Anim. Nutr., 15: 69-71.

Singh, M and Jackson, M.G. 1975. Improved utilization of Agricultural waste material and industrial by productsas livestock feed. Research Progress Report, 1969-74, G.B. Pant Univ., Pantnagar.

Singh, K.K., Pailan, G.H., Nag, S.K., Das, M.M. and Samanta, A.K. 2002. Nitrogen, sulphur and their ratio in some commonly used feeds and forages. Range Mgmt. and Agroforestry 23: 164-165.

Solaiman, S.G., Horn, G.W. and Owens, F.N., 1979. Ammonium hydroxide treatment of wheat straw. J. Anim. Sci., 49: 802-8Q8.

Sullivan, J.T. and Hershberger, T.V., 1959. Effect of chlorine dioxide on lignin content and cellulose digestibility of forages. Science, 30: 1252.

Sundstol, F. 1988. Feedstuffs. 5. Straw and other fibrous by products. Livestock Prod. Sci., 19: 137-158.

Sundstol, F., Coxworth, E. and Mowat, D.N., 1978. Improving the nutritive value of straw and other low-quality roughages by treatment with ammonia. World Anim. Rev., 26: 13-21.

Sundstol, F., Said, A.N. and Arnason, J., 1979. Factors influencing the effect of chemical treatment on the nutritive value of straw. Acta Agric. Scand., 29: 179-190.

Ulmer, D.C., Tongerdy, R.P. and Murphy, V.C. 1981. Solid state fermentation of steam treated feed lot waste fibre with Cheetomium cellulolyticum. Biotechnol. Bioeng. Sym. 11: 349-451.

Venkanna, P., Reddy, M.R. and Reddy, G.V.N. 1997. Rumen fermentation pattern in complete based diets based on dry grass and concentrate in crossbred bulls. Indian J. Anim. Nutr., 14: 245-249.

Wanapat, M., Sundstol, F. and Garmo, T.H., 1985. A comparison of alkali treatment methods to improve the nutritive value of straw.1. Digestibility and metabolizability. Anim. Feed. Sci. Technol., 12: 295-309.

Wanapat, M., Sundstol, F. and Hall, J.M.R, 1986. A comparison of alkali treatment methods to improve the nutritive value of straw. II. In sacco and in vitro degradation relative to in vivo digestibility. Anim. Feed. Sci. Technol., 14: 215-220.

Wilkinson, J.M., 1984a. Farm-scale dry treatment with sodium hydroxide. In: F. Sundst0l and E. Owen (Editors), Straw and Other Fibrous By-Products as Feed. Elsevier, Amsterdam, pp. 162-180.

Wilkinson, J.M., 1984b. Ensiling with sodium hydroxide. In: F. Sundstol and E. Owen (Editors), Straw and Other Fibrous By-Products as Feed. Elsevier, Amsterdam, pp. 181-195.

Yadav, K. K., Rathee, C.S. and Lohan, O.P. 1990. Effect of compaction of roughage based complete feed block on digestibility and rumen parameters. Indian J. Anim. Sci., 7: 27-30.

Zadrazil, F. 1980. Influence of ammonium nitrate and organic supplement on the yield of P. aajorcaju. Eur. J. Appl. Biotechnol. 2: 31-35.

Zadrazil, F. and Brunnert, M. 1982. Solid state fermentation of lignocellulose containing plant residues with Sporotrichum pulverulentum and Merulius tremellosus. Eur. J. Appl. Biotechnol. 16: 45-51.

Zadrazil, F., 1977. The conversion of straw into feed by Basideomycetes. Eur. J. Appl. Biotechnol. 4: 291-294.

CHAPTER - 13

CNCPS : A NEW TECHNIQUE FOR FEED EVALUATION

Sultan Singh, Ram Singh, S.S. Kundu and Ranjan Kumar

Ruminant utilize protein fractions of feed like, NPN, soluble peptide and soluble true protein as well as carbohydrate fractions like sugar, starch, non-starch polysaccharides and other structural carbohydrates at different rates and manner. Therefore, their estimation in animal feeds assumes significance. The development of the Cornell Net Carbohydrate Protein System (CNCPS) arose from the need to account for variation in forages, including environment and previous plans of nutrition effective dietary fibre, ruminally degradable carbohydrate and proteins, rate of digestion and passage and variable efficiency of use of absorbed energy and protein.

This is the recent method from Cornell University and seems to be most accurate method of evaluation that characterizes the carbohydrates and proteins of feeds and forage into different fractions as per their availability to the animals. This system characterizes the feeds into different protein (A, B_1, B_2, B_3 and C) and carbohydrate (Sugar, fructant; Fast degradation – starch pectin, oligosaccharides; slow degradable starch; Fast degradation fiber; and lignin) fractions. This method helps to rank the feeds based on carbohydrates and proteins degradability and thus guides in dietary formulation accordingly. This will assist in optimization and in efficient utilization of feeds/forages in nutrient feeding.

Conventional proximate analysis system of feeds and cell wall determination partially accounts for factors that influence animal requirements and feed utilization. The CNCP system meets these requirements as it characterizes the feeds and forages in to different protein and carbohydrate fractions. The protein fractions of feeds such as NPN content, soluble peptide, soluble protein as well as carbohydrate fractions like sugar content, starch, non-starch polysaccharides and other carbohydrates are utilized by different rumen microbes and animals differently, their estimation on feedstuffs is therefore important. The degradation rate of these stuffs and their passage rates are also one of the determinants of net energy and protein available from feeds. The CNCP system is an application model that uses a combination of mechanistic and empirical approaches and account to the effect of variation in animal factor as well as feed chemical entities.

The CNCPS (Chalupa *et al.* 1991 Sniffen *et al,* 1992) has developed the need for uniform procedures to partition feed nitrogen into A, B, C fraction (Pichard and Van Soest, 1977) while carbohydrate fractions are relatively standardized (based on NDF, ADF with corrections for ash, protein and lignin).

Protein and carbohydrate fraction of feed stuff

Based on chemical constituent as well as potential degradation rate Pichard and Van Soest (1977) and Krishnamoorthy *et al,* (1982, 1983) fractionated feeds into various categories

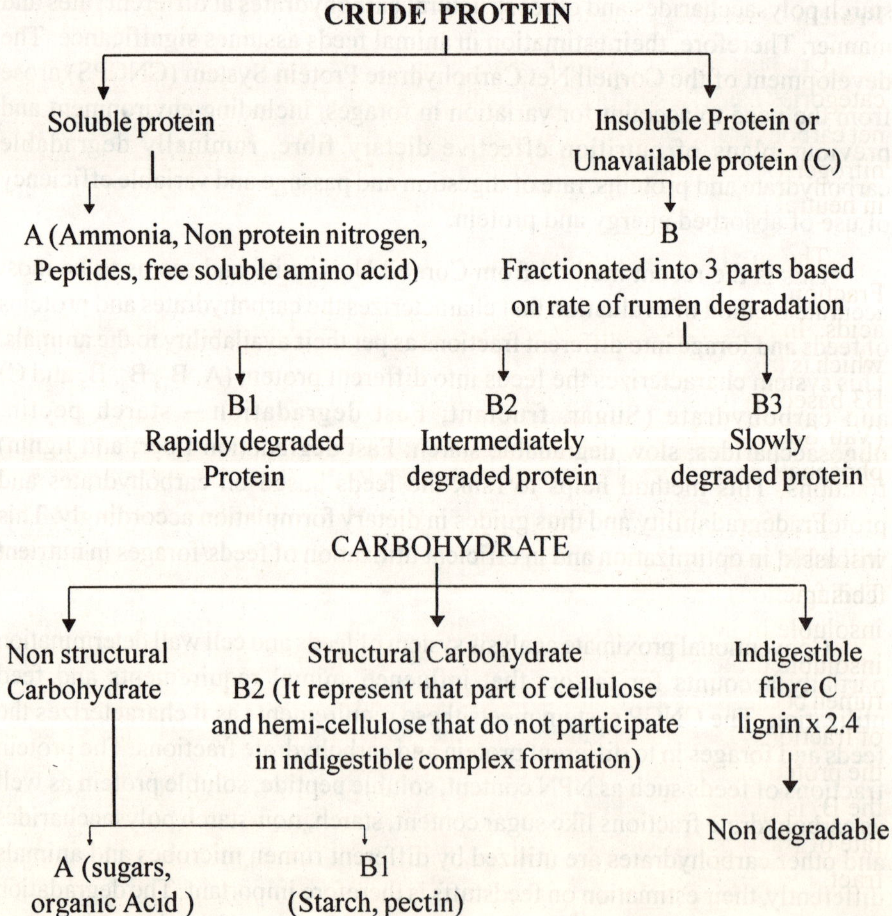

Fraction	Protein	Carbohydrate	Rumen Availability
A	Ammonia-N, Protein, soluble amino acids	Sugar, fructant, some starch	Soluble (generally highly available, 4% minimum to 2.0% maximum)
B1	Very rapid degrading protein, oligopeptides	Fast degrading starch pectin oligosaccharides	Insoluble, potential digestible 1-30 % hour
B2	Intermediate	Slow degrading starch	
B3	Slow degrading protein	Fast degrading fibre	
C	ADF bound protein	Lignin	Non digestible (0 % hour)

Although this gives some useful indication of degradation rate, the various B fractions are based on arbitrary criteria and vary considerably.

Protein fractions

Crude protein (CP) contents of feeds/fodders can be fractionated into two categories namely, soluble and insoluble based on their solubility. The Cornell net carbohydrate and protein system partition CP of the forages into non-protein nitrogen (A), soluble CP (B1), insoluble CP that was acid detergent but soluble in neutral detergent (B2) and CP insoluble in acid detergent (C).

The soluble protein is further fractionated into fraction A and fraction B. Fraction A of protein is mostly composed of NPN, peptides and free amino acids. In forage most soluble CP is actually NPN but contain a considerable N which is true protein. Fraction B of protein is further subdivided in to B1, B2 and B3 based on the degradation rate. Fraction B_1 is rapidly degraded in the rumen (Van Soest et al., 1981). Licetra et al. (1996) recommended use of borate-phosphate buffer for estimating soluble protein.

Fraction C of protein is the unavailable or bound protein fraction that is insoluble in the acid detergent (acid detergent insoluble protein, ADIP) fraction. This fraction (Fraction C) contains protein associated with lignin. Fraction B_3 is insoluble in neutral detergent but soluble in acid detergent [neutral detergent insoluble protein (NDIP) minus ADIP]. Fraction B_3 is slowly degraded in the rumen because it is associated with the cell wall. In the CNCPS, a high percentage of fractions B_3 escape degradation in the rumen. Buffer insoluble protein minus the protein insoluble in neutral detergent is used to estimate fraction B_2. Part of the B_2 is fermented in the rumen and remaining escapes to the lower gut. The fate of fraction B_2 depends on the relative rates of digestion and passage in GI tract.

Feed carbohydrate fractions

In CNCP system ccarbohydrates can be classified according to degradation rates, as with protein. Fraction A is fast and it is the sugar and organic acids, fraction B_1 is intermediate and is starch and soluble fibre, fraction B_2 is slow and is available cell wall (digestible fibre) and fraction C is unavailable cell wall (indigestible residue). These fractions are computed from feed content of non structural carbohydrate (NSC) and structural carbohydrate (SC) and indigestible fibre C. The fraction C is lignin x 2.4.

The NSC fraction represent the carbohydrates that are insoluble in neutral detergents and it can be estimated as 100 minus protein, NDF corrected for protein, lipids and ash in a feed stuff. The calculated NSC is usually in close agreement with direct measurements, but feed containing large amount of pectin will have a somewhat lower true NSC. Given proportion of starch and pectin in the NSC, the fraction A (Sugar and organic acids) can be determined by difference (Sniffen et al. 1992).

The CNCP system assumes that feedstuffs are composed of protein, carbohydrate, fat and ash. Protein and carbohydrate DM are further subdivided by chemical composition, physical characteristics, ruminal degradation and postiminal digestibility characteristics. NDF % of DM, lignin % of NSC, starch % of NSC are part of carbohydrate, while soluble % of CP, ADFIP% of CP, NPN% of SP, NDFIP% of CP are part of protein. Fractions of protein carbohydrate, ash fat in a feedstuff can be derived from the different method

1. NDF and lignin (Van Soest *et al.,* 1991)
2. Total nitrogen as assayed by macro and micro Kjeldahl (AOAC, 1980)
3. Soluble protein determined by the procedure of Krishnamoorthy *et al.,* 1983)
4. Nitrogen that is insoluble in neutral detergent and acid detergent (Van Soest *et al.,* 1991)
5. Ash (AOAC, 1980)
6. Solvent soluble fat (AOAC, 1980)
7. Determination of NSC from the determinations of NDF, protein, fat and ash (Van Soest *et al* (1991); 100 – [NDF- NDF protein + fat+ ash]
8. Starch by method of AOAC (1995)
9. NPN, soluble protein, NDFIN and ADFIN as per Licitra *et al.* (1996)

Estimation of Non Protein Nitrogen

Non protein nitrogen (NPN), soluble protein, Neutral detergent insoluble nitrogen (NDFIN) and acid detergent insoluble nitrogen (ADFIN) were estimated as per Licitra *et al.* (1996).

Apparatus

Erlenmeyer flask (125ml), What man filter paper 54 or 541, analytical balance, pH meter, filter funnels, Kjeldahl apparatus.

Reagents

1. Sodium tungastate ($Na_2WO_4 2H_2O$) (100 g/ L) solution in water, 0.30 M
2. 0.5 M sulphuric acid (H_2SO_4).

Procedure

Put about 0.5 g of dry ground sample into a 125 ml Erlenmeyer flask having 50 ml of cold distilled water. To it add 8 ml of 10 percent sodium tungstate solution. Keep flask at room temperature for 30 minutes. Adjust pH of the flask content to 2 with 0.5M sulphuric acid. Keep the flasks over night at room temperature. Filter the contents through Whatman filter paper 54 or 541. Wash the residue of filter paper twice with cold distilled water. Transfer this filter paper to kjeldahl flask and determine nitrogen Micro-kjeldahl procedure. Estimate NPN of the sample by subtracting residual nitrogen from total nitrogen.

SOLUBLE NITROGEN AND PROTEIN

Apparatus

One Erlenmeyer flask (125 ml) Whatman filter paper 54 or 541 analytical balance, water bath, vacuum source, filter manifold fitted with conical funnels (50 ml), Kjeldahl apparatus.

Reagent

1) Borate-phosphate buffer, pH 6.7-6.8 consisting mono sodium phosphate ($Na_2PO_4 . H_2O$) 12.20 g/l; sodium tetra borate ($Na_2B4O_7.10H_2O$) 8.91 g/l and tertiary butyl alcohol 100 ml/l
2) Sodium azide to 10 percent solution freshly prepared.

Procedure

Weigh approximately 0.5 gm ground sample and put into 125 ml Erlenmeyer flask. To the flask add 50ml borate-phosphate buffer. Then add one ml of sodium azide solution to the flask. Allow the flask to stand at room temperature for 3 hours.Filter the contents through Whatman filter paper 54 or 541 using mild vacuum. Wash the residue left after filtration with 250 ml cold distilled water. Estimate nitrogen in residue by Kjeldahl method. The N estimated in residue gives insoluble protein fraction. Calculate soluble protein by difference from total CP. The soluble true protein is obtained by subtracting the NPN estimated by tungstic acid procedure.

DETERMINATION OF ACID DETERGENT INSOLUBLE NITROGEN (ADIN)

Regents

1) Acid-detergent solution.
2) Acetone.

Procedure

For the estimation of ADIP, procedure of ADF estimation is used (Van Soest, 1973). For this 2 g of sample utilized for ADF refluxion is transferred to a Kjeldahl flask. Estimate the nitrogen in residue following the standard Kjeldahl procedure. Express ADIN of sample as percent of total nitrogen or as N x 6.25.

DETERMINATION OF NEUTRAL DETERGENT INSOLUBLE NITROGEN (NDIN)

The procedure is exactly same as that for acid detergent insoluble nitrogen except that neutral detergent solution was substituted for acid-detergent reagent (Van Soest *et al.*, 1991). Use of sodium sulphite is omitted. Pre-treatment with amylase is not done, although amylase may be used in the boiling step to aid filtration.

STARCH ESTIMATION

The starch is estimated as per the method of AOAC (1995)

Reagents

$CuSO_4$ Solution: Dissolve 34.639 g of 34.639g of $CuSO_4.5H_2O$ in water and make volume up to 500 ml and then filter through filter paper.

Alkaline Tartrate Solution

Dissolve 173 g of potassium sodium tartrate, $4H_2O$ (Rochelle salt) and 50 g NaOH in the H_2O and dilute to 500 ml. Allow solution to stand for 2 days and then filter.

Potassium Permagnate Standard Solution

Prepare approximately, 0.1573 N and containing 4.988 g/l and standardize with 0.35 g Na oxalate solution.

Ferric Sulphate Solution

Dissolve 135 g $FeNH_4 (SO_4)_2\ 12\ H_2O$ in H_2O and dilute to 1 liter. Acidify fifty ml $Fe\ (SO_4)_3$ solution with 20 ml 4 N H_2SO_4. Titrate with $KMnO_4$ solution and use this titre as '0' point correction.

Ferrous Phenanthrelene Indicator

Dissolve 0.7425 g O-phenanthrelene. H_2O in 25 ml, 0.025 M $FeSO_4$ solution (6.95 g $FeSO_4.7H_2O$)

Procedure

Stir about 2.5-3 g dry material for 1 hr in 250 ml beaker with 50 ml cold water. T Filter the contents and wash with 250 ml cold water. Heat the soluble residue for 2.5 hour with 200 ml water and 20 ml HCl (Specific gravity 1.13) in a flask provided with reflux condenser. Cool the flask contents and nearly neutralized with NaOH. Transfer to 250 ml volumetric flask, dilute to mark and filter. Weight of dextrose obtained x 0.925 = weight of starch. Add 25 ml each of $CuSO_4$ and alkaline tatrate solution to 400 ml beaker (alkaline resistant glass) and then add 50 ml reducing sugar solution. Heat the beaker on a pre-adjusted electric heater so that boiling begins in 4 minutes and continue boiling for two minute. Cover the beaker with watch glass during heating. Filter hot solution through sintered glass crucible using suction. Wash the precipitates of Cu_2O thoroughly with water and determine the amount of reduced copper by following method.

Transfer the crucible in the sample with glass rod and add 50 ml $Fe_2(SO_4)_3$ solution. Stir vigorously until Cu_2O is completely dissolved. Remove the crucible after making clear solution. Add 20ml, 4N H_2SO_4 solution and titrate with standard KMnO4. As end point approaches, add one drop of ferrous phenanthelene indicator. At the end point, brownish solution changes to green. The weight is reducing sugar equivalent to weight of copper obtained from the standard table (AOAC, 1995).

The equations that are used to estimate the different fractions of protein and carbohydrate of feeds are as:

CALCULATION

NON-STRUCTURAL CARBOHYDRATE (NSC)

NSC = 100-[(NDF-NDIP) + protein + fat +ash)]

Protein Fraction

PA (%CP)	=	NPN (%SOLP). 0.01. SOLP (%CP)
PB1 (%CP)	=	SOLP (%CP) – A (%CP)
PC (%)	=	ADIP (%CP)
PB3 (%CP)	=	NDIP (%CP) –ADIP (%CP)
PB2 (%CP)	=	100-A (%CP)-B1 (CP)-B3 (%CP) - C (%CP)
Where, CP (%DM)	=	Percentage of crude protein of the feedstuff;
NPN (%CP)	=	Percentage of crude protein of the feedstuff that is non–protein nitrogen x 6.25; SOLP (%CP) = Percentage of crude protein of the feedstuff that is soluble protein; NDIP (%DM) = Percentage of the feedstuff that is neutral detergent insoluble protein;
ADIP (%DM)	=	Percentage of the feedstuff that is acid detergent insoluble protein.
PA (%CP)	=	Percentage of crude protein in the feedstuff that is non-protein nitrogen;
PB1 (%CP)	=	Percentage of crude protein in the feedstuffs that is rapidly degraded protein.
PB2 (%CP)	=	Percentage of crude protein in the feedstuffs that is intermediately degraded protein.
PB3 (%CP)	=	Percentage of crude protein in the feedstuffs that is slowly degraded protein.
PC (%CP)	=	Percentage of crude protein in the feedstuffs that is bound protein.

Carbohydrate Fractions

CHO (%DM)	=	100-CP (%CP)-FAT (%DM)- ASH (%DM)
CC (%CHO)	=	100. (NDF (%DM). 0.01. LIGNIN (%NDF). 2.4)/ CHO (%DM)
CB2	=	100. {NDF (%DM) – NDIP (%CP). 0.01. CP (%DM) – NDF (%DM). 0.01. LIGNIN (%NDF). 2.4/ CHO (%DM)

CNSC (%CHO) = 100 – B2 (%CHO) – C (%CHO)

CB1 (%CHO) = STARCH (%NSC). {100 – B2 (%CHO) – C (%CHO)}/100

CA (%CHO) = {100-STARCH (%NSC)}. (100 – B2 (% CHO) – C (%CHO)}/ 100

Where, CP (%DM) = Percentage of crude protein of feedstuff; CHO (%DM) = Percentage of carbohydrate of the feedstuff; FAT (%DM) = Percentage of fat of the feedstuff; ASH (%DM) = Percentage of ash of the feedstuff; NDF (%DM) = Percentage of ash of the feedstuff; NDF (%DM) = Percentage of the feedstuff that is neutral detergent fiber; NDIP (%DM) = Percentage of neutral detergent insoluble protein of the feedstuff; feedstuff; LIGNIN (%NDF) = Percentage of lignin of the feedstuff NDF; STARCH (%NSC) = Percentage of starch in the non-structural carbohydrate of the feedstuff; SUGAR (%NSC) = Percentage of sugar in the non-structural carbohydrate of the feedstuff; CA (%CHO) = Percentage of carbohydrate of the feedstuff, that is sugar; CB1 (%CHO) = Percentage of carbohydrate of the feedstuff that is starch + NSP;CB2 (%CHO) = Percentage of carbohydrate of the feedstuff that is available fibre; and CC (%CHO) = Percentage of carbohydrate in the feedstuff that is unavailable fibre.

Bibliography

AOAC (1980). Official Methods of Analysis 13th Edition Association of Official Analytical Chemists Washington, DC: 1094

AOAC (1995). Official Methods of Analysis (16th Ed). Association of Official Analytical Chemists, Arlington VA

Chalupa, W. and Sniffen, C.J. (1996). Protein and amino acid nutrition of lactating dairy cattle today and tomorrow. Animal Feed Science and Technology, 58: 65

D' Mello, J. P. F. (Ed) 2000. Farm animal Nutrition and Metabolism CABI Publishing.

Fahey jr., G. C; Collin, M; Mertens, D. R and Moser, L. E. (Eds.) 1994. .Forage quality evaluation and utilization. CABI Publishing.

Fox, D.G., Barry, M.C., Pitt, R.E., Roseler, D.K. and Stone, W.C. (1995). Application of the Cornell net carbohydrate and protein model for cattle consuming forages. Journal of Animal Science, 73: 267-277

Fox, D.G., Sniffen, C.J., O'Connor, J.D., Russel, J.B. and Van Soest, P.J. (1992). A net carbohydrate and protein system for evaluating cattle diets. III. Cattle requirements and diet adequacy. Journal of Animal Science, 70: 3578-3596

Givens, D.I. Owen, E; Axford, R.F.E. and Omed, H.M. (eds) (2002). Forage Evaluation in Ruminant Nutrition. CABI Publishing.

Krishnamoorthy, U., Sniffen, C.J. and Van Soest, P.J. (1982). Nitrogen fractions in selected feedstuffs. Journal of Dairy Science, 65: 217

Krishnamoorthy, U., Sniffen, C.J., Stern, M.D. and Van Soest, P.J. (1983). Evaluation of a mathematical model of rumen digestion and in vitro simulation of rumen proteolysis to estimate rumen undegraded nitrogen content of feedstuffs. British Journal of Nutrition, 50: 555-568

Licitra. G., Harnandez, T.M. and Van Soest, P.J. (1996). Standardizations of procedures for nitrogen fractionation of ruminant feeds. Animal Feed Science and Technology, 57: 347-358

Pichard, D.G.R. and Van Soest, P.J. (1977). Protein solubility of ruminal feeds. In Proc, Cornell Nutrition Conference: 91-98

Sniffen, C.J., O'Connor, J.D., Van Soest, P.J., Fox, D.G and Russel, J.B. (1992). A net carbohydrate and protein system for evaluating cattle diets II carbohydrate and protein availability. Journal of Animal Science, 70: 3562-3577

Van Soest, P.J., Robertson, J.B. and Lewis, B.A. (1991). Symposium: Carbohydrate methodology, metabolism and nutritional implications in dairy cattle. Methods for dietary fiber, neutral detergent fiber and non starch polysaccharides in relation to Animal Nutrition. Journal of Dairy Science, 74: 3583-3597

Van Soest, P.J., Robertson, J.B. and Lewis, B.A. (1991). Symposium: Carbohydrate methodology, metabolism and nutritional implications in dairy cattle. Methods for dietary fiber, neutral detergent fiber and non starch polysaccharides in relation to Animal Nutrition. Journal of Dairy Science, 74: 3583-3597

Van Soest, P.J., Sniffen, C.J., Mertens, D.R., Fox, D.G., Robinson, P.H. and Krishnamoorthy, U. (1981). A net protein system for cattle: The rumen sub model for nitrogen. (Eds Owens F.N.). Protein Requirements for Cattle. Proceedings of an International Symposium, Mp-109: 265. Division of Agriculture, Oklahoma State University, Stillwater

Van Soest, P.J., Sniffen, C.J., Mertens, D.R., Fox, D.G., Robinson, P.H. and Krishnamoorthy, U. (1981). A net protein system for cattle: The rumen sub model for nitrogen. (Eds Owens F.N.). Protein Requirements for Cattle. Proceedings of an International Symposium, Mp-109: 265. Division of Agriculture, Oklahoma State University, Stillwater

CHAPTER - 14
CHEMICAL PRESERVATION OF FORAGES

S.S.Thakur

One of the main reasons for low productivity of Indian dairy animals is the lack of optimum feeding conditions during certain periods every year. These conditions can be overcome by developing fodder varieties that can be grown during scarcity period or by conserving surplus green fodder available during flush seasons so as to use it during scarcity periods. The importance of conservation of green fodder is felt in even those parts of country where assured irrigation facilities are available round the year as on one hand surplus green fodder is available during flush seasons whereas fodder scarcity exists for a considerable period of time.

For obtaining good quality conserved fodder, it is important to harvest green fodder at appropriate stage of maturity and to store it with minimum losses of nutrients. Undoubtedly, the stage of growth or maturity is a very important factor affecting the chemical composition and nutritive value of forages. At early stage of growth, plants have high moisture, protein and mineral contents with low fiber and lignin contents. With maturity, leaf growth slows down, stems elongate, floral parts appear and the products of photosynthesis accumulate. The cytoplasmic portion of each cell decreases and contents of proteins, lipids, soluble carbohydrates and soluble minerals decrease, whereas, the cell wall fibrous constituents increase and become more lignified. Total dry matter increases according to a sigmoid shaped curve but there occurs a progressive decline in nutritive quality particularly of digestibility of nutrients and digestible energy (Sullivan, 1973). Hay making and ensiling described as below are the two common methods of fodder preservation.

Hay making

The aim of hay making is to reduce the moisture content of the harvested green crop to around 15 to 20 per cent, which is low enough to inhibit the action of plant and microbial enzymes. A variety of losses occur due to chemical changes, action of plant enzymes, leaching, mechanical damage etc. Quality of hay is further deteriorated due to crop harvesting at a late stage of maturity, sun drying at high temperature and unwanted rains. Total dry matter (DM) losses have been estimated to be around 4-15 per cent due to plant respiration, 2-5 per cent from leaf shattering of grass hays, 3-35 per cent in legume hays and 5 to 14 per cent due to leaching of nutrients by rains. Thus heavy losses of nutrients occur during field preparation of hay whereas its preparation under controlled conditions may add to the cost in Indian context.

Silage making

Silage making is a good method of conserving green fodder under anaerobic conditions. For this the green fodder has to be harvested at around 30 percent DM and anaerobic conditions are achieved by compaction and proper packing. The main chemical changes occurring in the green fodder during ensiling involve soluble carbohydrates and nitrogenous compounds. Soluble carbohydrates such as glucose, fructose, sucrose and fructans are fermented to lactic acid and other products such as acetate, ethanol and CO_2. The main losses in ensiling occur during field wilting, respiration, fermentation and as effluents. The losses during field wilting take place due to loss of carbohydrates and DM. Losses in respiration occur due to plant enzyme activity on carbohydrates. The fermentation losses take place due to anaerobic microbial fermentation of carbohydrates and proteins. Dry matter losses ranged from 4.6 to 8.7 per cent for fresh and wilted herbage (Anderson and Jackson, 1970), fermentation and initial oxidation losses ranged from 3.6 to 7.4 per cent in grass herbage (McDonald and Whittenbury, 1973). On the basis of regression equations, it was observed that crops ensiled at 30 per cent DM produced little effluent. The characteristics of good quality silage are: pH value of < 4.2, butyric acid content < 0.1 percent and NH_3-N content < 11.0 percent of total N content (Carpintero et al., 1969). In order to control the microbial and biochemical changes leading to the nutrient losses and deterioration of quality of preserved green fodder, certain substances are added which can be classified into two categories viz. fermentation stimulants and fermentation inhibitors. These have been reviewed by Thakur and Sharma (1990 a & b).

Fermentation Stimulants

Additives rich is soluble carbohydrates such as molasses, cereal grains and sugar beet pulp have been used to encourage rapid lactic acid production which checks the growth of detrimental microbial activities. Another approach has been to add Lactobacillus inoculum along with molasses to stimulate growth of desirable microorganisms and produce good quality silage (Singh and Neelakanthan, 1982). Commercial microbial inoculants along with molasses decreased silage pH, NH_3-N and butyric acid contents of grass silage (Haigh et al., 1996) leading to production of good quality silage.

Fermentation Inhibitors

A large number of chemical compounds have been tested as potential fermentation inhibitors. One of the earliest was mixture of mineral acids proposed by A.I. Virtanan. Hydrochloric and Sulphuric acid were added during ensiling to lower the pH of the stored fodder below 4.0 leading to cessation of undesireable microbial activity, but due to corrosiveness of these acids, their use was discontinued. Sodium chloride applied @ 1 to 3 percent on fresh basis of fodder had inhibitory effect on undesirable butyric acid producing bacteria (Nash, 1978). Rape and barley silage preserved with sodium chloride @ 12 kg/ton produced good quality silage with improved digestibility and rumen fermentation characteristics in cattle(Fedulina et al., 1988).

To overcome the corrosive nature of mineral acids, organic acids such as propionic and formic acids have been extensively tested as silage additives. Propionic acid applied at 0.3 to 0.4 percent of cereal fodders improved quality, palatability and silage digestibility (Bolsen et al., 1996). Formic acid is probably the most widely studied chemical as fodder preservative. Formic acid applied at 0.3 to 0.6 percent level had given consistent results in terms of silage quality, palatability, aerobic stability, digestibility and utilization of nutrients (Riviere, 1986 & Haigh, 1996). Formaldehyde applied to silage made from cereal fodders @ 14 to 100 g/kg CP of fodder reduced carbohydrate and protein breakdown, thereby reducing ammonia-N and total organic acid contents (Kaiser et al., 1983), but feeding of such silage resulted in depressed DM intake and digestibility of all nutrients thereby reducing live weight gain (Kaiser et al., 1981). Formic acid and formaldehyde applied together in various combinations restricted silage fermentation, reduced NH_3-N, total and individual organic acids, increased water soluble carbohydrates but reduced its palatability, digestibility and nutrient intake.

Ammonia and urea have been extensively tried as silage additives as a source of adding non-protein nitrogen in addition to their antimicrobial and fungistatic effects (Thakur et al., 1997b). The literature establishes that preservation of high moisture green fodders after treating with urea at 0.5 to 1.0 per cent on fresh basis of cereal fodders holds potential as the ammonia liberated from its hydrolysis prevents the growth of undesirable bacteria, molds and yeasts. Urea treatment also increases the aerobic stability of silage (Adogla-Bessa et al., 1999). Animal performance in terms of growth and milk production has also been found to be better due to increased palatability, digestibility, and nitrogen availability. Certain antibiotics such as streptomycin, bacitracin, neomycin, penicillin, aureomycin, terramycin, nisin etc. have been tried as fermentation inhibitors but these appeared to be generally ineffective and no better than other conventional additives. Food preservatives such as sodium metabisulphite, potassium sorbate, sorbric acid etc. have been tried but the cost of these chemicals may rule out their commercial use as fodder preservatives. Certain other chemicals such as ammonium tetraformate, ammonium bicarbonate, ammonium carbonate, sodium formate, sodium diacetate, sodium acrylate, hexanoic acid, benzoic acids etc. have been tested for their efficacy in restricting the growth of undesirable microorganism but the results and the cost benefit ratio did not suggest their application on a wider scale. Based on the above information available in literature, the techniques for chemical preservation of a few cereal fodders was standardized and the preserved fodders were evaluated for their nutritional quality..

a. Preservation of maize fodder:

At National Dairy Research Institute, Karnal looking into the cost and effectiveness of various green fodder preservatives on the basis of available literature, sodium chloride at 0, 0.5 and 1.0 percent; formic acid at 0, 0.3 and 0.4 percent and urea at 0, 0.25 and 0.5 percent on w/w basis of fresh fodder, in a factorial experimental design, were used to preserve chaffed maize fodder harvested at soft dough stage (24% DM) for two months in double layered polythene bags. The DM loss was least in urea + common salt treated samples, 3.91 (urea 0.5+common salt 1%), 4.78 (urea 0.5 +common salt 0.5%) and 6.25 percent (urea 0.25 + 1% salt) as compared to 14 percent observed in control. pH was highest (4.36 to 4.57) at higher urea application levels and lowest (3.71 to 3.93) in 0.5 percent common salt alone and formic acid + common salt treated samples. Lactic acid concentration varied between 6.87 to 9.61 per cent. Lowest lactic acid concentration, 6.87 and 7.39 per cent were observed in urea plus common salt treatment combinations. Butyric acid content was less than one percent in all samples. Effective DM degradability was highest 55.55 to 60.55 per cent in urea + common salt treatment combinations as compared to 46.81 in control (Thakur and Sharma, 1995).

Table 1 Chemical composition and nutritive value of different maize silages

Parameter	Control silage	Urea 0.5% +Salt 1% treated silage	Urea 0.5%+ Salt 0.5% treated silage
pH	3.46	3.82	3.79
Total N (g/100g DM)	1.27	2.04	2.23
NH_3-N (g/100g DM)	0.03	0.40	0.43
Lactic acid (g/100g DM)	10.19	7.23	8.15
TVFA (mM/100g DM)	32.26	20.83	22.63
Butyrate (mM/100g DM)	1.29	0.83	0.68
CP content (%DM basis)	7.91	12.74	13.98
TDN (%DM basis)	54.81	58.27	57.96
DCP (% DM basis)	3.15	8.76	10.28

Salt : Common Salt

b. Rumen fermentation study

On the basis of results of laboratory evaluation of 27 treatment combinations, the best two treatments were selected on the basis of DM loss, lactic acid, total VFA, butyrate content and DM degradability. Nine crossbred rumen fistulated male calves of 1 to 2 years age were divided into three groups of three animals each. Animals of control group were fed untreated silage prepared from maize fodder harvested at soft dough stage (23% DM), animals of group I were fed maize silage preserved after treating with urea 0.5 + common salt 1%, those of group II were fed maize silage preserved after treating with urea 0.5% + common salt 0.5%. The concentration of ruminal total N (mg/litre SRL) was increased from 542.13 in control group to 645.33 in group II calves, whereas that of TCA precipitable-N varied from 299.47 in control to 348.13mg/litre SRL in group I calves. Concentration of ammonia-N varied between 87.27 in control to 182.60 mg/litre SRL in group I calves. Total volatile fatty acid concentrations (mg/litre SRL) were 68.67 in control and 94.77 in group I calves (Thakur and Sharma, 1993).

Table 2 Intake and digestibility coefficient of nutrients of maize silage in calves

Parameter	Control silage	Urea 0.5% +Salt 1% treated silage	Urea 0.5% +Salt 0.5% treated silage
DM intake (kg/d)	2.93[a]	4.44[b]	3.61[b]
DM intake (kg/d/100 kg BW)	1.63[a]	2.01[b]	1.82[a]
Digestibility coefficient (%)			
DM	59.39[a]	62.78[b]	61.24[a]
OM	65.33	66.82	65.87
CP	37.79[a]	68.74[b]	73.83[b]
CF	67.20[a]	74.95[b]	75.67[b]
NFE	79.99[a]	87.98[b]	87.76[b]
N balance g/d	1.40[a]	16.88[b]	10.65[b]

The production rate of VFA's were the highest in group II (10.64) and lowest in control (6.92 moles/day). Bacterial production rate (g/day) were 154.06, 256.80 and 254.67 in calves of control, I, and II groups, respectively (Thakur and Sharma, 1998).

Table 3 Concentration of nitrogen fraction (mg/litre) and total volatile fatty acids in strained rumen liquor (mM / litre) of crossbred calves

Parameter	Control silage	Urea 0.5% +Salt 1% treated silage	Urea 0.5% +Salt 0.5% treated silage
Total N	542.13	639.33	645.33
TCA precipitable N	299.47	348.13	313.07
NH_3-N	87.27	182.60	139.00
TVFA	68.87[a]	94.77[c]	75.67[b]
Acetate	44.30[a]	59.18[b]	49.88[a]
Propionate	11.78	17.62	16.54
Butyrate	12.79	15.71	9.00

b. Growth study on crossbred calves

Eighteen male crossbred calves of 5-8 month of age were randomly divided into three similar groups on the basis of body weights. All the calves were fed concentrate mixture @ 1.5 kg/calf /day and silage ad lib. The control group calves were fed control maize silage, group I calves fed maize silage preserved after treating with 0.5% urea +1% common salt and those of group II were fed maize silage preserved with 0.25% urea + 0.5%, common salt. After a growth period of 120 days, total increase in live weight was 42.50, 67.50 and 50.83 kg with a corresponding average daily gain of 354.2, 541.7 and 423.6 g in control, group I and II calves, respectively. Calves of group II exhibited higher growth and I by 46.0 and 17.4 percent that of control (Thakur and Sharma, 1993).

Table 4 Body weight gain, feed conversion efficiency and feed cost per unit gain in crossbred calves.

Parameter	Control silage group	Urea 0.5% +Salt 1% treated silage fed group	Urea .25% +Salt 0.5% treated silage fed group
Growth performance			
Initial body weight, kg	111.00	111.67	111.00
Total gain in BW, kg in 120 days	42.50[a]	67.50[b]	50.83[b]
Average daily gain, g/d	354.[a]20	541.70[b]	423.60[a]
Feed consumed/ kg gain in BW, kg	10.55[c]	7.49[a]	8.97[b]
Feed cost (Rs)/ kg gain in BW	16.88[b]	11.58[a]	13.94[a]

DM consumption (kg) per kg body weight gain was minimum in group I (7.49) and the highest in control group (10.55). The feed cost (Rs) per kg gain in liveweight was 16.88, 11.58 and 13.94 in calves of control, group I and II calves. Digestibility co-efficients of DM, OM, NFE and cellulose were lower in calves of control group than those of group I and II. Intake of CP (g/d) was higher by 35.6 and 26.0 percent, that of DCP (g/d) by 53.5 and 29.6 per cent and TDN (kg/d) by 17.2 and 11.6 percent in group I calves than that of control and group II calves. Similarly, Nitrogen balance (g/d) was 15.57 in control group and 81.2 and 58.8 percent higher in calves of group I and II over that of control. N retained as percent of N-intake was 17.46, 23.93, 25.61 and as per cent of N absorbed was 26.51, 32.42 and 35.31 percent in control, group I and II calves, respectively. Better growth performance in calves fed urea + common salt treated silage was mainly due to higher CP, DCP, TDN intakes and better nutrient utilization.

c. Evaluation of sorghum fodder preserved after treating with urea plus common salt

Green sorghum fodder was harvested at inflorescence stage (21.8% DM) and after chaffing was preserved in underground uncemented pits after treating with a mixture of 0.5% urea + 1% common salt as fresh basis of fodder. After 2 months, this silage was fed ad lib to ten buffalo calves (2-2.5 years of age), divided into two similar groups of five in each. Group I calves were fed untreated silage ad lib and supplemented with urea and salt to make common salt and nitrogen content similar to that of group II whereas group II animals were fed sorghum silage prepared after treating with urea + commons salt. Total DMI, DMI per 100 kg body weight and digestibilities of DM, OM, CF and NFE were similar in two groups whereas that of CP and EE was improved by 13.5 and 15.0 units in group II calves than that of group I. Similarly, digestibility of NDF, ADF, hemicellulose, cellulose and cell contents were enhanced by 12.76, 11.62, 12.47, 11.24 and 10.19 units in group II calves as compared to that of group I. Nitrogen balance (g/d) and N retained as percent of N intake were improved by 22.88 and 16.93per cent in group II calves over that of group I calves (Thakur et al., 1997).

Table 5 Nutrient intake digestibility and N-balance in buffalo calves fed sorghum silage

Parameter	Untreated silage fed group	Urea 0.5% +Salt 1% treated silage
Nutrient intake		
DM (kg/d)	6.43	6.24
DM (kg/100kg BW)	2.33	2.29
CP (g/d)	720.94	777.65
DCP (g/d)	372.75[a]	506.92[b]
TDN (kg/d)	3.26	3.55
Digestibility coefficients (%)		
DM	53.34	61.49
CP	51.68[a]	65.14
NDF	46.18[a]	58.94[b]
ADF	56.74[a]	66.93[b]
Hemicellulose	46.96[a]	59.43[b]
Cellulose	60.64[a]	71.88[b]
N Balance (g/d)	19.04[a]	41.92[b]

A lactation trial on ten crossbred cows yielding around 10-14 litres milk per day was conducted to evaluate sorghum fodder(harvested at 22% DM) after treating with 0.5% urea + 1% common salt. Cows in first quarter of lactation were divided into two groups yielding 12.50 ± 0.96 kg and 12.35 ± 1.15 kg milk per day with 4.1 ± 0.29 and 3.98 ± 0.38 percent fat. Cows of control group were fed concentrate, wheat straw and control sorghum silage, whereas those of group I were fed concentrate mixture, wheat straw and sorghum fodder preserved after treating with 0.5% urea + 1% common salt. The experiment was conducted for a period of 90 days. The overall average milk yield was 12.15±1.05 kg and 13.95±1.75 kg/day in control and group I, respectively exhibiting an increase of 13.08 percent in group I over that of control. Average fat percent was 4.28 ± 0.39 and 4.13 ± 0.28 which was statistically similar. Dry matter intake (kg/d) was similar in the two groups i.e. 11.95 ± 0.85 and 12.19 ± 1.05.

Based on the literature and work conducted on preservation of green cereal fodders at National Dairy Research Institute, Karnal and other institutions in India, the following recommendations are made:

1. Maize, sorghum and oats fodder harvested at 22-24% DM levels, can be preserved by treating with 0.5% urea and 1% common salt on fresh basis of fodder at organised dairy farms. The two chemicals need to be just broadcasted/dusted on the chaffed fodder layer by layer. The improved silage thus obtained, will improve the productivity of dairy animals.
2. The progressive farmers of areas where surplus green fodder is available, should be encouraged to adopt the technique of chemical preservation of cereal fodders. This work can be taken by national and state level extension agencies, which will definitely sustain milk production during periods of lean fodder availability.

Bibliography

Adogla Bessa,T.; Owen, E. and A Desogan, A.T. 1999. Ensiling of whole crop wheat with cellulose and hemicellulose based enzymes comparing effects of urea or enzyme treatment on forage composition and quality. Animal Feed Sci. and Techol. 82:51-61.

Anderson, B.K. and Jackson, N.1970. Conservation of herbage of varying dry matter content in air tight metal containers with reference to the carbohydrate fraction. J. Sci. Food Agriculture 21:228-241.

Bolsen, K.K.; Ashbell,G. and Weinberg, Z.G.1996. Silage fermentation and silage additives - Review. Asian Australasian Journal of Animal Sciences. 9:483-493.

Carpintero, M.C.; Holding, A.J. and McDonald, P.1969. Fermentation studies on silages. J. Sci. Food Agric. 20:677-684.

Fedulina, N.N.; Mukhamedyanov, V.M. and Coldatova, V.V.1988. Rumen microflora, feed digestibility and productivity of heifers fed on silage preserved with sodium chloride. Nutr. Abst. & Rev. (Series B) 58:2967A

Haigh, P.M.; Chapple, D.G. and Powell, T.L.1996. Effect of silage additive on big bale grass silage. Grass and Forage Sci. 51: 318-322.

Kaiser, A.G. ; Osbourn D.F. and England,P. 1983. Intake, digestion and nitrogen retention by calves given ryegrass silages: influence of formaldehyde treatment and supplementation with maize starch or maize starch plus urea. J.Agric.Sci. Camb. 100:63-74.

Kaiser, A.G.; Tayler, J.C.; Gibbs, B.G. and England, P. 1981. The effect of the quantity of formladehyde applied at ensiling and of urea supplementation at feeding on utilization of red clover silage by young growing Cattle J.Agric. Sci. Camb, 97:1-11.

McDonald, P. and Whittenbury, R. 1973. The Ensilage Process. In: Chemistry and Biochemistry of Herbage vol.3 edited by Butler, G.W. and Bailey, R.W., Academic Press, London pp 33-58.

Nash, M.J. 1978. Pretreatment for Conservation and Storage. In: Crop Conservation and Storage, Pergamon Press, Oxford, U.K. pp 274-314.

Rieviere,F. 1986. Perennial rye grass silage: as good as maize. Dairy Science Abstract 48:4900A.

Singh, K. and Neelakantan, S. 1982. Note on the effect of inoculum and additive on hybrid Napier grass silage. Indian J. Anim. Sci. 52:685-687.

Sullivan, J.T.1973. Drying and storing Herbage as hay. In: Chemistry and Biochemistry of Herbage vol 3 edited by Butler, G.W and Bailey R.W., Academic Press, London pp 1-31.

Thakur, S.S. and Sharma, D.D. 1998. Bacterial and volatile fatty acid production rates in crossbred calves fed chemically preserved maize fodder Indian J.Anim. Nutr. 16(1): 24-29.

Thakur, S.S. and Sharma, D.D. 1993. Nutrient utilization and rumen fermentation in crossbred calves fed maize fodder preserved with urea plus common salt treatment . Indian J.Anim. Sci., 63:456-460.

Thakur, S.S.; Kundu, S.S.; Oberoi, P.S and Sharma, D.D. 1997a. Nutritional evaluation of Sorghum fodder preserved after treating with urea and common salt. Indian J.Dairy Sci. 50(2): 88-93.

Thakur, S.S.; Kundu, S.S.; Oberoi, P.S and Sharma, D.D. 1997b. Chemical characteristics and degradability of green Jowar fodder preserved by urea plus common salt treatment Indian Anim. Nutr. 14:65-69.

Thakur, S.S. and Sharma D.D. 1990 a. Chemical preservation of high moisture fodders-A Review. Agric. Rev 11(1):44-60.

Thakur, S.S. and Sharma, D.D. 1990 b. Chemical preservation of green fodder-A new approach. Indian Dairyman 42:270-272.

Thakur, S.S.and Sharma, D.D. 1995. Influence of urea, formic acid and common salt on preservation and in situ rumen degradability of maize fodder at dough stage. Indian J.Anim,Sci., 65:1018-1025.

CHAPTER - 15
FEED BLOCK TECHNOLOGY FOR ARID REGION

R.C. Jakhmola and M.K. Tripathi

The most common feed resources of arid regions are often low in energy and digestible proteins (Table 1). By supplementing and balancing deficient nutrients like energy and nitrogen, a better utilization can be possible. The nutritive value of the crop residues and agro-industrial by-products is taken into consideration, while using these feed sources for livestock feeding. The supply of additives may balance these feed for protein, minerals and vitamins. Therefore, appropriate mixture of crop residues and by-products with deficient nutrients would provide cost-effective feed option, which is balanced for optimized digestion and thus, low quality feedstuff may be utilized by animals for improved production of meat and milk.

A major part of ruminants diets in arid areas are also derived from shrubby vegetation. Numerous tree and shrub species contain high level of protein. Supplementation of tree leaves/ leaves of leguminous shrubs to livestock diet is expected to promote fermentation of roughage in the rumen for improved animal performance. However, tannins are the most widely occurring anti-nutritional factors found in these resources. Tannins form soluble and insoluble complexes mainly with proteins and also have affinity towards carbohydrates, amino acids and minerals, rendering them unavailable for rumen microflora and host animal (Makkar, 2003). These phenolic compounds also reduce the nutritive value of numerous feedstuffs thereby decrease livestock production and reproductive performance (Waghorn et al., 1999). Recently, it has been shown that tannins can be used in ruminant's diet to decrease the ruminal degradation of protein thus providing more amino acids for intestinal absorption. Another potential for complete feed block supplementation that can be exploited through the use of medicated blocks, for example to control parasites. The supplementation strategy can improve to be highly advantageous especially in the extensive system of animal production where there is little control over the animals. Several authors (Mc Beath et al., 1979; Bogan and Marriner, 1983) proved the success of feed blocks in controlling nematode parasites of sheep. Anindo et al., (1998) confirmed this success when medicated molasses–urea blocks have been distributed to grazing ram for strategic prophylactic treatment against nematodes.

Benefits of complete feed block technology

- Simple and efficient technique for long term conservation of crop residues and agricultural by-product
- Ease in transport and feeding stall-fed and/ or free grazing animals
- Reduced use of conventional concentrate feeds, thereby feeding cost would be alleviated.
- Allows a synchronous and fractionated supply of essential nutrients for ruminants fed on low quality roughages.
- Used as carrier of several chemicals and anthelmatic medicines.
- Environment pollution may be reduced.

Impact of feed block technology on the utilisation of agrowastes available in arid region

The use of mustard straw (MS) as crop residue for animal feeding is fairly a recent development. It is woody in nature thus unpalatable to ruminants with lower voluntary feed intake (Mishra et al., 1995) and poor digestibility. Based on straw and seed ratio of 3:1, it is estimated that 20.16 MT of MS is produced annually in India. An advantage of MS is that the large amount of unutilized straw can be used in animal feeding with the methodologies that are developed to improve its intake as well as nutritive value. Optimum inclusion level of mustard straw in sheep diets in the form of complete feed block was reported to be 60 % (Dhuria et al., 2004). Using complete feed block technology mustard straw can be used in animal feeding during scarcity. Rumen fermentation of poor quality roughages could be improved by their incorporation in complete diets (Reddy et al., 1993). This improves nutrient availability to animals when fibrous crop residues fed to ruminants in the form of complete feed block. Groundnut straw and mesquite straw 20 and 30 parts can effectively we used in 70 % roughage based feeding in sheep if fed in block form (Arya et al., 2004). Sheep had good health and positive nitrogen balance on the block feeding which contained groundnut straw 20, mesquite straw 30, DORB 20 and groundnut cake 13 parts with mineral and salt supplementation. A restricted quantity of unconventional and unpalatable feeds (Table 2) can be used in animal feeding using feed block technology. The animal responses on block feeding were variable with animal species (Table 3) but favorable demonstrating the feasibility of the technology.

Table 1: Potential availability of straws and stovers (million tones) 2003.

Name	Straw grain ratio	Present availability 2001-2002	Tenth Five Year Plan 2002-2003	2003-2004	2004-2005	2005-2006	2006-2007
Jowar kadbi	1:4	41.60	41.60	42.00	42.24	42.44	42.64
Bajra kadbi	1:4	26.40	26.52	26.68	26.80	26.92	27.08
Maize stover	1:3	34.20	35.22	36.27	37.38	38.48	39.66
Ragi straw	1:4	8.80	8.84	8.88	8.92	8.96	9.00
Millet straw	1:4	2.56	2.56	2.60	2.60	2.64	2.64
Pulses straw	1:4	72.80	74.80	77.20	79.60	82.00	84.00
Oilseed stalk	1:2	56.00	58.20	60.20	62.60	65.00	67.60

Table 2: Maximum incorporation level of unconventional feed in feed blocks

Feed Resource	Maximum leve in block %	Reference
Prosopis juiliflora pods	30	Ravikala et al., 1993
Arachis hypoegis hulls	40	Prasad et al.., 1986
Arachis hypoegis straw	20	Arya et al., 2004
Citrullus lanatus seed cake	20	Sharma et al., 2001
Ailanthus excelsa leaves	30	Raghuvansi, 2004
Azadirachta indica leaves	30	Raghuvansi, 2004
Albezia lebbek leaves and pods	30	Raghuvansi, 2004
Brassica campestris straw	70	Chaturvedi et al., 2004Mishra et al., 2004
Cicer arietinum straw	70	Nagpal et al., 2004
Fallen tree leaves	70	CSWRI, 2004

Supplementary feeding VS Block feeding

Supplementation of deficient nutrients to animals is an often-used feeding system to improve animal productivity. Grazing animals are generally supplemented with a small amount of concentrate before or after grazing. Supplementary feeding even of dry fodders is required during monsoon season, when vegetation contains very high moisture. Although green fodder availability and intake is quite adequate during lush green seasons but grazing animals are not even able to meet their dry matter requirement, high moisture of the pastures

Table 3: Feed resources used in block formation and its impact on utilization by animals

Feed	Animal	Physiological stage	% of feed in CFB	Impact of CFB feeding	Reference
Guar phalgati (GP), Tree leaves (TL)	Camel calves	Growing	GP 70-75; TL-8-15	Higher DMI, better nutrient utilization and moregain	Nagpal and Arora, 2002
Wheat straw	Buffalo calves	Adult	70	DMI, Digestible DMI, Digestible OMI and TDN values were improved between 17-20%. Nitrogen, Ca and P retention were higher.	Verma et al., 1996
Natural grasses	Goat kids	Growing	60	Improved anaerobic fungi population	Samanta et al., 2003
Mustard straw	Cows	Lactating	70	Improved milk yield	Chaturvedi et al., 2002
Mustard straw	Sheep	Ewes, Pregnant, growing	60-70	Improved reproduction in ewes, lambing in pregnant and growth in kids	Chaturvedi et al., 2003
Mustard straw	Sheep	Adult	70	Improved DMI, nutrient digestibility . and N balance.	NATP, 2003
Bajra Kadbi	Sheep	Adult	70		Chaturvedi et al., 2004
Mustard straw	Sheep	Adult	40-60	Mustard straw can be included in the diet of sheep upto 60%, this will have optimum intake nutrient utilization and nutrient balance	Dhuria et al., 2004
Wheat straw, Ground nut fodder	Calves	Growing	30 parts each	Compaction of mash diet in block form improved DMI and nutrient digestibility, which resulted a 55 g higher ADG	Sharma et al., 2004
Mesquite pods, Ground nut straw, Sewan straw	Sheep	Adult	GN Straw, 20, Mesquite pods, 30, Sewan straw, 20	Sheep can be effectively maintained with high nitrogen balance and good health if fed in block form.	Arya et al., 2004

Feed	Animal	Physiological stage	% of feed in CFB	Impact of CFB feeding	Reference
Natural grasses	Cow	Lactating	70	Improved DMI, DM and OM digestibilities, Nitrogen balance, milk production, milk fat and protein contents	Sharma, 2004
Natural grasses	Heifers	Growing	60	Block feeding to heifers improved growth performance by 25% over mash feeding due to 30% higher dry matter intake.	Singh et al., 2004
Natural grasses	Sheep	Adult	70	Improved DMI & digestibilities of DM, CP and fiber fractions. N balance also improved due to compaction in block.	Sharma, 2004
Gram Straw	Camel Calves	Growing	70	Growth improved by 11 % due to block feeding	Nagpal et al., 2004
Guar palgati & Moth chara	Camel Draft	Adult	70,30	Block feeding improved dry matter and water intake	Nagpal et al., 2004a

also causes diarrhea and weight loss. Complete feed block technology may play an important role in maintaining an optimum nutritional status of animals by providing proportionate concentrate and roughage dry matter intake, which promote rumen function thereby better nutrient utilization and absorption. Block supplementation will reduce the concentrate wastage and will increase concentrate consumption time, which will be beneficial for rumen microbial ecosystem. High quality feed block was also recommended by Wanapat et al. (1999) as a strategic supplement in lactating dairy cows especially when fed on low-quality roughages or crop residues. Poor quality roughage based complete feed block feeding improved milk production in native animals over concentrate supplementation based feeding system (Chaturvedi et al., 2004). Raghuvansi (2004) shown that the feeding of complete feed block with comparable nutrient density to grazing plus concentrate supplementation in hoggets improved intake and nutrient digestibility, increased fibrolytic enzymes activity in rumen resulting a better energy availability and growth (Table 4, 5 & 6). The effect of complete feed blocks on feed intake in ruminants is inclusive on the basis of available data. Houmani and Tisserand (1999) and Bensalem et al. (2001) did not observe any improvement of straw intake when concentrate was partially or totally replaced by feed block, however Nyarko-Badohu et al., (1994) reported an increase of straw intake by sheep supplemented with molasses containing feed blocks, similarly Raghuvansi (2004) also observed higher dry matter intake in sheep fed roughage based complete feed blocks. Overall, block supply did not reduce feed intake suggesting an absence of substitution effect of feed block supplement.

Table 4: Nutrient intake in hoggets fed complete feed block (CFB)and grazing plus supplementation (G+CON).

Attributes	**Feeding system**	
	CFB	**G+CON**
$W^{0.75}$ (kg)	12.19	11.99
Dry matter intake (day^{-1})		
g/day	1547.79	1079.49
g/kg $W^{0.75}$	126.97	89.95
Digestible organic matter intake (day^{-1})		
g/kg $W^{0.75}$	69.63	28.08
Crude Protein intake (day^{-1})		
g/kg $W^{0.75}$	16.23	7.38
DCP Intake (day^{-1})		
g/kg $W^{0.75}$	10.58	3.06
DE intake (day^{-1})		
Mcal/100 kg BW	16.34	9.89
Mcal/ kg $W^{0.75}$	0.38	0.071
Nutritive value of diets		
DCP (%)	8.31	3.43
DE (kcal/g diet)	2.98	2.51

Table 5: Nutrient Digestibility and Rumen microbial enzymes (units/100ml RL) activities in hoggets fed CFBF and GCON.

Attributes	Feeding system	
	CFB	**G+CON**
Dry matter	55.06	50.00
Organic matter	64.11	44.87
Gross energy	66.32	54.77
Rumen microbial enzymes Activity		
Cmcase	91.49	89.26
CMCase (EC)	42.79	59.88
CMCase (C)	48.69	29.38
Mccase	24.54	7.68
MCCase (EC)	9.83	2.49
MCCase (C)	14.72	5.18
Esterase	62.28	35.85
Acetyl Esterase (EC)	21.79	6.07
Acetyl Esterase (C)	40.49	29.78
Xylanase	484.09	527.88
Xylanase (EC)	210.57	325.18
Xylanase (C)	273.52	202.70
Protease	309.02	525.40
Protease (EC)	67.03	189.23
Protease (C)	241.99	336.17

NS= non-significant, *= P<0.05, **= P<0.01, c= Cellular, ec= extracellular

Table 6: wool yield, growth performance and feed efficiency of hoggets.

Attributes	Feeding system	
	CFB	G+CON
Total wool yield (g)	855.63	782.13
Growth performance		
Initial BW (kg)	22.13	22.09
Final BW (kg)	31.02	28.98
Total gain (kg)	8.89	6.89
Average daily gain (g)	101.0	78.3
Feed intake		
Total feed intake (kg)	117.76	94.84
Average daily feed intake (g)	1338.12	1077.70
Feed efficiency	13.40	14.16

Complete feed mash Vs block feeding

Complete feed mashes are comprised of roughage, concentrate and other supplementary nutrients and non-nutrients compounds in a proportionate balanced form that are desired for optimum animal productivity. When these mashes are offered to animals a significant amount of micronutrient get settled at the bottom in the feeding trays. The animal consumes the feed ingredient rapidly that they like more while others they consume later and slowly. A significant part of the feed mash remained in the feeding trays because of selectivity habits of the animals. These makes this feeding method uneconomic due to wastage of feed and unbalanced intake resulting in poor animal performance. Complete feed block technology seems promising as this technology checks all above disadvantages of complete feed mashe feeding. Feeding of complete fed block did not reduce forage dry matter intake (Nyarko-Badohu et al., 1994; Raghuvansi, 2004) but an increased organic matter and crude protein digestibility of feed block containing diets has been reported (Houmani and Tisserand, 1999; Nyarko-Badohu et al., 1994; Verma et al., 1996; Raghuvansi, 2004) and the improvement of microbial protein synthesis was also reported (Ben Salem et al., 2001). Higher concentration of rumen anaerobic fungi was noticed (Samanta et al., 2003) in goat kids fed complete feed block over mash diet. Therefore, block feeding promotes rumen microbial growth and activity thereby increase in digestion of poor quality roughage. Block feeding synchronises the supply of nutrients to rumen microorganism which enhance rumen fermentation. Consumption of hard feed block over the day ensures a synergy between nutrient demand of rumen microorganism to degrade low quality feedstuffs and the provision of adequate level of these nutrient without exceeding the transformation capacity of microbial population. When mash diets are fed, concentrate portion of the diet is generally consumed rapidly by animals and degrade in the rumen, therefore nutrient flux in the rumen peaks in the first few hours of post feeding, great part of which is lost in urine and feces. Consequently, availability of nutrients for rumen microorganism over the rest of the day will be limited and degrading capacity of fibrous feeds will be compromised. Verma et al. (1996) showed that feeding straw based complete feed compressed diets improved dry matter and digestible dry matter intake by 20 and 17 %, respectively over the mixed mash diet. The improvement in digestible organic matter intake and TDN were 20 % due to block feeding. Nitrogen, Ca and P balances were also higher in block fed animals. A 55 g higher daily gain in growing calves was reported by Sharma et al. (2004) due to block technology, in their study 30 parts of each wheat straw and groundnut fodder were used as roughage source with 40 parts concentrate, these were fed either in mixed mash or block form. The improvement in gain was explained to be the response of higher dry matter intake and nutrient utilization as noticed by higher nutrient digestibility in block fed animals. Improved dry matter intake and nutrient digestibilitis, and higher milk production was reported in lactating cows fed block diets as compared to mash (Sharma, 2004). Block feeding resulted 30 % higher dry matter intake and 25 % higher daily gain over mash feeding in crossbred growing heifers fed a diet with 60:40, roughage to concentrate ratio (Singh et al., 2004). Improved intake, nutrient utilization, nitrogen retention and better animal performance are the bright features of block feeding over mash (Table 7). Higher protein digestibility on block feeding might be due to the better protease activity (Vaithianathan et al., 2003) in block fed animals.

Table 7: Improvement noticed on feed utilization and performance of animals

Variable	Experimental animal	Improvement		Reference
		Mash	Block	
Dry matter intake (%BW)				
	Cow	2.97	3.90	Sharma, 2004
	Calves	2.60	209	Sharma, 2004
	Sheep	4.12	4.38	NATP, 2003
	Sheep	3.5	3.95	Chaturvedi et al.., 2004
	Sheep	3.15	3.63	Sharma, 2004
	Camel draft	2.15	2.48	Nagpal et al., 2004a
	Cattle calves	3.83	4.04	Sharma et al., 2004
Digestibility %				
DM	Cow	56.31	58.63	Sharma, 2004
	Sheep	54.93	59.48	Sharma, 2004
	Sheep	50.1	51.0	NATP, 2003
	Sheep	54.6	57.6	Chaturvedi et al., 2004
	Calves	62.04	64.14	Sharma et al., 2004
CP	Cow	60.34	64.50	Sharma, 2004
	Calves	60.68	61.94	Sharma, 2004
	Sheep	58.2	61.1	NATP, 2003
	Sheep	65.10	71.50	Chaturvedi et al.., 2004
	Sheep	62.16	67.13	Sharma, 2004
	Calves	58.69	61.24	Sharma et al., 2004
Daily gain (g/d)				
	Camel calves	377.55	420.92	Nagpal et al., 2004
	Cattle calves	233.33	288.88	Sharma et al., 2004
	Heifers	332.89	448.89	Singh et al., 2004
Milk production				
(l/d)	Cows	3.96	6.93	Sharma, 2004
Milk protein %	Cow	3.11	3.82	Sharma, 2004
Milk Fat %	Cow	4.50	5.50	Sharma, 2004

Complete feed block technology during feed scarcity in arid region

Arid ecology has the harshest conditions for animal production due to lack of feed and fodders during all seasons of the year except monsoon. Dry matter availability to animals is also not adequate even during monsoon. Grazing, top feed and a little fibrous feed are the major components of ruminant feeding systems. Availability and nutritional values of grazing pasture and top feeds varies due to there harvesting time etc. These feed resources can be harvested at an optimum nutritional state; various combinations can be made to prepare a balance or required diet types. After compaction in the block form these can be stored and/or can be fed as and when required. The block making process will facilitate to reduce wastage of feed resources and nutrient, because a few top feeds and grasses possesses very higher nutritive value thus using block technology nutrient looses due to imbalance of nutrient in diet can be minimized. Blocks can be prepared when availability of feed is adequate and can also be stored for scarcity feeding. Scarcity of feed and fodders will continue because of shrinkage in grazing lands, over stocking density of animals on grazing lands will further degrade the quality and quantity available for grazing. Whereas top feed resources are available only in limited quantity and a short period of time.

Role of complete feed bock technology for Drought management

Various unconventional feed resources that have reasonable nutritional value are either under utilized or goes waste. Improper feeding practices also results in under utilization on available feed resources. In hot arid reasons of India particularly in Rajasthan dry fodders such as Jawar, Bajra Kadbi etc are being fed to animals with out chaffing, this practice looses about 20% feed which is mainly the stem portion. A huge quantity of mustard straw is also available in arid regions of India, is not being used in animal feeding. Whereas, mustard straw has nutrient density similar to rice and wheat straw (Mishra et al., 2000). Mustard straw together with other crop residues either being burned or left in the field for natural decomposition, this practice either adds to environmental pollution and or loosing the biomass. Using complete feed block technology, experimental results suggested that mustard straw (Chaturvedi et al., 2004) and fallen tree leaves (CSWRI, 2004) can be used in ruminants feeding after blending with other feeds. These can we used as high as 70 % of the total diet to sustain the adult stock. Using the block technology fallen tree leaves and forest grasses, which are available approximately 300-350 million MT annually (Jakhmola and Misra, 2000) can be added to the feed resource of the country. A 22 million MT mustard straw can also be used in feed block technology (Mishra et al., 2004). A large number of unconventional feeds have been identified to be used in animal feed formulations, however poor palatability and imbalance nutritional state remained a problem in utilization of these feed. To mitigate the feed scarcity these unconventional feeds can be added in the diets, palatability can be enhanced by addition of some palatability enhancers and nutrients balancing compounds. Complete feed block technology could be an effective way of feeding livestock to improve and sustain animal productivity. It helps in developing low cost feed formulations, better nutrient utilization due to proportionate intake of roughage and concentrate, avoiding refusal of unpalatable portion of plant residues, improves utilization of NPN compounds that enables use of locally available ingredients (Jakhmola, 2003; Reddy and Reddy 1998; Reddy 1987) in an efficient manner.

To combat the feed scarcity situation during famine such as drought, which is regular feature in arid regions of India. The surplus feed and fodders from different parts of country can be used following block technology and be transported to required places. For example-plenty of rice straw is being produced in northern and central part of India, states likes UP, Haryana and Punjab. Rice straw is not being used for animal feeding in these states because good quality fodder availability in states are adequate. After grain harvesting rice straw is being burned in the field to grow rabi crops. This rice straw can be collected free of charge and blending with some concentrate ingredients complete feed blocks can be prepared, easily transported and stored for future use. Since keeping quality of these blocks is quite good hence fodder banks can be developed.

Complete feed block technology seems promising for intensive animal production systems through the improvement of rumen function by optimal rumen fermentation via the adequate and continuous supply of desired nutrients thereby minimizing fermentation and absorptive losses. Block based feeding strategies are pertinent under urban animal production systems due to less space requirement for storage and ease in transportation and feeding. Low transportation cost, high keeping quality and less residual losses will make the technology economically viable for animal producers. Block technology will helps in developing low cost feed formulations, avoiding refusal of unpalatable portion of plant residues, improves utilization of NPN compounds that enables use of locally available ingredients in an efficient manner.

Bibliography

Anindo, D., Toé, F., Tembely, S., Mukasa-Mugerwa, E., Lahlou-Kassi, A. and Sovani, S. 1998. Effect of molasses– urea-block (MUB) on dry matter intake, growth, reproductive performance and control of gastrointestinal nematode infection of grazing Menz ram lambs. Small Rum. Res. 27: 63–71.

Arya, R. S., Sharma, T., Dhuria, R. K., Harish, M. and Purohit, G. R., 2004. Nutritional evaluation of non conventional mesquite (Prosopis juliflora) and groundnut (Arachis hypogea) straw based complete feed block in sheep. Proc: XI Animal Nutrition Conference, Jabalpur, Jan 5-7, 2004. p80.

Bagon, J.A. and Marriner, S.E. 1983. Uptake of fenbendazole by grazing sheep with access to feed blocks containing fenbendazole. Br. Vet. J. 139, 223–227.

Ben Salem, H., Nefzaoui, A. and Messaoudi, L. 2001. Les blocs alimentaires à base de grignons d'olive. Une aleternative economique pour la complémentation des fourrages pauvres. Annales de I'INRAT 74, 187-203.

Bhat, P.N. 2000. Livestock revolution as a sustainable alternative for poverty alleviation: A strategic approach. In: 2nd International Conference on Sustainable Agriculture, Water Resources Development and Earth Care Policies, New Delhi, 18-20 Dec., 2002, pp 1-13.

Chaturvedi, O H., Prasad, R., Tripathi, M K., Mishra, A. S., Vaithianathan, S., Raghuvansi, S K S., and Jakhmola, R. C. 2004. Nutrient utilization in sheep fed sodium hydroxide or alkaline hydrogen peroxide treated millet (Pennisetum typhoides) kadbi based complete feed diet in mash or block. Proc: XI Animal Nutrition Conference, Jabalpur, Jan 5-7, 2004. pp81.

Chaturvedi, O.H.; Santra, A.; Mishra, A.S.; Prasad, R.; Misra, A.K.; Vaithiyanathan, S. and Jakhmola, R.C. 2002. Titrabale groups and soluble phenolic compounds in untreated and treated mustard (Brassica campestris) straw and their relation with cell wall constituents and dry matter disappearance in situ. Indian J. Anim. Sci. 72:75-79.

CSWRI. 2004 Studies on utilization of dead litter based complete feed block as scarcity feed. Annual report 2003-2004. Central Sheep and Wool Research, Institute, Avikanagar.p 34-35.

Dhuria, R. K., Purohit, G. R., Sharma, T., Arya, A., and Garg, D. D., 2004. Nutiritional evaluation of mustard straw based feed in sheep. Proc: XI Animal Nutrition Conference, Jabalpur, Jan 5-7, 2004. p49.

Houmani, M. and Tisserand, J.L. 1999. Complémentation d'une paille de blé avec des blocs multinutritionnels: effects sur la digestibilité de la paille et intérét pour des brebis et des agneaux en croissance. Ann. Zootech. 48, 199-209.

Jakhmola, R C. and Misra A K., 2000. Feeding strategy using unconventional feeds. Proc. III Biennial Conference of animal Nutrition association. Nov 7-9, Hisar pp83-98.

Jakhmola, R. C., 2003. Principle and preparation of complete feed block. Short course on recent advances in Animal Nutrition Research., March 5-25, CAS, IVRI, Izatnagar. Pp139-143.

Makkar, H.P.S. 2003. Effects and fate of tannins in ruminant animals, adaptation to tannins, and strategies to overcome detrimental effects of feeding tannin-rich feeds. Small Rum. Res. 49: 241–256.

Mc Beath, D.G.; Preston, N.K. and Thompson, F. 1979. Studies on the efficacy of fenbendazole administered via feed-block carrier. Br. Vet. J. 135, 271–278.

Mishra, A S., Chaturvedi, O H., Khali Ananta, Prasad, R., Santra, A., Misra, A K., Parthasathy, S. and Jakhmola, R. C., 2000. effect of sodium hydroxide and alkaline hydrogen peroxide treatment on physical and chemical characteristics and IVOMD of mustard straw. Anim. Feed Sci Technol., 84, 257-264.

Mishra, A S., Misra, A K., Tripathi, M K., Santra, A., Prasad, R. and Jakhmola, R. C. 2004. effect of sodium hydroxide plus hydrogen peroxide treated mustard (Brassica campestris) straw based diets on rumen degradation kinetics, fermentation pattern and nutrient utilization in sheep. Asian-Aust. J. Anim. Sci., 17, 355-365.

Misra, A.K.; Karim, S.A.; Patnayak, B.C. and Verma, D.L. 1995. Growth performance of lambs on mustard (Brassica compestris) straw based complete feed. Indian J. Small Rumin., 1: 31-34.

Nagpal, A K., Arora, M. and Roy, A K., 2004a. Utilisation of guar palgati (Cyamposis tetragonoloba) and Moth chara based complete feed in camel diets. NATP progress report Oct 2002 to Jan 2004, Evaluation of locally available feed and fodder to improve quality and formulate complete economic rations with high roughage diets. CSWRI, Avikanagar. Pp 47.

Nagpal, A K., Arora, M. and Singh, G P., 2004. Nutrient utilization of gram straw (Cicer arietinum) based complete feed blocks in camel calves. Indian J. . Anim. Sci. (In Press)

Nagpal, A.K. and Arora, Manju. 2002. Utilization of guar phalgati and groundnut haulms based complete feeds in camel calves, Indian J. Anim. Nutr., 19: 69-72.

NATP. 2003.Comparative utilization of mash and block diet in sheep feeding. Annual Report 2002-03, National Agricultural Technology Project. CSWRI, Avikanagar. P3-6.

Nyarko-Badohu, D.K., Kayouli, C., Bâ, A A. and Gasmi, A. 1994. Valorisation des pailles de céréals en alimentation des ovins dans le nord de la tunisie: trairement à l'urée et à l'ammoniac et complementation par des blocks mélasse-urée. Options Méditerranéennes. Série B: Etudes et Recherches 6, 129-141.

Prasad, J. Rama; Krishna, N.; Parthasarathy, M. and Prasad, D. Anjaneya. 1986. Daily gain and nutrient utilization in nellore rams on feeding complete diets at three protein levels, Indian J. Anim. Nutr., 5: 212-217.

Raghuvansi, SKS., 2004. Preparation and testing of high roughage based complete feed block in sheep feeding. Ph.D. Thesis, Udai Pratap Autonomus College, Varanasi, UP, India.pp 115.

Ravikala, K.; Patel, A.M.; Murthy, K.S. and Desai, M.C. 1993. Rumen metabolites in growing lambs on feeding Prosopis juliflora based complete feeds. Indian J. Anim. Nutr., 10: 177-180.

Reddy, G.V.N. and Reddy, M. R., 1998. Utilisation of expander-extruder processed complete diet containing sunflower heads on Ongole bull calves. Indian J. Anim. Nutr. 15, 272-275

Reddy, M R.. 1987. Processing of crop residues for formulation of complete feeds. Proc: V Animal Nutrition Research Workers Conference, Feb 14-16, Udaipur.

Reddy, N.M.; Reddy, G.V.N. and Reddy, M.R.(1993. Effect of fodder based complete diets on the rumen fermentation pattern in crossbred bulls. Indian J. Anim. Nutr., 10: 7-12.

Samanta, A.K.; Singh, K.K.; Das, M.M.; Maity, S.B. and Kundu, S.S. 2003. Effect of complete feed block on nutrient utilization and rumen fermentation in barbari goats, Small Rumin. Res. 48: 95-102.

Sharma V K., 2004 Preparation and evaluation of CFBs for milk production in university dairy farm. NATP progress report Oct 2002 to Jan 2004, Evaluation of locally available feed and fodder to improve quality and formulate complete economic rations with high roughage diets. CSWRI, Avikanagar. Pp 35-40.

Sharma, S. 2001. Efficiency of utilization of non-conventional protein sources in complete feed of Marwari goat for growth and lactating performance. Ph.D. thesis RAU, Bikaner.

Sharma, T., Arya, R.S., Dhuria, R.K., Garg, D.D., Harish, M. andPurohit, G.R. 2004.Comparative utilization of wheat straw and groundnut fodder based complete feed in loose and block form in calves. Proc: XI Animal Nutrition Conference, Jabalpur, Jan 5-7, 2004. p79.

Singh, K. K., Das, M. M., Samanta, A. K., Kundu, S. S. and Sharma, S.D., 2004. Effect of complete feed block on nutrient utilization and growth in crossbred heifers. Proc: XI Animal Nutrition Conference, Jabalpur, Jan 5-7, 2004. p79-80.

Vaithiyanathan, S.; Raghuvansi, S.K.S.; Mishra, A.S.; Tripathi, M.K.; Misra, A.K.; Prasad, R. and Jakhmola, R.C. 2003. Effect of feeding chemically treated mustard (Brassica campestris) straw on rumen fiber degrading enzymes in sheep. Asian-Aust. J. Anim. Sci., 16 : 1610-1613.

Verma, A.K.; Mehra, U.R.; Dass, R.S. and Singh, A. 1996. Nutrient utilization by murrah buffaloes (Bubalus bubalis) from compressed complete feed blocks. Animal Feed Science Technology, 59: 255-263.

Waghorn, G.C.; Reed, J.D. and Ndlovu, L.R. 1999. Condensed tannins and herbivore nutrition. In: Buchanan-Smith, J.G., Bailey, L.D., McCaughy, P. (Eds.), Proceedings of the 18th International Grassland Congress, vol. 3. Association Management Centre, Calgary, pp. 153–166.

Wanapat, M.; Petlum, A. and Pimpa, O. 1999. Strategic supplementation with a high quality feed block on roughage intake, milk yield and composition, and economic return in lactating dairy cows. Asian-Aus. J. Anim. Sci., 12: 901-903.

CHAPTER - 16
ROUGHAGE PARTICLE SIZE

K.K Singhal And Raman Malik

Roughages, the integral part of ruminants ration, are required to fulfill their physiological and nutritional requirements. Numerous processing techniques are generally used to enhance the fibre utilization by ruminants. The technique includes particle size reduction by grinding, with or without combining other techniques such as pelleting and steam treatment. The roughages being utilized are straws and stovers and for their improvement, in addition to these physical techniques some chemical and biological treatments have also been attempted with varying degree of success. Here, the major emphasis will be given on particle size reduction technique and its impact on animal performance. Reduction in particle size of roughages is accomplished by cutting and/ or grinding, which increases the surface area and bulk density of stem and leaf fractions of roughages. Pelleting of roughages results in increased bulk density, reduction in dustiness and increased ease in handling. Increased fiber level and roughage particle size, effectively increase chewing activity and also increase saliva flow, rumen pH, acetate-to-propionate ratio and milk fat levels (Cassida and Stokes 1986; Mertens, 2000). The most consistent responses of particle size reduction of roughages are increase in feed intake, body weight gain and feed efficiency, particularly with poor quality roughages. However, significant increases in voluntary intake of processed poor quality roughages often occur only in conjunction with nitrogen (N) supplementation of the diet (Westen, 1967). Greenhalgh and Reid (1973) demonstrated that intake response to particle size reduction were greater for sheep (45%) than cattle (11%) and greater in young animals (38%) than the mature ones (17%).

Less extreme processing of roughages, such as chopping or wafering, does not result in consistent improvements in animal performance although these result in reduced diet wastage and sorting (Beardsley, 1964). Cattle fed rations lacking in physical structure, excessive amounts of long, coarse forage may impaired rumen fermentation and function and can result in limit intake and digestibility, ultimately affecting energy balance of the animal (Allen 1997). Current NRC guidelines have proven useful in defining animal requirements and feed composition but do not provide recommendations on physical form of the ration. The concept of effective fiber was created to unite the chemical and physical nature of the forage, and to quantify its value for rumen function. Chewing and rumination are known accurate measurements of the roughage characteristics for ruminant diets (Sudweeks et al., 1981). Physically effective NDF is defined as that dietary fiber source which effectively stimulates rumination and salivation (Mertens 1997). Poppi et al. (1983) found that particles those were retained on a sieve measuring 1.18-mm pass out of the rumen slower than those, which are not retained. Mertens (1997) found that particles size of > 1.18-mm to pass out of the rumen they would have to be reduced through pulverization and as a result these particles would stimulate more saliva secretion as compared to < 1.18-mm particles size. Routinely measurement of feed particle size has been difficult until the introduction of The Penn. State Particle Separator (PSPS). Using the PSPS, a particle distribution is determined by separating particles according to size in the following categories (Lammers et al., 1996).

1. Particle size more than 19.0-mm
2. Particle size between 19.0 and 8.0-mm
3. Particle size less than 8.00-mm.

With the construction of an additional screen (1.18-mm), the PSPS is capable to making physically effective NDF measurements proposed by Mertens (1997). Additionally the third screen may be useful in measuring the particle size of non-forage fiber sources as well as processed grains.

Effect of Forage Particle Size on DMI

Effect of particle size on DMI is not clear with conflicting results being reported (Kennedy and Doyle 1993). Few researchers have observed effects of alfalfa silage on intake when well balanced rations are fed to mid-lactation cows (Table 1). DMI have been reported to improve with reduced particle size in some studies feeding corn silage of different particle size (Stockdale and Beavis, 1994) but have not been observed in others (Table 2). Difficulty in interpreting the response of particle size on DMI may be partially due to digestibility and specific gravity, factors independent of the individual feeds particle size (Dulphy and Van Os 1996, Singhal 1996). Furthermore, the degree of particle size reduction varies depending upon processing and particle size determination methods and making treatment types difficult to quantify in relation to others. When, intake is increased on reduced particle size, it is usually positive with the magnitude depending upon the extent of particle reduction, the type of forage, digestibility of forage and the energy status of the animal. Increased DMI with reduced particle size have been noted in many poor quality forages containing high cell wall contents (Singhal et al., 1992; Kusmartono et al. 1996). Several authors have reported increased DMI with reduced particle size while feeding high quality forages (Beauchemin et al., 1997), however, most authors report that no effect on DMI occurs when good quality forage is fed (Tables 1 and 2). It appears that DMI is increased by particle size reduction only when a poorly digestible feed with high cell wall content is fed and had no effects when good quality/highly digestible forages are fed.

Eating and Rumination Activities

Best and accurate measurements to assess the roughage quality for ruminant diets are chewing and rumination (Balch 1971). When total ration NDF level and ration particle size is increased, it resulted in the more rumination time, which resulted in increased saliva flow, which buffers rumen pH (Cassida and Stokes 1986). When chopped and dehydrated alfalfa was fed, there was no difference between different chop lengths but differences in rumination times between chopped hay and finer dehydrated forage products were observed (Beauchemin et al., 1997 and Shaver et al. 1988). Increased mean rumen pH levels accompanied increased rumination times. Differences between eating, ruminating, and total chewing activity (TCA) are often not observed when feeding corn silage of different particle size. Both Bal et al. (2000) and Schwab and Shaver (2001a, 2001b) observed that fine cut length reduced eating activity but did not effect rumination or total chewing time per kilogram of DMI. As a result of these observations it seems likely that increasing corn silage particle size increases the sorting tendency of dairy cattle, ultimately resulting in the diets being consumed being different than those formulated.

Rumen pH and Digestion

Digestibility of ground roughages generally is depressed relative to that of the parent material fed in long or chopped forms. The average depression in DM digestibility was found greater for grasses (up to 150 g/ kg) than legumes (30 to 60 g/kg). Decreased digestibility as a result of reduction in particle size of roughages increase in magnitude as feed intake increases i.e. enhanced GE and DE intake. Increases in GE and DE intakes as a result of roughage processing were found greater as forge maturity advances, reflecting the increased intake response as forage quality decreased. Fine grinding reduces the time that ruminants spend eating and ruminating and consequently, saliva production is reduced significantly with manifestation of decreased buffering capacity and depressed ruminal pH. Low ruminal pH and increased rate of passage of digesta result in deceased cell wall carbohydrates fermentation resulting depressed TVFA production as a proportion of GE. In addition, reduction in acetate: propionate ratio often results with particle reduction of roughages, which has been implicated as a reason for the milk fat reduction. Rumen escape protein generally increased by particle size reduction as it is achieved by grinding and in the processes of grinding heat is generated to increase the rumen escape protein (Beever et al., 1981) without any affect on microbial protein synthesis efficiency. Though processing increases the rumen escape protein to the small intestine, it should be recognized that intestinal availability of this additional protein may be reduced if heat damage is excessive (Thomson and Beever, 1980).

It is difficult to understand the complete dynamic nature of ruminal fermentation because a number of factors affect rumen fermentation. A number of studies have been designed in an attempt to understand the effects of particle size on ruminal fermentation (Tables 1 and 2). The effects of different theoretical cut lengths of alfalfa silage and corn silage do not have significant effect on dry matter digestibility. Dry matter digestibility does not appear to be linked to different cut lengths of silages however differences have been observed when forages are excessively ground or when dehydrated forms are fed (Beauchemin et al. 1997; Rodrigue and Allen, 1960). Although increasing particle size often results in increased rumination activity and higher mean rumen pH values, the resulting effects are usually quite small. Grant et al. (1990) harvested alfalfa silage at three different cut lengths, although increasing chop length increased milk fat percentage by 0.8%, with a slight change in mean rumen pH. Singhal and Grant (1998) observed that rate of fiber digestion decreased when fine wheat straw particle was used. Fine particles though giving large surface area showed the lowest rate of fibre digestion due to higher rate of passage as a result of higher functional specific gravity. Apparent extent of NDF digestion was highest for chopped straw and decreased on reducing in to fine particles. They conclude that processing of wheat straw into fine particles does not favor the efficient utilization of its fiber content. Overall there is little evidence showing that different cut length of roughages effects rumen pH and volatile fatty acid level. In future work should be carried out to determine how particle size might be important when feeding cows in different stages of lactation.

Table 1: The effect of alfalfa physical forage form on digestive parameters and milk production in lactating cattle.

Reference	DMI	Rumination (min/d)	DMD	Acetate: Propionate	pH	Milk Yield	FCM	Milk Protein%	Milk fat%
Beauchemin et al. 1997	S	S	N/A	N/A	N/A	NS	NS	S	S
Yang et al. 2001	NS	N/A	NS	NS	NS	NS	**	*	S
Fischer et al. 1994	S	NS	N/A	NS	NS	NS	*	NS	S
Grant et al. 1990	NS	S	N/A	S	S	NS	S	NS	S
Woodford & Morphy 1985	NS	NS	NS	NS	NS	NS	NS	NS	NS

Table 2: The effect of corn silage physical forage form on digestive parameters and milk production in lactating cattle

Reference	DMI	Rumination (min/d)	DMD	Acetate: Propionate	pH	Milk Yield	FCM	Milk Protein%	Milk fat%
Bal et al. 2000	NS	NS	NS	NS	NS	NS	NS	NS	NS
Schwab & Shaver, 2001a,b	NS	NS	N/A	N/A	N/A	NS	NS	NS	NS
Stockdale & Beavis, 1994	N/A	N/A	NS	NS	NS	NS	NS	NS	NS
Clark and Armentano 1999	NS	NS	N/A	N/A	N/A	NS	NS	NS	NS
Odai et al. 1988 a, b	NS	N/A	NS	NS	N/A	NS	NS	NS	NS

N/A= parameter not measured in study
S=significant, NS = not significant ($P < 0.05$)
DMI= dry matter intake; DMD=dry matter digestibility; FCM=fat corrected milk

Milk Production and Composition

In alfalfa based diets, forage particle size has been shown to significantly affect both yield and composition however most differences are reported when forage is in the dehydrated form (Woodford and Murphy 1988; Shaver et al. 1986; Beauchemin et al., 1997). When forage is in silage form and different lengths of cut are fed improvements have been observed in milk fat (Grant et al., 1990; Fischer et al., 1994; Shaver et al., 1986) and protein percentage (Beauchemin et al., 1997). No any studies have reported a significant response in milk production or fat corrected milk by feeding corn silage of different particle size (Clark and Armentano, 1999, Bal et al. 2000). Increased milk protein with decreasing corn silage particle size has been observed in some studies and believed to be a result of increased starch and energy availability (Clark and Armentano, 1999). Fat percentage is not affected by feeding rations of different corn silage particle size (Clark and Armentano, 1999; Stockdale and Beavis, 1994) however, in one study coarse particle size resulted in excessive amounts of sorting and NDF refusal and milk fat was actually higher in rations of shorter particle size (Bal et al., 2000). Although significant effects of particle size on rumen pH differences are small and are not likely to result of the large shifts in volatile fatty acid production required to alter milk fat composition (Grant et al., 1990; Shaver et al., 1986; Woodford and Murphy, 1988). Observed differenced in milk composition are seems to be a result of other factors such as passage and rate or extent of total tract digestion.

Particle Size and Effective Fiber

Fiber is defined as the slowly digestible or indigestible organic fraction of the feed, which occupies space in the gastrointestinal tract (Mertens 1997). When an animal is deficient in fiber there is decreased chewing, resulting in less salivary buffer and a lower rumen pH and low acetate to propionate ratio, all factors affecting the rumen fermentation, milk production and over all performance of animals. Yang et al. (2001) however noted that contribution of increased total daily saliva output due to increased total chewing activity on rumen pH is often overestimated. Although reduced particle size may decrease total chewing activity, changes in total saliva production are small (approximately 4%), as resting saliva secretion will increase. It seems likely that the physical characteristics of the feed have a smaller influence on rumen pH than originally believed. In comparison, level of starch in the ration may have a greater effect on rumen pH and management of the ration nonstructural carbohydrate level may be more useful in identifying high-risk rations that lead to ruminal acidosis. It should be noted that to date there have been no controlled studies designed to evaluate the effects of feeding rations of different particle size on long term reproductive performance or health status (displaced abomasums, laminitis etc.) and that continually feeding high energy diets of fine particle size may have other negative effects.

Particle Size Recommendations

So far research published on feed particle sizes distributions are only beginning, thus making recommendations difficult. Present NRC recommendations indicate that a minimum mean particle length of 3 mm for alfalfa diets is necessary to maintain rumen pH, chewing activity, and milk fat percentage. Additionally, diet NDF content should be increased if excessively fine forages or high amounts of rapidly fermentable starch are fed. It is apparent that when silages are too coarse, sorting by the animals occurs at the feed bunk, this problem is further exacerbated when drier silages are fed. Listed below are the current particle size recommendations (Table 3.) for haylage and corn silage (Lammers et al., 1996). These may change in the future as more research data becomes available.

Table 3: Particle size recommendations for haylage and corn silage based rations

Particle Fraction	Haylage TMR	Corn Silage TMR
> 19.0-mm	5-10	3-5
19.0-8.0-mm	30-35	45-55
< 8.0-mm	53 - 65	42-50

Grinding of roughages does not enhance structural carbohydrate utilization by ruminants. Increases in animal performance primarily are consequences of increased DE intake. Although low quality roughages is enhanced by grinding and pelleting, the processed material still has a relatively poor NE content. Particle reduction followed by chemical treatment of roughages may enhance the feeding value of such low quality roughages.

Bibliography

Allen, M.S. 1997. Relationships between fermentation acid production in the rumen and the rumen and the requirement for physically effective fiber. J. Dairy Sci. 80: 1447-1462.

Bal, M.A. Shaver, R.D., Jirovec, A.G., Shinners, K.J., and Coors, J.G. 2000. Crop processing and chop length of corn silage: Effects on intake, digestion, and milk production by dairy cows. J. Dairy Sci. 83: 1264-1273.

Balch, C.C. 1971. Proposal to use time spent chewing as an index of extent to which diets for ruminants posses the physical property of fibrousness characteristics of roughages. Br. J. Nutr. 26:383-389.

Beardsley D.W. 1964. Symposium on forage utilization: Nutritive value of forages as affected by physical form. Part II. Beef cattle and sheep Studies. J. Anim. Sci. 23: 239-245.

Beauchemin, K.A., Rode, L.M., and Eliason, M.J. 1997. Chewing activities and milk production of dairy cows fed alfalfa as hay, silage, or dries cubes of hay or silage. J. Dairy Sci. 80: 324-333.

Beever, D.E., Osbourn, D.F., Cammell, S.B. and Terry, R.A. 1981. The effect of grinding and pelleting on the digestion of Italian ryegrass and timothy by sheep. Br. J. Nutr. 46:357-370.

Cassida, K. A., and Stokes, M.R. 1986. Eating and resting salivation in early lactation dairy cows. J. Dairy Sci. 69: 1282-1291.

Clark, P.W., and Armentano, L.E. 1999. Influence of particle size on the effectiveness of fiber in corn silage. J. Dairy Sci. 82: 581-588.

Dulphy, J.P., and Van Os, M. 1996. Control of voluntary of precision-chopped silages by ruminants: a review. Reprod. Nutr. Rev. 36: 113-135.

Grant, R.J., Colenbrander, V.F., and Mertens, D.R. 1990. Milk fat depression in dairy cows: role of silage particle size. J. Dairy. Sci. 73:1834-1842.

Greenhalgh, J.F.D. and Reid, G.W. 1773. Long and short term effects on intake of pelleting a roughage on sheep. Anim. Prod. 19:77-86.

Kennedy, P. M., and Doyle, P.T. 1993. Particle size reduction by ruminants- effects of cell wall composition and structure. pp 499-528. In: Forage cell wall structure and digestibility. Eds; H.G. Jung, D.R. Buxton, R.D. Hatfield, and J. Ralph.

Kusmartono, A., Shimada, A., Stafford, K.J., 1996. Intra-ruminal particle size reduction in deer fed fresh perennial ryegrass (Lolium perenne) or chicory (Cichorium intybus). J. Agric. Sci. 127: 525-531.

Lammers, B.P., Buckmaster, D.R., Heinrichs, A.J. 1996. A simplified method for the analysis of particle sizes of forage and total mixed rations. J. Dairy Sci. 79:922-928.

Mertens, D.R. 2000. Physically effective NDF and its use in dairy rations explored. Feedstuffs. pp 11-14.

Mertens, D.R. 1997. Creating a system for meeting the fiber requirements of dairy cattle. J. Dairy Sci. 80: 1463-1482.

Odai, M., Takahashi, T., Itabashi, H., Kobayashi, T., Ono, M. 1988, b. Effects of length of cut of whole crop corn silage and milk production in dairy cows (2) effects on milk yield, milk composition and ruminal VFA during long term feeding. Bull. Nat. Inst. Anim. Ind., No 45. 7-12.

Poppi, D.P., Hendrickson, R.E., and Minson, D.J. 1983. The relative resistance to escape of leaf and stem particles from the rumen of cattle. J. Agric. Sci. 105: 9-14.

Rodrigue, C.B., and Allen, N.N. 1960. The effect of fine grinding of hay on ration digestibility, rate of passage, and fat content of milk. Canadian. J. Anim. Sci. 40: 23-29.

Schwab, E.C., and Shaver, R.D. 2001a. Crop processing and chop length effects in brown midrib corn silage on dry matter intake and lactational performance by dairy cows. J. Dairy Sci. 79: Suppl 1 pp 197.

Schwab, E.C. and Shaver, R.D. 2001b. Crop processing and chop length effects in brown midrib corn silage on chewing activity and mean particle size of silage and masticates. J. Dairy Sci. 79: Suppl 1 pp 349.

Shaver, R.D., Nytes, A.J., Satter, L.D., and Jorgenson, N.A. 1986. Influence of feed intake, forage physical form, and forage fiber content on particle size of masticated forage, rumen digesta, and feces of dairy cows. 71: 1566-1572.

Singhal, K.K., Kundu, S.S. and Sharma, D.D. 1992. Effect of chopping of oat hay on nutrient utilization and rate of passage of liquid digesta through gastrointestinal tract in crossbred calves. Indian J. Dairy Sci. 45:123-125.

Singhal K. K. 1996. Dynamics of particle breakdown in rumen. Proc. Summer Institute on Dietary manipulation of rumen fermentation for increased productivity and reduced methane production in ruminant conducted at NDRI, Karnal.

Singhal K. K. and Rick Grant 1998 Effect of particle size reduction on fibre digestion kinetics of wheat straw. Indian J. Anim. Nutr., 15(1)1-5.

Stockdale, C.R., and Beavis, G.W. 1994. Nutritional evaluation of whole plant maize ensiled at three chop lengths and fed to lactating dairy cattle. Aust. J. of Exp. 34: 709-716.

Sudweeks, E.M., Ely, L.O., Mertens, D.R., Sisk, L.R. 1981. Assessing minimum amounts and form of roughage in ruminant diets: roughage value index system. J. Dairy Sci. 53: 1406-1411.

Thompson, D.J. and Beever, D.E. 1980. The effect of conservation and processing on the digestion of forages by ruminants Pp[291-308 In Digestive physiology and metabolism in ruminants. Ed. Y. Ruckebusch and P. Thivend. MTP Press Ltd. Lancaster, England.

Weston, R.H. 1967.Eactors affecting the intake of feed by sheep. II Studies with wheaten hay. Aust. J. Agric. Res. 18: 983-1002.

Woodford, S.T., and Murphy, M.R. 1988. Effect of physical form of forage on chewing activity, dry matter intake, and rumen function of dairy cows in early lactation. J. Dairy Sci. 71: 674-686.

Yang, W.Z., Beauchemin, K.A., and L.A. Rode. 2001. Effects of grain processing, forage to concentrate ration, and forage particle size on rumen pH and digestion by dairy cattle. J. Dairy Sci. 84: 2203-2216.

CHAPTER - 17
PROCESSED FEED FOR GOAT

Nawab Singh and T. K. Dutta

In animal production enterprise, feed alone costs about 60-70% of total production cost depending on the system of production. The shortage in feed and poor quality of available feed resources have been one of the major constraints. There is deficit of about 44% in concentrates, 38% in green fodder and 45% in dry roughage. India possesses about 123.5 million goats i.e. 21% of World's 743.0 million goats population (FAO, 2000). They produce about 0.48 million tones meat, 1.684 million tones of milk, 0.085 million tones Pashmina and 0.109 million kg skin in addition to 390 thousand metric tones of manure. The estimated value of different types of produce from goat is about Rs. 2612.00 million per year and it also generates about 4.2% rural employment per year. (C.I.R.G. Annual report 2002-2003). Indian farmers generally maintain their goats under zero input system in which the productivity of goats is low due to lesser availability of nutrients and seasonal variation in herbage biomass quality. Secondly, grazing land is also shrinking year after year as more land is brought under crop cultivation and there has been introduction of commercial goat farming with the result inputs in terms of feed is the main constraint. Available concentrate mixture ingredients like grains, oilseed cake and by-product of pulses and cereals are generally being fed to large ruminant like cattle and buffaloes. The tree leaves which are available in plenty and if lopped scientifically, can be utilized easily by goats in the form of complete feed blended with other cheaper feed ingredients and thus sparing costly concentrate mixture ingredients for large ruminants. To cope with increasing future demand of meat in the country as well as export demand, the present system of goat rearing which can hardly supports 3 goats per hectare will have to be replaced by intensive system where 100 goats per hectare can be maintained by providing feed from outside sources. The available feed and fodder in the country are not enough to sustain the existing population size of livestock. The available fodder in India is poorer in nutritive value and seldom grown and managed scientifically (Bhattacharya, 1989). The Himalayan region presents a typical example of poverty amidst of plenty of forage. Therefore, concerted efforts are required to use these vast green resources in a sustainable manner for enhancing the income generation of rural mass of these regions.

The training programme organized at CIRG every year for commercial goat rearing has promoted commercial goat production at national level. Several trainees of this programme have already started goat farming on commercial scale. At present 26 trained farmers from nine states have started commercial goat farming on large scale. Therefore, commercial goat farming is growing fast, which will be considered as new entrepreneur for resource poor and unemployed people. To cope with the nutritional requirement under intensive and semi-intensive farming condition proper feeding strategy must be developed for higher growth within a short time, higher milk production, quality skin and wool production from this entrepreneur. The balance scientific feeding schedule to maximize the return from the goat enterprise under intensive feeding system has some advantages over extensive or semi-intensive system of management, due to scarcity of grazing land. Processed feed in the form of feed pellets, hay and silage has greater role to play specially for stall-fed goats in intensive system of production on commercial lines.

Tree leaves as source of fodder

In India, about 20 per cent area is under forest cover. Goats are basically browsing animal. Browsing, a term, which is used to refer to the tender shoots, twigs and leaves of shrubs and trees, is a very valuable feed for goats. Many tree species have good potential for supplying quality fodder to goats throughout the year. Tannin is one of the limiting factors of top leaves. However, lopping from following tree species can be used as a source of fodder if it is used up to certain level, i.e., up to 30 per cent of the total dry matter intake. Most of the tree species are available in different parts of India under natural forest cover or in improved silvi-pastoral system.

· Acacia lecophloea (Safed Kikar, Reru Rinj, Raunj, Rhea)
· Albizia procera (Safed Siris, Gurar, Karak)
· Anthocephalus chinensis (Kadam)
· Bauhinia malabarica (Amli, Amlosa)
· Bauhina purpurea (Khairwal, Karar, Koliar, Kaniar)
· Bauhinia semla (Semal, Semul)
· Gauga pinnata (Kharpat, Kaikar)
· Mangifera indica (Aam)
· Populas ciliata (Pahari pilal, Chalaun, Chalni, Safeda, Ban Pipal)
· Schleichera oleosa (Kusum)
· Syzygium cumini (Jamun)
· Tamarindus indica (Imli)

- Terminalia belerica (Bahera)
- Ziztyphus mauratiana (Ber)
- Acacia catechu (Khair)
- Acacia nilotica (Babool)
- Aegle marmelos (Bel)
- Ailanthus excelsa (Ardu)
- Albizia lebbek (Siris)
- Artocarpus heterophyllus (Kathal)
- Azadirachta indica (Neem)
- Diospyros melanoxylon (Tendu)
- Ficus glomerata (Gular)
- Ficus lacor (Pakar)
- Ficus religiosa (Pipal)

These are commonly found tree species in different parts of India. Scientific lopping of such trees may fill some gap of demand and supply of green biomass for traditional and commercial goat production.

Feed containing 20% subabul was found to be best for growth when compared to other feed. Similarly, subabul feed gave high carcass yields. Lactation studies revealed that feed containing 15% desi babul leaves was found to be most suitable. The feed containing 20% lucerne leaves was found to be more cheaper followed by feed containing 15% Desi babul leaves and control (conventional concentrate mixture). Thirty per cent replacement of concentrate mixture (or 9.80% in total diet consumed) with Leucaena leaves in pellet has given comparable DM, TDN and DCP intake with that of mash fed control. Values of ruminal pH, total VFA, total-N, ammonia-N and protozoa counts were also similar with that of control. Hence, pelleted feed with 30% Leucaena leaf can be fed easily to adult Marwari goats for their maintenance. Addition of Leucaena leaves in the mash reduced the cost of pellet by 21.43 per cent (Dutta, et al. 2002).

Silvi-pastoral system

Silvi-pasture farming approach is based on the integration of woody species with agricultural crops; pasture grasses and legumes in the same piece of land. It has been recognized as the low input technology for useful utilization of wasteland. Under poor soil-water and nutrient situation where cropping is not possible the programme of silvi-pastoral management serves the twin purpose of forage and wood production with system conservation. Integrated silvi-pastoral system of farming has been recognized as low input technology for the utilization of wasteland for meeting the shortage of forage and leaf fodder besides fuel wood and timber. During the establishment year, pasture production is normally low and varies from 1.8 to 2.5t/ha. Normally tree seedlings are planted during the monsoon after a good shower followed by sowing/planting of pasture grasses and legumes. Pasture production of two Cenchrus ciliaris and C. setigerus is association with different tree densities of Acacia tortilis and Leucaena leucocephala. The results revealed that dry foarge production varies from 3.1 t/ha in the second year and 5.1 t/ha in the 5th year giving a mean forage production of 4.2 t/ha/year (Deb Roy, 1989), and with improved management, forage production could be further increased from 5.6 to 7.8 t/ha.

Two-tier silvi-pastoral system is maintained with ground grass cover and tree species as top layer. Whereas, three-tier silvi-pastoral system is maintained with three layers i.e., trees species as top, shrubs as middle and grasses as ground cover. Few examples of silvi-pastoral systems with plant species are mentioned in following tabular form.

Table 1: Constituent plant species of natural and silvipasture combinations.

Systems	Tree (top layer)	Shrubs (middle layer)	Grasses (ground layer)
Natural rangeland	Acacia nilotica Prosopis cineraria	-	Saccharum munja Heteropogon contourtus Cenchrus ciliaris S. setigerus etc.
Silvipasture system (2-tier)	-	Leucaena leucocephala (3,250)	Cenchrus ciliaris Cynodon dactylon Stylosanthes hamata
Silvipasture system (2-tier)	-	Dichrostachys nutan (1, 250)	Cenchrus ciliaris Cynodon dactylon Stylosanthes hamata
Silvipasture system (3-tier)	Acacia nilotica (312)	Leucaena leucocephala (1, 250)	Cenchrus ciliaris Cynodon dactylon Stylosanthes hamata

Source: Sharma, K. (1989) The figures in parenthesis indicate plants density/ha

Following models may be adopted for development of silvi-pasture or agro-silvi-pasture in different agro-climatic zones of India.

Table 2: Agro-forestry systems for different agro-climatic regions of India.

Regions	Agro-forestry systems
Arid zone	Silvi-pasture
Agro-silvi pasture	
Semi-arid	Silvi-pasture
Agro-silvi pasture	
Humid	Agro-silvi pasture
Sub-humid	Agro-silvi pasture
Temperate	Silvi-pasture
Agro-silvi pasture	
Alpine	Silvi-pasture

Source: Sharma, K. (1989)

Development of tree leave based complete feeds for goat production

At CIRG, Makhdoom different types of tree leaves based complete feed have been developed for intensive goat rearing purposes (Singh and Rao, 2004).

1. Neem (Azadirachta indica) leaves based pelleted feed consisting of 30% mature dry fallen neem 37% maize grain, 10% groundnut cake, 20% wheat bran, 1% salt and 2% mineral mixture was tried in Jakhrana goats. DCP and TDN content of the feed was 9.43 and 67.42% respectively. The dry matter intake/100 kg body weight was 2.41 kg/day.
2. A complete feed consisting of 30% dry shesham leaves, 30% arhar straw and 40% concentrate mixture (maize 50%, groundnut cake 30%, wheat bran 17%, min. mix. 2% and common salt 1%) was prepared and fed to Jakhrana goats. The DCP and TDN content of pelleted feed were 9.47 and 56.77 respectively.
3. Nutritive value of complete feed made of 50 parts Ardu (Ailanthus excelsa) leaves and 40 parts barley grain for goats came to be 12.92% DCP and 68.93% TDN. The dry matter intake per 100 kg body weight was 2.54 kg.
4. A complete feed in pelleted form was prepared and evaluated for goats, by mixing 60 parts dry leaves of Anni (Chlordendron phlemides) and 40 parts concentrate mixture. Dry matter intake per 100 kg body weight was 3.11kg. The DCP and TDN content on dry matter basis were 13.02 and 62.94 per cent, respectively.

5. A complete pelleted feed was prepared by mixing 40% dried and ground Heens (Capparis horrida) leaves, 20% Arhar (Cajanus Cajan) straw and 40% concentrate mixture. Nutritive value was determined by using Sirohi male goats. DCP and TDN content was 13.67 and 64.59 per cent, respectively.
6. Green leaves of Neem trees (Azadirachta indica) were collected and dried in winter season. A pelleted complete feed was prepared by mixing 40 parts neem leaves, 30 parts arhar straw, 10% til cake, 17% barley grain, mineral mixture 2% and salt 1%. Nutritive value was determined using adult male goats. The DCP and TDN contents were 8.89 and 58.90% of the complete feed. The DMI was 3.63 kg/100 kg body weight.
7. A complete ration was formulated by using urea (3%) treated barley straw 46.7 parts, Arhar straw 20 parts, barley grain 32.3 parts and mineral mixture with salt 1 part. Feeding trial in Jakhrana goats revealed intake of DM, DCP and TDN was well above the suggested requirements. The DCP and TDN of the complete ration was 5.51 and 64.78 per cent respectively.
8. A complete pelleted feed using dried mulberry leaves 40%, arhar (Canjanus cajan) straw 10%, til (Sesamum indicum) cake 10%, barley (Hardeum vulgare) 28.5%, wheat bran 10%, mineral mixture 1%, and salt 0.5%. Feeding trial was conducted using adult Barbari goats. The DCP and TDN content of the complete ration were found to be 9.65 and 64.51 per cent respectively. The dry matter intake was 2.71kg/100 kg body weight.
9. In a traditional system of SANI preparation straw (wheat or barley) is moistened with fresh water and mixed with concentrate. A substantial increase was noted in dry matter (DM), crude protein (CP) and TDN intake values after addition of common salt+ mineral mixture to the sani of goats. Addition of 15 g common salt and 15 g mineral mixture to sani of adult male goats can increase the availability of protein and energy by about 16 per cent.
10. A complete feed in pelleted form was prepared and evaluated for goats, by mixing 60 parts dry leaves of Anni (Chlordendron phlemides) and 40 parts concentrate mixture. Dry matter intake per 100 kg body weight was 3.11kg. The DCP and TDN content on dry matter basis were 13.02 and 62.94 per cent, respectively.
11. Complete diets in pelleted form having non-traditional feed ingredients were tested using male kids of Jakhrana breed for meat production. Feeding was done for about six months slaughter studies indicated that non-traditional feeds can be used for meat production from goats.

12. Ber (Zizyphus mauritiania), Pipal (Ficus religiasa) and Siras (Albiza lebbek) tree leaves based pelleted feeds were evaluated for its nutritional potential for goats.

Cultivated fodder as raw material for processed feed

Feeding of cultivated fodder for goat rearing purposes is not commonly practiced in India under traditional system of management at farmers door. However, there is growing concern to feed quality fodder to improve production traits of goats under commercialized condition. Cultivation of quality fodder with scientific crop rotation can be one such solution for large commercial farm entrepreneur. Cultivated fodder contains good amount of major nutrients like, CP and TDN. However, fodders like, barseem, alfalfa, oat, barley, bazra (pearl millet), maize, sorghum, cowpea and guar fodders are mainly used for milk production from cows and buffaloes. In any organized farm proper schedule of legume and non-legume fodder crops should be maintained to enhance the total green biomass production and total protein content of any particular cropped area. Barseem, lucerne (alfalfa), oat and barley are the major fodder crops during winter season in most of the parts of India, whereas, lobia (cowpea), pearl millet, maize, guar, sorghum are the major fodder crops during kharif season. Time of harvesting is one of the major parameter to maintain the quality of fodder. Harvesting at pre-flowering to 50% flowering stage is considered to be the best. Introduction of intercropping of leguminous and non-leguminous crops may enhance the biomass productivity per unit area and may enhance the soil fertility. Area under fodder production in India is very marginal, hence judicial and scientific use of available area may increase the biomass productivity without deteriorating the soil fertility. Therefore, proper crop rotation should be followed throughout the year in any organized goat farm. Hay and silage prepared from surplus fodder may overcome the problem of green biomass during the lean part of the year in any organized farm.

Intake of Nutrients and Digestibility as influenced by conserved fodder

At CIRG, Makhdoom, few experiments have been conducted with silage and hay feeding to goats. Dry matter intake per 100 kg body weight and TDN intake per 100 kg body weight was almost similar in berseem fresh and hay fed groups of post-weaned Barbari kids and less in silage fed groups (Table 3). Adult Barbari can relish berseem silage fodder easily than Jamunapari and Marwari breed (Table 4).

Table 3. Intake pattern of post weaned Barbari kids under different feeding system

Parameters	Berseem Fresh	Berseem Hay	Berseem Silage
DMI(kg)/100 kg b. wt.	5.21^{a}	4.03^{ab}	2.88^{b}
DMI from Berseem (kg)/100 kg b. wt.	3.87^{a}	3.20^{ab}	1.83^{b}
TDN Intake (kg)/100 kg b. wt.	3.80^{a}	2.44^{a}	1.51^{b}
DCP Intake (g)/100 kg b. wt.	681.79^{a}	442.25^{b}	403.35^{b}
TDN % in Feed	71.89	67.64	51.87
DCP % in Feed	12.84	10.89	13.91

Table 4. Intake pattern of adult Barbari, Jamunapari and Marwari goat under berseem silage based feeding

Parameters	Barbari	Jamunapari	Marwari
DMI (kg)/100 kg b. wt.	2.86^{a}	2.43^{b}	2.18^{c}
DMI from silage (kg)/100 kg b. wt.	2.22^{a}	1.78^{b}	1.66^{c}
TDN Intake (kg)/100 kg b. wt.	1.51^{b}	1.11^{a}	1.31^{ab}
DCP Intake (g)/100 kg b. wt.	383.35	345.27	408.32

DM intake/kg $W^{0.75}$ was significantly ($P<0.01$) lower in silage fed Barbari kid group than in fresh or hay fed group, which was mainly due to reduction of fodder intake through silage. Digestibility co-efficient of DM, CP and CF were significantly ($P<0.01$) reduced in hay or silage fed groups. Narrowing the ratio of TDN and DCP in silage fed group could be the reason for lower DMI. Hence, supplementation of energy is required to improve nutrient availability in berseem silage based feeding to post-weaned Barbari kids. TDN intake by fresh and hay fed groups were sufficient to meet the nutrient requirement, whereas, kids under berseem silage feeding required energy supplementation. pH value, ammonia-N and total volatile fatty acids concentration were similar in all three treatments (Fresh, hay and silage). Total ciliate counts were also similar among treatments. Although rumen fermentation pattern was not influenced by different form of berseem fodder but its lower tract digestibility was affected.

Hence, during scarcity period of green fodder, berseem hay can be fed to post-weaned Barbari kids without affecting its TDN and DCP intake. Berseem silage as a sole source of fodder was not so effective as it reduced DCP and

TDN intake. Comparatively N-balance was also affected due to berseem silage feeding. Berseem fodder can be mixed with non-leguminous fodder like oat, and silage can be prepared. This will give better silage quality than leguminous fodder alone.

Utilization of crop residues

Crop residues are the remains of harvested crops primarily planted for man or agriculture-based industries. In most cases, crop residues are roughages that have an appreciable level of ADF, NDF and lignin. With the growing ruminant animal production industries in general and the goat in particular, there is a need to appraise all avenues of cheaper but readily available feed sources and feedstuffs of lesser dietary importance to man and mono-gastric animals. This will reduce the competition for feed resources between ruminants and mono-gastric animals. In spite of their availability, by-products are still not extensively used in goat diet.

The availability of crop residues, which was to the tune of 267.53 million tones in 1985-86, has increased to 373.21 million tones in 1999-2000 registering an increase of 39.50% over a period of 15 years (Khub Singh et al., 2004). The increase in the availability of crop residues over the years has largely been due to increase in production of paddy and wheat crops. Utilization of crop residues with maximum level in the rations of small ruminants is an essential step in enhancing livestock productivity. The availability of coarse cereals straw (like maize, pearl millet, sorghum) was 145 million tones during 1999-2000. However, goats do not generally prefer these straws. Straws of pulses like pigeon pea (arhar), gram, mung, musur etc are preferred by Indian goats. This could be due to higher nutrients (total protein and total digestible nutrient) content in pulses straws as compared to cereals straws. Nutritive value of arhar (Cajanus cajan) straw for goats was determined and the DCP and TDN contents came at to be 5.99 and 56.89% respectively. Moreover, the competition between small ruminants and large ruminants is very marginal in respect of these pulses straws. Pulses straw is available in many parts of India, like Madhya Pradesh, parts of Maharastra, Uttar Pradesh, Andhra Pradesh, Rajasthan etc. The availability of pulses straw was 34.16 million tones during 1999-2000. These pulses straw is not used scientifically to improve the production potential of small ruminants in particular goats. Commercial goat farms, which is emerging as a new entrepreneur in many parts of the country can store and use these straws for supplying bulky dry fodder throughout the year. The potential availability of fine straws (rice, wheat, barley, oats and ragi) was 194.65 million tones during 1999-2000 (Khub Singh et al., 2004). Fine straws can be utilized by goats up to certain level or after treatment. Pulses straws have been frequently used for preparation of pelleted feeds for goats.

Newly developed processed feed for commercial goat farming

a) Milk Replacer and by-products based starter ration for kids

Mother's milk is the ideal food for newborn kids. Kids are fed with normal milk after having sufficient colostrum. The mother's milk can not only meet newborn kids' nutritional requirement but also match kids' taste and their body fluid type. However, for some practical reasons, such as multiple births or milk shortage for weak dams, large scale finishing system, industrialized rearing, early weaned or super-early weaning of kids, fine quality does producing two litters in one year, etc, the availability of mother's milk is either limiting or diverted for human consumption.

Kids can be reared economically using milk replacers or substitute, and weaned at an early age, if adequately grown. Alternative rearing systems are a high priority to producers of goat milk. The advantages include reduced costs, increased production, and breaking disease cycles. The success of rearing kids using milk replacers requires strict adherence to correct management practices, particularly in ensuring good hygiene in rearing facilities, and cleanliness of feed and feeding equipment. Artificial methods of rearing goat kids are widely used overseas, and are gaining acceptance in other countries. They have a number of advantages over natural methods, particularly in more intensive commercial goat-breeding operations.

Most of the goat breeds in India are poor yielder of milk and have higher multiple birth rates. Milk replacer/substitute supplies essential nutrients to multiple birth kids (twin, triplet and quadruplet) and the kids having no or minimum accessibility of mother' milk. Therefore, the life of multiple birth kids may be saved and growth rate may be maintained. Surplus milk of high yielding goats may be diverted for human consumption, if milk replacer is used for rearing the kids of high yielding does. Thus, the gap of demand and supply of milk for human consumption may be shortened to some extend.

The milk replacers developed at CIRG were found palatable, as the total DM intake was >3 per cent of body weight in Barbari kid (pre-weaned) tested under replacer/substitute feeding. Feed conversion efficiency was found better in milk fed kids, however, kid were maintained well in replacer fed groups and up to 24% CP containing replacer (Table 3) was cheaper than mother's milk. Among replacer fed groups of kids the group having 24% CP milk replacer was found best in terms of FCE, as this is directly related to the economics of feeding. No mortality of kids was recorded under different milk replacer feeding regimen. The replacer/substitute having 24% CP may be used safely for early growing kids (2-13 weeks old) having minimum accessibility to mother's milk (poor yielder) and having higher multiple birth rate. For high yielding does surplus milk may be diverted for human consumption, if this milk replacer is used for rearing such kids. Thus, the gap of demand and supply of milk for human consumption may be shortened up to some extent (Dutta et al., 2003).

Table 5: Formulation of milk replacers.

Ingredients	Ratio	CP	Energy (kcal/100g)
Skim milk powder	47	16.44	168
Soybean Meal	9	3.92	39
Wheat flour	35	3.85	121
Coconut oil	7	0.00	63
Mineral mixture	2	0.00	0
Total	100	24.21	390

Followings are additional per kg of milk replacer powder

Butyric acid 0.2 ml, citric acid 2 g and Probiotics $1.5X10^8$ spores (Sporolac) per kg milk replacer powder.

Kids of early stage of growth are susceptible to many diseases if they are undernourished or maintained with poor nutrition. Therefore, under present system of management mortality of kids is recorded high. Henceforth, a suitable ration schedule is the need of the hour to maintain and improve growth performances of kids at an early stage of growth under intensive system of production. The energy and protein are thought to be basic nutrients. The diet must contain adequate protein, no other nutrient can substitute for it and efficient utilization of nutrients depends on an adequate supply of energy, which is of paramount importance in determining the productivity of goats (NRC, 1981).

Starter ration with CP 20% and TDN 65% was found most suitable for Barbari kids ageing 50 to 175 days during summer season in northern part of India. This ration contained maximum level of agro-industrial by-products (92%, pulses chunnis + wheat bran + oilseed cakes), thereby lowest cost of feed/kg body weight gain (Rs. 29.76) was obtained in this treatment groups. ADG (58.06g) and FCE (15.56%) were recorded better in early growing kids fed with this starter ration. During winter season, early growing Barbari kids (30 to 150 days) responded in a better way with starter ration having CP18% and TDN 65%. This ration also contained maximum level of agro-industrial by-products (91%, pulses chunnis + wheat bran + oilseed cakes), thereby, lowest cost of feed/kg weight gain (Rs. 38.10) was calculated in such kids with higher ADG (44.35) and FCE (13.81%) values. However, due to extreme winter stress kids attained little lower body weight gain as compared to summer trial. For early growing kids, protein was found most limiting factor, as 18-20 per cent CP in starter ration gave highest ADG and FCE and lowest feed cost/kg weight gain. Whereas, 65 per cent TDN in starter ration was sufficient to maintaining such growth rate in Barbari kids (Dutta et al., 2003).

b) By-products based broiler ration for goats

Development of economic ration schedule for faster growth of post-weaned kids is one of major important factor in commercialized farming. Pulses by-products based complete ration consisting CP 12% and TDN 60% was found economical to give >50 g ADG and > 10% FCE in post weaned (4-9 months of age) Barbari kids during summer season. Carcass traits and quality of value added product Kofta was found best under this ration schedule. The ration contained 72.8% agro-industrial and agricultural by-products (pulses chunn + wheat bran + rice polish + arhar straw) and had minimum proportion of conventional cakes (6%) and cereal grains. During winter season also, the pelleted feed having CP 12% and TDN 60% gave >50 g ADG in Barbari kids (5-10 months), whereas, FCE was found little less (7.63%) in this group. This could be attributed to diversion of feed energy for body temperature maintenance during extreme winter at Northern part of India. The ration was recorded economical in terms of cost of feed/kg weight gain. Quality of carcass parameters and sausage was found best under such treatment in Barbari kids. The goat breed under study had the potential for meat production as they attained about or above 10 kg carcass weight at the age of 9 or 10 months when proper combination of CP and TDN was maintained in their ration (Dutta et al., 2003). It was inferred that, complete feed pellets of 60:40 roughage concentrate ratio could fulfill the maintenance and growth requirement of goats (Rekhate et al, 2004).

Effect of feed processing on goat production

Eight crop residues (maize straw, sorghum straw, maize cobs, cottonseed hulls, cotton straw, sunflower, ground nut hulls and sugarcane baggasse) and 2 forest dry grasses (Sehima nervosun and Hetoropogan contortus) were subjected to physical processing like chopping grinding and stream pelleting, and the effect of processing on the nutritive value, rumen fermentability and rumen degradability on goats and sheep reported. The buck density of chopped residues increased considerably (32 to 88%) on grinding and further improved (54 to 236%) on pelleting. Pelleting improved dry matter intake and total digestible nutrients in all the feed compared to chopping and grinding. Goats utilized fibrous residues more efficiently than sheep, however sheep consumed more dry matter than goats. With complete rations containing crop residues even to the extent of 40 to 50% optimum growth rates were recorded amongst local goats and sheep (Reddy and Raghavan, 1992).

Hydrolyzed feather meal (HFM) was identified as a less rumen degradable protein source for lactating dairy goats. Meat and bone meal has been suggested as an excellent source of less degradable protein in the rumen, calcium and phosphorus. Integration of meat and bone meal with urea is an effective strategy for economical milk production. Christopher and Rubino, (1992) further stated that hydrolyzed feather meat (HFM) can be utilized as a less rumen degradable protein source for lactating dairy goats. Feeding of hydrolyzed animal fat can be considered as an alternative to alleviate problems associated with fat depression and off-flower in goat milk. Urea can be fed at 33% of dietary N in diets containing 65% TDN without adversely affecting milk yield and should be considered as an important nitrogen source for goats. Potential for altering composition of meat and milk by nutritional manipulation is tremendous and merit further research investment. Mechanical processing or chemical treatment can enhance utilization of crop residues.

The apparent comparable performance observed in goats fed sun dried pelleted Leuceana foliage suggests the possibility of its long-term use as a sole feed for the maintenance of stall fed goats without any significant deleterious effect (Srivatava and Sharma, 1998).

There is very little scope for enhancing the area under fodder production keeping in view of the priority of food grains and pulses production. It is therefore necessary to increase the productivity of available forage resources per unit area by incorporating the improved variety and also to introduce the suitable forages on fallow lands in the existing grain crop rotation of small and marginal farmers, and to increase the biomass production in community wastelands, village common grazing lands, and marginal lands owned by farmers. Stress should be given to incorporate maximum amount of by-products (agricultural and agro-industrial by-products) in the diet of goats as conventional ingredients are generally used for milk production from cows and buffaloes. Conservation and processing of feed and fodder is also the need of the hour to improve the utilization of existing resources for getting higher productivity of goats in terms of milk, meat, wool and skin. The intensive system provides shelter in hygienic conditions, ensuring provision of controlled feeding, medical care and clean drinking water. Goat farming under intensive system can be more economical if the locally available feed ingredients are subjected to cost effective processing.

Bibliography

Annual Report. 2003-2004. Central Institute for Research on Goats, Makhdoom, Farah, Mathura, UP (281 122).

Bhattacharya, N. K. 1989. Management of fodder resources for sustainable livestock production. In: Proceedings of lecture notes on "Fodder and pasture resource development in the sub-tropical and remperate region of the eastern Himalayas" Documentation Centre, Himalayan Pasture and Fodder Research Network, Krishi Bhawan, New Delhi. pp 14-19.

Christopher, D.L.U. and Rubino, R. 1992. Recent advancements on availability and utilization of feed resources for goats in north America and Europe; In: Pre-conference proceedings of Invited Papers. Vol II, pt I, pp 105-120. V International Conference on goats, New Delhi, India.

Deb Roy, R. 1989. Pasture and forage production strategies under silvi-pastoral systems. In: Proceedings of lecture notes on "Fodder and pasture resource development in the sub-tropical and remperate region of the eastern Himalayas" Documentation Centre, Himalayan Pasture and Fodder Research Network, Krishi Bhawan, New Delhi. pp.51-62.

Dutta, T.K., Sahoo, P.K., Singh, Nawab, Rao, S.B.N. and Chaudhary, U.B.2002. Partial replacement of concentrate mixture with Leucaena leucocephala leaves in pelleted feed of goats. Indian Journal of Animal Science, 72:820-822.

Dutta, T.K., Singh, Nawab, Rekhate, D.H., Reddy, T.J and Shettar, V.B. (editors). 2003. Final Report of NATP/RPPS-12 entitled "Utilization of by-products of pulses, oilseeds along with coarse cereals grains for intensive goat production" published by NATP/RPPS-12, CIRG, Makhdoom, Farah, Mathura, UP.

FAO, 2000. Food and Agricultural Organization of the United Nation, Statistical Databases.

NRC, 1981. Nutrient requirements of domestic animals, No. 15, Nutrient requirements of goats: Angora, Dairy and Meat Goats in Temperate and tropical countries. National Academy Press, Washington DC, USA.

Reddy, M.R. and Raghavan, G.V. 1992. Recent advances on feed processing and complete feed for goats. Pre-conference proceedings of Invited papers. Vol II pt. I pp. 168-182, V International Conference on goats, New Delhi, March, 1992.

Rekhate, D.H., Raut, R.G., Dhok, A.P., Deshmukh, V.B. and Ambulkar, D.R. 2004. Utilization of complete feed pellets by goats. In: Proceedings of National Seminar on "opportunities and challenges in nutrition and feeding management of sheep, goat and rabbit for sustainable production", held at CSWRI, Avikanagar,Raj.

Sharma, K. 1989. Potentialities of silvi-pasture development in problem soils. In: Proceedings of lecture notes on "Fodder and pasture resource development in the sub-tropical and remperate region of the eastern Himalayas" Documentation Centre, Himalayan Pasture and Fodder Research Network, Krishi Bhawan, New Delhi. pp 23-33.

Singh, Khub and Ramachandra, K.S. 2004. Feed and fodder availability, requirement and strategy for meeting the shotage in different regions for small ruinants. In: The proceedings of National Seminar on "Opportunities and challenges in nutrition and feeding management of sheep, goat and rabbit for sustainable production" held at Central Sheep and Wool Research Institute, Avikanagar-304 501, Rajasthan (India) on 10-12 February 2004.

Singh, Nawab and Rao, S.B.N. 2004. Development and evaluation of tree leave based pelleted feeds for goats. In: Proceedings of workshop on "Design and development of indigenous equipments for feeding, manufacturing, preservation and packaging of animal feed", organized by College of Veterinary Science and Animal Husbandry, UP Pt. Deen Dayal Upadhyaya Pashu Chiktsa Vigyan Vishwavidyalaya evam Gau Anusandhan Sansthan, Mathura. pp 80-84.

Srivastava,S.N.L. and Sharma, K. 1998. Response of goats to pelleted diets containing different proportions of sun dried Leucaena leucocephala. Small Ruminant Research, 28:139-148.

CHAPTER - 18
FEED PROCESSING FOR CAMEL

G.P Singh and A.K Nagpal

Camel is known as ship of desert. It has become the symbol of Rajasthan as pride and prestige of people living in villages and attracts a large number of tourists from European and North America, which boost tourism in Rajasthan. Normally camels are used for drought; ploughing and many others to day-to-day work. In desert, above works are possible by camel only. Other animals are not able to do such type of work. Camels can survive in desert condition where climate is harsh and there is always limited food as well as water scarcity, but camels have certain special feature which enable them not only to survive but also work for carrying camel cart, ploughing of field, bringing water and woods etc. without any difficulties. Therefore, it will be essential to know the special features of camels to decide the camel feed and its processing for improved productivity.

Special quality of Camel

1. Camel's body structures are such that it can tolerate heat and dust very comfortably.
2. Pads on camel feet help in carrying load on sand of desert which other animal can not do.
3. Camels changes body temperature according to climate. It has been reported that camel body temperature rises with the rise of day temperature and temperature change is 3 to 5°C.
4. Camel digestion system specially rumen digestion is very efficient than other ruminant (Singh etal. 2004). This is the reason that camel can eat and digest the most rubbish feed which other animal can not consume. Therefore, camel can survive in very adverse situation.
5. Digestion of fibre and highly lignified feed are very high as compared to other ruminants. Due to this camel digest Guar Chara (highly lignified feed) and extract maximum nutrients for maintenance and milk production.

6. Reason for digestion of fibre and lignified crop residue is higher cellulytic bacteria and fungi. Protozoa has almost no role to play in rumen of camel (Singh et al., 2004 and Singh and Saini, 2002)
7. Camel utilizes nitrogen of feed very efficiently because of highly efficient nitrogen recycling. Losses of nitrogen in urine is very minimal. Nitrogen of diet recycled in rumen in the form of urea through saliva, which act as a source of nitrogen for microbial protein synthesis in rumen of camels.
8. pH in rumen of camels is reported to the higher that other ruminants, this is indication of efficient ammonia utilization, secondly variation in rumen pH during 24 hrs is minimum.
9. Volatile Fatty Acids production is higher in camels than in other ruminant.
10. Acetate in volatile fatty acids is always higher than other animals because of fibrous feed in diet.

Camels are usually fed legume straw, cereal straw, grasses, shrubs and tree's leaves in addition to grazing. Camels move considerable distance in search of forages. In dry period, camels are taken for forages outside the village or even beyond the regular foraging area. However, when rain comes camels are brought to their original village. Camels move about 50 to 70 km daily to obtain their feed under poor range condition and camels graze slowly when plants are thorny (Gauthier- Pilters, 1979). On the other hand grazing land has deceased with very fast rate (Danik Bhaskar, 2001) and competition for grazing and fodder with other productive animals like cattle sheep and goat in arid zone has created problems for camel. Stall-feeding of camels is increasing and the most rubbish fodder is used for camels feeding. Frequent drought is another problem, which affects the camel population of India. Due to poor quality of feed and fodders, processing/treatment will increase nutrient utilization.

Common feed and fodder for camels

Deciduous shrubs, trees, grasses and crop residues are the major components of camel diets in both dry and wet seasons. During winter season, all the grasses are dried and shedding of tree leaves take place. Therefore, practically nothing is available for the grazing except thorny stem, pala, Moth chara, groundnut straw, guar phalgati are the main feeds available for feeding of camel. During summer season camel consumes Phog, Khess, Bawali, Khejri, and Jal. Neem Species are also available. Scenario is changing fast. During last 10-15 years lands are being brought under cultivation, therefore, availability of trees and shrubs is decreasing and crop residues are now becoming main fodder for camel. During Rainy season if rains are sufficient, green grasses like Bhusal,

Anjan, Sevan, Bokoria, Gokhru etc. are available for a short period in addition to moth, guar, bajra fodder. Dried leaves of Khejeri, pala, Phog is also kept for feeding during the rainy season.

Major feed and fodder of camels contain high crude fibre and lignin, which is a major source of energy in ruminants including camel (Table 1). Camels are classified as browsers as they prefer to browse and they are more efficient users of woody shrubs and tree leaves containing higher fiber content.

In General, camels diets have a large fibre and lignin content. The level of neutral detergent fibre (NDF) and acid detergent lignin (ADL) are slightly lower in overall average of forages in the dry season than in wet season (Elmi, 1989). Fibrous materials are retained longer in rumen thereby subjected for fermentation for longer period and produce more fermentation product (Heller etal, 1986) accompanied by efficient nitrogen utilization through urea recycling.

Rumen fermentation: Very limited literature is available on camels rumen fermentation. In India too only preliminary work is initiated. The pH of rumen fluid in dromedaries was 6.7 which is higher than that of sheep (6.5) and goats (6.4). Similar results were found by (Maloiy, 1972; Kayoulli et al., 1991; and Kayouli et al., 1992). The pH of rumen fluid seemed to decrease during day for all animal species. However, such a decrease was less marked in dromedaries. The pH stability observed in dromedaries might be explained by the higher buffering capacity of camelids than that of sheep and goats (Vallenas and Stevens, 1971).

The volatile fatty acids (VFA concentration (m mole/1) was significantly higher in dromedaries than in other species. The VFA concentrations were 81.8, 65.7, and 69.1 in dromedaries, goats, and sheep, respectively, which agrees with the levels reported by Kayouli et al., (1991, 1992). Molar proportions of more important VFA were similar among all animal species except for acetic acid. For the butyric acid, molar concentration were 9.3, 8.1, and 8.0 for dromedaries, goats and sheep, respectively. Similar levels were reported by Ali A. (1990), Lechner-Doll et al., 1991) and Kayouli (1991). The butyric acid molar concentration variation among species were attributed to the low activity of the 3 hydroxybutyrate dehydrogenase in dromedaries, responsible for transforming butyrates to ketone bodies in the rumen wall (Emmanuel. 1979; Chandrasena, 1979).

The concentration level of (NH_3-N) in the rumen liquor was significantly lower ($P<0.05$) in dromedaries than in sheep and goats. This result confirms previous reports (Ghosal et al. 1981 and Kayouli et al 1992). These finding might be explained by:

- A more important microbial activity in dromedaries than in sheep and goats.
- A rational utilization of ammonia.

It was also shown that camelids recycle indigenous ammonia mainly when diets are based on forages poor in soluble ammonia (Emmanuel, 1976; Von Enghelhardt and Schneider, 1977 and Von Engelhardt, 1978).

The cellulolytic activity in dromedaries evaluated by the ADF degradability is comparable to that of other ruminant. Furthermore, the DM degradability is more important in dromedaries than in other animal species. These results are in agreement with finding reported by Hintz et al. (1973) Farid et al (1979) and Gihad et al (1988).

The DM intake is lower in dromedaries than in other ruminant species. The pH level of the rumen liquor in dromedaries was in a range favourable for important microbial activities. Even through the concentration of (NH_3-N) was low and that of VFA was high, the PH values were relatively stable. This might be explained by the buffering capacity of rumen content of dromedaries. The ability of dromedaries to digest DM and ADF as compared to goats and sheep shows a better utilization for forages rich in cell wall.

Fungi play a very significant role in the digestion/fermentation of fibrous feed, while fungi is absent or low in number on low fibre diet. Protozoa has no role to play for the digestion of fibrous feed. Therefore, it is only bacteria and fingi responsible for the fermentation of fibre in the rumen. Fungi play a leading role in initiating the process of fermentation of fibre by partially digesting and creating sites for bacterial attachment to fibre. Therefore, it is beyond doubt that fungi plays a very important role in camel productivity and role of fungi can be increased by suitable biotechnological studies and by introducing the different combination of fungi for maximum digestibility. (Singh and Saini 2002).

Nutrient Requirements of Camels

Camels are most neglected animals because of their today utility and no serious attempt has been made to determine the nutrient requirement of camel. Requirements of growing and adult camel have been calculated based on NRC data and compared to requirement published in ICAR 1985 publication. The published requirement (Table -2) seems to be very high than cattle and buffaloes. It is quite obvious that due to climatic condition, efficient microbial digestion and efficient rumen ammonia utilization, the requirement of camel is comparatively lower than other species. Therefore, serious efforts are required in this direction.

Table-1: Common feed and fodder of camel and their composition (on % DM basis)

FEED & FODDER	CP	CF	Total CHo
STRAWS			
Moth Chara (Phaseouls aconitifolius)	9.7	25.2	73.6
Missa Bhusa/Gram Straw (Cicer arietinum)	6.5	43.0	76.5
Groundnut Straw (Arachis-Hypogea)	8.11	40.63	80.5
Guar Phalgati (Cyamopsis tetragonoloba)	8.9	34.1	75.7
Bajra Straw (Pennisetum typhoideus)	2.6	37.15	85.6
GRASSES			
Seven grasses (Lasiurus sindicus)	3.1	36.8	84.4
Bakari (Indigofera Cordifolia)	6.2	28.3	73.1
Anjan /Dhaman (Cenchrus ciliaris)	8.3	30.54	73.8
Lampa (Aristrida depressa)	2.1	42.25	84.2
Gokhru (Tribulles terrestris)	19.11	16.75	62.1
Dub (Cynodon dactylon)	10.47	28.17	75.4
Gramna/Blue pannic (Panicum antidotale)	7.26	40.47	83.6
Makara (Dactyocternium-aegyptiancum)	8.32	35.19	75.7
Bui (Aerua-tomentosa)	5.0	32.1	82.4
Karad (Dichanthium annulatum)	7.39	28.17	75.8
Grass mixture	4.39	32.43	84.0
SHRUBS AND TREES			
Babool (Acacia arabica)	7.04	33.32	84.5
Phog (Calligonum-Polygaonoides)	6.5	30.2	79.1
Siras/Sares (Albizzia Lebbeck)	16.81	31.52	67.7
Peepal (Ficus religiosa)	9.02	15.93	68.3
Khejeri (Prosopis spicigera)	14.7	15.14	68.1
Khejri Loong	11.9	17.5	61.0
Jhar beri (Zizyphus nummularia)	11.53	33.82	80.6
Pala	13.8	30.7	74.9
Kikar (Accacia nilotica)	15.98	7.50	69.7
Khair (Acacia catechu)	13.03	22.55	73.5

Source:- Singh & Saini (2002)

Table-2 : Nutrient Requirement of Dromedary Camels

GROWTH								
Body wt.	**ICAR (100 g/d)**				**NRC (300 g/d)**			
(kg)	**DCP (g/d)**	**TDN (kg)**	**Ca (g/d)**	**P. (g/d)**	**DCP (g/d)**	**TDN (kg)**	**Ca (g/d)**	**P. (g/d)**
175	-	-	-	-	210	1.90	15	10
200	250	2.00	80	30	242	2.15	18	13
250	325	2.50	100	35	283	2.43	22	15
300	350	3.00	120	50	308	2.76	27	19
350	470	3.80	145	60	321	3.07	31	22
400	600	4.70	185	80	332	3.36	36	27
450	650	5.00	200	100	328	3.69	45	33
Maintenance								
500	500	5.50	300	100	241	3.42	24	18
550	525	6.00	225	120	259	3.70	25	19
600	550	6.50	250	150	275	3.99	27	20
650	600	7.00	250	150	295	4.18	28	22
750	650	7.50	250	150	312	4.46	30	23

Protein and Urea supplementation and nutrient utilization in camels

The digestibility coefficient of CP were 54.36, 60.13 and 51.10 percent and the CP digestibility was significantly higher ($P<0.05$) on supplementation of khejri leave. Crude fibre digestibility was significantly ($P<0.05$) lower in camel fed on guar chara as sole ration. Low fibre digestibility could be due to high lignification of guar stem which was major part in guar chara. However, supplementation of urea and khejri leaves increased the CF digestibility because of catalytic effect of supplements on the ruminal microbial activity due to proper ammonia level in rumen and other factors. On the contrary NDF digestibility of ration was higher in group fed guar chara as sole ration than those observed on supplementation of 2% urea or khejri leaves to guar chara. ADF showed similar trend in respect of digestibility due to supplementation of 2% urea or khejri leaves. (Singh & Saini. 2001).

The cost of feed per day in three groups were Rs. 13.56, 16.77 and 12.00 respectively on urea supplementation, khejri leaves supplementation and guar chara as sloe feed. Cost of per kg DOM in respective groups were Rs. 3.52, 3.73 and 3.38. Cost of DOM was lower on urea supplementation (Rs. 3.64) than as khejri leaves feeding (Rs. 4.00) and guar chara as sole feed (Rs. 3.96).

Although cost of feed was lower on guar chara but the cost of available nutrient was cheaper on urea supplementation. Thus urea feeding was economical in terms of nutrient availability which is main index for productivity than a simple DM.

Feeding of ground nut chara in combination with guar chara has shown beneficial effect in camels. The total dry matter intake of camels were 6.84, 7.20 and 7.80 kg/day ($P<0.05$), respectively in guar chara, guar + GN chara and ground nut alone fed camels. Similarly, dry matter intake as percent of body weight in respective grounds were 1.65, 1.71 and 1.86 ($P<0.05$). Total increase in body weight were 4.50, 7.30 and 10.20 kg in 27 days, respectively while, increase in body weight per day basis were 166.67, 270.37 and 377.78 in groups in respective groups.

Intake of digestible organic matter in groups I, II and III were 3.62, 3.96 and 4.54 kg/day, respectively and CP intake was 240,327 and 447 g/day indicating significantly higher intake of DOM and DCP due to groundnut chara.

In this experiment 56.52, 45.64 and 36.22% of dietary nitrogen recycled on guar chara, guar chara + groundnut chara and groundnut chara, respectively. Requirement of feed per kg body weight gains were 41.42, 26.63 and 20.65 in three groups respectively. Requirement of feed/kg gain was significantly lower on feeding of guar chara + groundut chara and groundnut chara as sole feed. Amount of DCP required per kg gain was also significantly lower on groundnut chara (1183g) as compared to guar chara + groundnut chara (1209g) and guar chara as sole feed (1440g). Similarly DOM requirement was 21.72, 14.64 and 12.02 kg/kg gain ($P>0.05$). Thus the feed utilization efficiency was higher on groundnut chara feeding as sole ration (Singh etal., 2004).

While cost of feeding was Rs. 12.41, 16.23 and 21.03 per day in groups I, II and III respectively, the cost per kg gain were 74.50, 60.04 and 55.67, respectively on feeding of guar chara, guar chara + groundnut chara and groundnut chara as sole feed.

Processing method

More than 30 processing methods have been developed experimentally or practically to improve the intake and digestibility of crop residues. The processes include physical, chemical and biological methods. Most of the methods have been technically successful and have been exhaustively investigated.

Physical methods

Physical treatments include soaking, grinding, pelleting, extrusion, steam under pressure and irradiation. Chaffing is a less severe processing method and aimed at preventing selection of highly nutritive tender parts by animals, to reduce the particle size and to prevent feed wastage. In comparison to grinding and chopping, expander-extruder processing and steam pelleting markedly increased the digestibility of nutrients and nutritive value of many crop residues (Reddy, 2003). In order to utilize crop residues in the rations of ruminants, they have to be subjected to at least a minimum processing like chopping/grinding. However, in camels most of crop residues like, guar chara, groundnut chara, moth chara etc. are fed after chaffing but a comparative study on particle size is very much required.

Complete diet system

The concept of complete feeds for feeding ruminants is relatively recent origin in India. In this system, all feed ingredients inclusive of roughages are proportioned, processed and mixed into a uniform blend, which is freely available to the animal to supply adequate nutrients. The product is fed as sole source of nutrients. This system ensures the supply of balanced nutrients, controls the ratio of concentrate to roughage, helps in improved utilization of low grade fibrous agricultural residues, reduces feed wastage and also reduces feeding costs. It also promotes better consumption and avoids refusal of the unpalatable portions of the feedstuffs.

One of the main advantages of this system is that it provides a uniform mixture of processed feedstuffs (concentrates and roughages) in a form, which prevents selection. Consumption is more widely spread in this system than with the conventional one. Higher intake of feed is also associated with less fluctuation in the release of ammonia so that non-protein nitrogen may be more efficiently utilized and may lead to an increase in the . general utilization of protein. This system gives good scope for increasing the use of home grown fibrous agricultural residues, by-products and animal wastes.

Mash: The different ingredients of the diet have to be proportioned as per formula into l00 Kg batches using appropriate scales. The ingredients required for grinding is ground in a hammer mill. The ground ingredients come to mixer through screw conveyor and bucket elevator and hopper above the mixer. The ingredients that do not require mixing is directly added to mixer. The micro-ingredients (vitamins, antibiotics etc.) are made into a premix by diluting with ground grain or bran and required quantity of premix are added to the mixer directly. Later, all the ingredients are mixed for about 10 minutes and then is collected in sacs and stored.

Pelleting : For pelleting, the mash from the mixer is dropped into bucket elevator and is lifted and conveyed into a hopper over the pellet mill. The mash feed is conveyed from the hopper into conditioning chamber of the pellet mill through a screw conveyor. The rate of flow of the feed into the conditioning chamber is controlled by wheel valve. The steam produced from a boiler attached to pellet mill is supplied into the conditioning chamber of the pellet mill. Required quantity of steam with 97-98°C temperature is supplied into conditioning chamber through control valve. Conditioned mash at 90-92°C temperature and with 16-17% moisture is conveyed to pellet mill.

Complete ration based on Ardu leaves, Neem leaves and Khejri leaves ware tried and compared with guar phalgali. The DCP intake /kg $W^{0.75}$ was significantly ($P<0.01$) lower in control diet (C) obviously due to lower crude protein content in guar phalgati. No statistical difference in TDN intake /kg $W^{0.75}$ was observed among 4 rations indicating even guar phalgati could be comparable to complete rations for providing energy but not sufficient protein requirements. Comparison with recommendations revealed that all the 3 rations could very well meet the energy and protein requirements of camel calves. It may be inferred that all the complete rations could meet the protein and energy requirements of camel calves CR I (Ardu) being the cheapest at Rs. 270/q and CR III (Khejri) being the costliest at Rs. 290/q due to inclusion of higher

quantity in the complete ration.

Similarly in another experiment, camels were fed guarphalgati and Ardu leaves (84:15), guarphalgati and neem leaves (80:20) and guarphalgati and khajri leaves (76:24). Nutrient intake in terms of DM/ kg $W^{0.75}$ was higher ($P<0.05$) in khejri than Ardu and neem fed groups and in terms of DCP and TDN it was similar among 3 camels weighing 350 kg growing @ 100 g recommended, the daily intake of DCP (g) in the present experiment varied between 382 g in neem to 413 g in khejri whereas TDN intakes varied between 3.36 kg in neem and 2 and 3.63 kg in kharji with growth rates ranging from 241 to 311 g/d. It is seen that in the present experiment both protein and energy intakes were lower coupled with higher growth than the ICAR (1985) recommendations.

Chemical methods

Most popular chemical treatments that have been undertaken in the country include use of sodium hydroxide, anhydrous ammonia treatment. aqueous ammonia treatment, hydrogen peroxide and urea-ammoniation techniques.

One of the oldest chemical methods was soaking straw in a 1.5 % NaOH solution over night and then rinsing in fresh water (Homb et al., 1977). This technique was further modified later in which straw was placed in a chamber/ container and NaOH solution was sprayed over straw for 7 to 8 hr (Homb,

1984).

The technique of treating straws with the anhydrous ammonia includes use of NH3 at

30g kg-I of straw followed by 8 weeks (Sundstol and Coxworth, 1984). Aqueous ammonia treatment was also extensively tried using 120g of 25 % NH3 solution per kg straw with a storage period of 8 weeks (Sundstol and Coxworth, 1984). Anhydrous ammonia and aqueous ammonia have the same effect in improving the nutritive value of cereal straws

Urea - ammoniation

Urea is generally the most practical and often the cheapest chemical available for treating straws in many areas of the tropics. The mechanism of urea-ammonia treatment is basically the reaction of ammonia released from urea with the fibres of the straw. Ammonia in the presence of water produces the alkali, ammonium hydroxide, which in turn reacts with the fibre components in the cell wall. The recommended method for urea treatment of paddy straw is 4kg urea dissolved in 65 liters of water has to be sprayed with a garden sprayer on 100 kg of paddy straw (lCAR, 1985) and the treated material has to be stored for 21 days in anaerobic conditions either by ensiling or by stacking.

Feed Block for camels

The camels were given complete ration consisting of chane ki khar, groundnut chara and concentrate in the ratio of 70.3:15.0:14.7 in loose form (L), complete ration in the form of feed blocks (FB) and also given complete ration consisting of chane ki khar, groundnut chara and concentrate in the ratio of 60.3.:25.0:14.7 in feed block form (FB-2) for 98 days. Growth rate as well as feed gain ratio were similar among L, FB-I, FB-2. Average daily gain ranged from 377.55 g/d in L to 420.92 g/d FB and feed/gain ratio from 13.15 in FB-1 to 14.30 FB-2. Feed intake was 5.60% higher in FB-1 and 8.42% in FB-2 compared to L due to higher feed density of complete feed blocks. Feeding cost (Rs./d) was higher in FB-2 because of higher feed intake but because of higher feed efficiency in FB, cost/kg gain was minimum (Nagpal et al., 2005)

The results indicated that chane ki khar alone is not sufficient for camel calves. No significant difference was observed among loose and complete feed block feeding to camels in respect of place of nutrition, growth and feed efficiency. Feeding of complete feed blocks consisting of change ki khar, groundnut chara and concentrate may be used for optimum nutrient utilization and growth of camel calves of getting additional advantage of less storage space, low transport cost and ease in feeding managent.

Thus, it is concluded that camel's feeds are of very poor quality with low protein and high fibre but camel can digest these poor quality roughages because of efficient rumen fermentation. Although processing technology of poor quality roughage are available but these technologies were applied to very limited scale. Processing of camel's feed (Crop residue) will prove very effective under efficient rumen digestion of camels.

References

Ali, A. 1990. Comparison des parametres physicohimiques de la digestion et de retention des particules alimentaires dans le rumen du dromadaire et du mouton nourris a base de 3eme cycle, I.N.A. Tunisie.

Chanddrasena, L.G., B. Emmanuel D.W., Hamar, and B.R. Howard. 1979. A comparative study of ketone body metabolism between the camel (Camelus dromedarius) and the sheep (Ovis aries). Comp. Biochem Physiol. 64b : 109.

Danik Bhasker 2001. SIKUDTE CHARAGAHAN SE PARESHAN HAIN OOT PALAK, BIKANER, 13-12-2001.

Elmi. A.A. 1989. Management, foraging behavious, composition and forage quality of free ranging but herded camels in Ceeldheer District, Central Somalia. Ph.D. Disserattion. Utah State University, Logan, 221 pp.

Emmanuel, B. 1979. Comparative metabolic studies between ruminant species. Ann. Rech. Vet. 10:480.

Emmanuel, B., B.R. Howard, and MM. Emady. 1976. Urea degradation in the camel. Chand J. of Anim. Sci. 56:595.

Farid, M.F.A., S.M. Shawket, and M.H.A. Abdelrahman. 1979. Observations on nutrition of camels and sheep under stress. In the camelid and all purpose animal., volume 1, Rosscockrill. W. ed. Scandinavian institute of africam studies. Uppsala. 29.

Gauthier – Pilter, H 1979. Some ecological aspects of the camel in the Eastern Sahara. In: Corkrill, W. (ed). The Camelid: An All Purpose Animal. Proc. Worksop on Camels. Khartoum. December, 1979. The Scandinavian Institute of African Studies. Uppsala. PP 412-430.

Gihad EGA., T.T. El-Gallad A.E. Sooud, H.M. Abou-A1-Nasr, and M.F.A. Farid. 1988. Feed and water intake digestibility and nitrogen utilisation by camels compared to sheep and goats fed low protein desert by-products. In – J.L. Tisserand (ed digestion, la nutrition et l'aimentation du dromadaire. Ouergla (Algerie): 28 fev-1 mars, CIHEAM, A2:75.

Ghosal., A.K., R.K. Auner, and P.K. Dwaraknath , 1981. Note on rumen microorganisms and termination pattern in dromedary. Ind. J. anim. Sci. : 1011. India.

Hintz, H.F.H.F. Ryver, and M. Halbert. 1973. A note on the comparison of digestion by new world camels, sheep and ponies. Anim. Prod. 303.

Holler, R., M.Lechner-Dool, H. Weyreter, and W. Von Engelhardt. 1986. Forestomach fluid volume and retention of fluid and particles in the gastrointestinal tract of dromedarious (camelus dromedarius) J.Vet. Med., A33:3 396.

Homb T., Sundstol F. and Arnason. 1977. Chemical treatment of straw at commercial and form levels. Proc. Tech. consult. New Feed Resources. FAO Rome. pp. 25-37.

Homb T. 1984. Wet treatment with sodium hydroxide. In straw and other fibrous by products as Feed. (Ed: F. Sundstal and F. Owen). pp. 106-126.

ICAR (1989) Nutrient requirement of livestock and poultry. Indian Council of Agricultural Research, New Delhi.

Kayouli, C., J.P. Jouany, and J. Ben Amor. 1991. Comparison of microbial activity in the forestomach of dromedary and sheep measured in vitro and in sacco on mediterranean roughages. Anim. Feed. Sci. Tech., 33:237-245.

Kayouli, C., J.P. Jouany, ALI., Demeyer, A. All-All, H.Taoueb, and C. Dardillat, 1992. Comparative Studies on the degradation and mean retention time of solid and liquid phases in the forestomach of dromaderies and sheep fed on low quality roughages from Tunisia. Anim. Feed. Sci Tech.

Lechner-Doll, M., M. Kaske, and W. Von Englhardt. 1991. Factors affecting the mean retention time of particles in the forestomach of ruminants and camels. In : Physiological aspect of digestion and metabolism in ruminant. T. Tsuda, Y. Sassaki and R. Kkawashimma (eds). Academic press, Tokyo (Japan), 455-483.

Maloiy, G.M.O. 1972. Comparative studies on digestion and fermentation rate in the forestomach of the one-humped camel and the zebu steer. Res. veto Sci. 13:476.

Nagpal A.K, Manju Arora and G.P. Singh 2005. Indian J. Anim. Sci., 75 59-63.

Reddy G.V.N. 2003. Effect of processing of crop residues on physical characteristics and nutritive value. In short course on "Feed Processing Technology' held at ANGRAU, Rajendra Nager, Hyderabad. pp. 199-214.

Singh G.P. and Saini. N. 2001 Annual Report of National Research Centre on Camel, Bikaner (2000-2001).

Singh G.P. and Saini. N 2002. Role of anaerobic fungi in fibre digestion and its special significance Camel Nutrition. Indian Dairyman, 54. 64-68.

Singh G.P., A.K. Nagpal and N. Saini 2004. Proc. VI Animal Nutrition Conference Jabalpur. pp. 216-222.

Sundstal F. and Coxworth E.M., 1984. Ammonia treatment. In: Straw and Fibrous By – products as Feed. Elsevier Science Publishers, Amsterdam, The Netherlands.

Vallenas, A. and C.E. Stevens. 1971. Motility of the llamas and guanaco stomach. Ann. J. Physiol., 220:275.

Von Engelhardt, W. 1978. Adaptation to low protein diets in some mammals. In : Zodiac symposium on adaptation, pp 110- 115. Center for agricultural publishing and documentation, Wageningen.

Von Engelhardt, W. and W. Schneider. 1977. Energy and nitrogen metabolism in the llama. Anim. Res. develop., 5:68.

CHAPTER - 19
FEED PURCHASE

Nand Kishore

Economical and balanced feed formulation for livestock feeding is the main objective of the feed-manufacturing units. The purchase of feed ingredients forms the major part of the investment in working capital and determines the margin of profit in the feed plants. In recent years, the purchasing function has gained significance and has become an important component of the total management. In general the principles of purchasing any item remains almost the same, but the purchase of feed ingredients involves much more complexity and poses many problems and challenges because of frequent price fluctuation, erratic market behaviour, susceptibility to adulteration and quality deterioration. Therefore, a close co-ordination among purchase executive, production manager and quality control officer is highly essential for efficient purchasing. Ingredient variation and lack of adequate standards poses a real problem in purchasing. Tremendous opportunities exist for feed industry to save ingredient and finished feed costs without sacrificing quality. This can be done by establishing purchasing guidelines based on nutritive value rather than on test weight, moisture, colour, texture and foreign material. Some of the important factors which would serve as guidelines for efficient purchases of livestock feed raw materials are:

Planning for the purchase of raw materials

Planning is highly essential to provide timely supplies of materials at a relatively cheaper rate and thus ensure a regular supply of low cost finished products to the customers. The organizations that do not plan are likely to face many problems such as, non-availability of critical feed materials, which either renders the production facilities idle or forces the management to make off-season purchases at higher rates. Sometimes, there is overstocking of raw materials which also takes away the margin of profit due to heavy capital investment and storage cost.

The following information must be collected and analysed properly which would help in planning for purchasing different raw materials on sound lines for good results:

i) Annual requirements

For making a successful purchase strategy, the total annual requirement of different feed ingredients should be carefully estimated based on the previous year's consumption and the current year production requirements. It is also important for the purchase executive to know the peak season of availability of different feed ingredients so that the purchase can be made at a cheaper rate.

ii) Finance

Purchase programme cannot be effectively implemented without providing adequate funds to meet the total cost of feed ingredients required to meet the total annual requirement of the production manager. The purchase executive, therefore, has to estimate the total funds required and split up budget at different times of the year based on average prevailing market rates, well in advance. This will help the management not only to provide necessary funds at the proper time, but will help to utilize extra funds somewhere else for better returns. This is important because a number of critical raw materials are of seasonal nature and such items are to be purchased in the peak season of their availability to purchase them at reasonable price. Therefore, off season purchase of seasonal raw materials, which cost heavily in off season, can be avoided if sufficient funds are made available at the disposal of the purchase department.

iii) Market information and market analysis

For arranging timely and competitive purchase of raw materials, it is highly essential to know the major sources of availability of materials in the market and the price trends in different areas. There is a great variation in the price of raw material from one region to the other due to many reasons. It is necessary for the buyer to locate the key market production centers of various raw materials to affect best purchase. This can be achieved by making suitable arrangements with different trade associations. It is, therefore, necessary for the purchase executive to establish a net work of suitable channels to get all such information in time regularly so that he can properly plan to effect purchases to the best advantage of the organization. It may also be pointed that the quality of raw materials differs from place to place and for buying the best quality raw materials, the purchaser should know the regional characteristics of all the raw materials. The entire process of purchase is the critical analysis of the market information and finally the common sense of the purchase executive.

iv) Identifying and developing of suppliers

It is extremely important to identify reliable and honest supplier and developing good relationship with them for making best purchases. You can identify a good supplier by personal contacts enquiring from different sources and based on previous performance records of the suppliers. A supplier should be identified keeping in view, honesty and reliability potentials to supply raw materials at competitive rates. Once good suppliers are identified, the purchase executive should win their confidence and good will by good behaviour, honest and fair dealing and making timely payments. It is emphasized that it is only good supplier relations that ensure emergency supplies of raw materials when sales and production forecasts go wrong. Moreover, when there are chances of minor disputes due to quality or shortage/excess of raw materials, it is only supplier good will that will help the buyer in amicable settling these matters. So proper care and attention need to be given for identifying, developing and retaining good suppliers.

v) Transport

Due to high costs the transport has become costlier. There is a scope to save some money in the transportation of raw materials. In an attempt to save transportation cost, there is need to locate purchase centers near the feed factory. Railway transport is relatively cheaper. Therefore, purchase centre having rail head facilities may be preferred. Possibilities of having their own trucks or having rate contract with private transport company may also be explored.

vi) Government Policies

It has been observed that government restrictions/controls on interstate movements of raw materials and also government policies with regard to export or import of certain commodities, affect the market rate of many feed ingredients. It is, therefore, highly desirable to collect information on these aspects, which would help to modify the purchase requirements of different feed raw materials.

vii) Substitute development

Due to government policies, natural calamities like drought and floods, some critical feed ingredients are not available and if available, their price goes very high. Therefore, some alternate feed ingredients should be identified. Moreover, it is also desirable to use new non-conventional feed ingredients in feed formulation, without sacrificing the quality of the finished products. It is, therefore, wise to include such exercises in the planning processes.

Quality Control

Low cost feed formulation may be the main objective of feed industry but not at the cost of quality of finished feeds. It is, therefore, highly essential to develop a sound and strict quality control programme for effecting the purchase of feed raw materials.

The first step in this direction should be to formulate simple and effective specifications for all the feed materials to be purchased. The specifications will not only help to purchase good quality material, but would also help to minimize the chances of adulteration, legal problems etc. It should be emphasized that the specifications of raw materials should have minimum parameters, which should be acceptable to the suppliers. As prices of raw materials are fixed on the basis of quality standards, quality analysis of the materials immediately on receipt is a must to safe guard the interest of the purchaser to purchase good quality material at competitive prices.

The quality control function should be independent of the purchase department. Close co-operation between quality control and purchase department is necessary for conducting smooth purchase operations. The quality control man should be encouraged to give his unbiast and independent opinion on the quality of all incoming materials.

Procedures of Purchases

Like industrial purchases, the purchase of feed raw materials does not confirm to any set pattern of buying procedures. Here one has to adopt different methods of purchase depending upon the necessity, availability, market trends, nature of the materials etc. Some of the commonly used procedures are given below:

i) Purchase through commission agents

Some of the grains like maize, barley, jawar and bajra arriving in grain markets during peak season of their production can be purchased at lowest cost through commission agents. Deal through an agent is necessary because feed industry men cannot enter into direct bidding without having proper licence. In this procedure the middle man profit is eliminated and the purchase is made direct from the farmers through the agent at nominal commission. In addition, purchase can be affected from grain stockiest through brokers. In this procedure, rates are negotiated and deals are finalized either across the table or on telephone.

In any case the purchaser should not entirely depend on the commission agents; he should enquire the prevailing market rates to ensure good purchase.

The most important thing to be looked into in this procedure is the selection of an honest and reliable commission agent.

ii) Spot purchase direct from manufacturers

It is cheaper to purchase feed materials like wheat bran, rice polish, deoiled rice polish, guar churi and different kinds of cakes, direct from the manufacturers of these raw materials. In this procedure also the middle men profit is eliminated. But the purchaser should have the skill of negotiating to reduce the rates. As a matter of fact the buyer should know the prevailing market rates and the expected market trend. Negotiation is an art and the purchaser learns only through experience.

iii) Purchase through tenders

This procedure may be convenient and useful when bulk quantity of feed raw materials are to be purchased It involves inviting of supply tenders advertising through news papers and mailing tenders direct to the main suppliers of these materials. While inviting tenders, the following points should be kept in view:

a) Exact quantity of raw materials to be purchased should be mentioned along with clear cut specifications.

b) The date, time and place of opening the tenders should be given.

c) Delivery schedule may be clearly mentioned giving the last date of supply.

   It may be pointed out here that regular supply of materials is highly essential for optimum production of finished products. The purchase executive, therefore, should carefully plan the delivery schedule of different raw materials.

d) The terms and conditions of payment should be clearly mentioned to avoid any further complication.

e) It may be clearly mentioned that the chairman of the purchase committee does not find himself to accept the lowest quotation and reserve the right to reject any or all quotations without assigning any reason.

f) In case of any dispute, the jurisdiction of the court may also be specified.

g) The validity period of atleast 7 days from the date of opening tenders should also be mentioned.

h) The security or earnest money in the shape of "Deposit at Call" in the name of chairman of purchase committee from any nationalized banks must be supplied along with tenders. The amount of security should be clearly mentioned.

i) Conditions of rebate/discount or premium should be clearly mentioned. In case of inferior quality percentage of deduction and in case of better quality premium will be given.

j) It should be clearly mentioned that if the material supplied is considered below standard (unfit for animal consumption) the supply will be rejected.

iv) Public Auction

Many a time damaged food grains as well as certain other materials unfit for human consumption are disposed off by different Government/Semi Government organizations by way of open public auction. Before participating in the public auction, random samples should be drawn from such lots of damaged food grains and tested to ascertain the quality and the suitability of the material. At times, this proves to be a good source of bulk purchase of quite a few important raw materials.

Legal aspects of purchasing

Because of the rapid fluctuations in the prices of a number of raw materials and also their susceptibility to adulteration and quality deterioration in transit, there is a possibility of dispute in many of the deals. Therefore, to protect the interests of the buyer, certain precautions are to be taken essentially, some of which are mentioned below:

i) Whenever deals are struck on telephone/telex or telegraphically. They should be immediately regularized and confirmed by placing written purchase orders stating therein all the terms and conditions including specifications. Purchase orders should be send to supplier in duplicate asking the party to return the duplicate copy duly signed by them in token of their acceptance.

ii) When purchase of any of the raw material is made as per sample, identically sealed samples should be kept with the supplier as well as the buyer.

iii) While taking delivery through truck, in case of any loss or damage, a written statement from the driver concerned should be obtained. While in case of delivery through wagons, same may be obtained from railway staff.

iv) As and when any material is found to be damaged or in shortage or inferior quality, the supplier should be intimated immediately.

v) 10 to 20% of the bill amount should be retained till the material is tested and found to be acceptable.

vi) In case it is intended to accept low standard material on rebate/discount or premium and the minimum acceptable quality limit should be clearly stated in the purchase orders.

CHAPTER - 20
FODDER AND FEED TRANSPORTATION

B.D. Shukla and S. Mangraj

The main feed resources in India include native grasslands, cultivated fodders and trees, crop residues, agro-industrial by-products, and non-conventional feed resources. There has been a gradual reduction in grazing resources due to the increasing use of these lands for crop cultivation. Pasture supplies about one-third of the total feed resources used by ruminants. As a result of a breakdown in the traditional management of common property resources, these areas are currently in a state of degradation throughout the country with low vegetation cover and a high vulnerability to erosion. The grazing intensity on these rangelands has been estimated as 2–6 adult cows/ha per year with very wide variations depending upon the density of ruminants in an area.

In areas where intensive cropping is common, native pasture and cultivated fodder contribute little to ruminant (animals) diets. In such areas, crop residues are the major feeds and are increasing in demand. The principal residues are the straws of cereals and pulses. However, estimates of the annual availability of different crop residues vary greatly, for example in wheat 55–74 million tones; rice 74–152 million tones; sorghum 23–47 million tones; and groundnut 14–31 million tones. The data on extent of use of these feeds are not available. In the Gangetic plains, characterized by the highest intensification of grain production and with a significant ruminant population, the availability of crop residues is higher than calculated requirements. However, in state like Punjab, about 80% and 50% of rice and wheat straw, respectively, are burned in the field and only about 17.3% and 51.5% of the total rice and wheat straw, respectively, are used as feed for ruminants. Type and quality of crop residues vary according to the crops and adopted cropping patterns

In the arid/semi-arid areas, crop residues are mostly used for the animal maintenance requirements. In these areas, sorghum and millet are very important sources of roughage for ruminants. Sorghum is the preferred source of fodder for dairy animals. However, fodder production from sorghum and millet has declined significantly mainly due to decrease in the cropped area and the use of new varieties having higher harvest index. These decreases have resulted in a shortfall in roughage availability, and have led to a significant increase in the price of these crop residues resulting hardships for landless milk producers. A tenfold increase in fodder price coincided with only a fourfold increase in grain price with the grain-stover ratio price declining from 6:1 to 3:1. The value attached by rural and landless farmers to sorghum and millet stover has slowed down the decrease in the area of these crops for grain.

In the irrigated areas, small farmers fed crop residues such as wheat and rice straw, and sorghum and millet stover in mixtures with green fodder. The increasing land area under cash crops like, oilseeds, medicinal and aromatic plants, flowers, etc, in Rajasthan, Madhya Pradesh, Maharashtra, Andhra Pradesh, Karnataka and Tamil Nadu has reduced the availability of fodder for livestock. The use of short-straw and dwarf cultivars of traditional crops such as pearl millet, sorghum, rice and wheat also reduces fodder availability and farmers balance the planting of these cultivars with traditional landraces according to their need for animal fodder. Considerable component research has been conducted on the biological and chemical treatment of cereal straws. However, largely the research institutions have conducted these studies and there has been little adoption at farm level.

Cultivated fodders account for only 3.3% of the total cropped area in the country. Over the years, fodder cultivation has increased in irrigated areas and has declined in rainfed areas. In the states like Uttar Pradesh, Madhya Pradesh, Gujarat, Rajasthan and Haryana, cultivation of fodders such as berseem and oats in the cool season and fodder sorghum and maize in the summer are common in irrigated areas, which account over 75% of the total cultivated fodders in the country. There has been a discernible shift to the growing of fodder crops in areas covered by milk procurement schemes either by government, co-operatives or privately owned agencies. Sugar-cane tops used as fodder are also available in irrigated areas.

Agro-industrial by-products (AIBP) include rice and wheat bran; groundnut, rapeseed, soybean, sunflower and cottonseed cakes; and cereal and brewers grains in that order of importance. In addition, bran and cakes are used more in semi-urban and urban areas than in rural areas. In addition to sugar-cane tops, the major by-products of the sugar industry are molasses and bagasse. Most of the bagasse is used as fuel in the sugar factories. Also, about 10% of the 3.0 million tones of molasses produced are available for animal feeding. Some important recommendations for utilization of these feeds have been made under the All-India Coordinated Research Project (AICRP) on Agro-Industrial By-Products Utilization and AICRP on Post Harvest Technology. These projects have several cooperating centers at ICAR institutes as well as State Agricultural

Universities having many components including feed and fodder development and utilization. Somehow, the research out puts has not percolated to users. A number of multi-purpose trees like Leucaena leucocephala, Gliricidia sepium and Prosopis species are used by smallholder farmers for fodder purposes. In some States, these sources of fodder represent 80% of the diet of goats during 5–8 months of summer and about 40% during the rest of the year.

An estimated 12 million tones of non-conventional feed resources (NCFR) are produced annually, which represent a large potential source of supplements for ruminants. They include rubber seed cake, spent annatto (Bixa orellena) seeds, cassava starch waste, sal (Shorea robusta) seed meal, panewar (Cassia tora) seed, mango seed kernels, babul (Acacia nilotica) seeds, vilayati babul (Prosopis juliflora) pods, warai (Panicum milliceum) bran, tamarind (Tamarindus indica) seed, Niger (Guizotia abyssinica) seed cake, ambadi (Hibiscus cannabirus) cake, kokam (Garcinia indica) cake, damaged apple waste, karanj (Pongamia glabra) cake, coconut pith, mahua (Madhuca indica) seed cake, kosum (Schleichera oleosa) cake, cocoa pods and castor bean meal. The utilization of these feeds can be increased significantly with improved processing technologies reducing the negative effects like anti-nutritional factors (hydrocyanic acid, tannins, mowrin and karanjine) present in the same.

There are huge numbers of fodders and feeds used for feeding huge varieties of animals but storage of these fodders has not been addressed sincerely for suitable solution. Fodder transportation from one place to other even in local areas is very common. In the case of natural calamities, like earthquake, drought, flood and tsunami, fodders are transported from one State to another and most nearest place of natural calamities. Modes of transport are in very poor. Mostly traditional systems of transports are followed. The losses of nutrients as well as quantity of fodders during storage and transportation are unavoidable in traditional system.

Appropriate transport system for fodder is essential. The requirements and suggested modes of transport and storage systems for fodders and feeds have been discussed in this paper. Scientific methods for reduction of volume of fodders facilitating long distance transportation have been covered. Besides, measures to check the loss of nutrients during transportation have also been covered in this paper.

Animal health hazards

There are several bacteria and fungus, which are very dangerous to health of animals. They may affect the quality of fodders especially during transportation and storage. Some important factors related to health hazards of animal are following:

Mycotoxins

Mycotoxins are secondary metabolites produced by fungi various generations when they grow on agricultural products before or after harvest or during transportation or storage. Both intrinsic and extrinsic factors influence fungal growth and mycotoxin production on a given substrate. The intrinsic factors include water activity, pH, and redox potential whereas extrinsic factors, which influence mycotoxin production, are relative humidity, temperature and availability of oxygen. Many mycotoxins with different chemical structures and biological activities have been identified. Much of the published information on toxicity concerns studies in experimental animals and these may not reflect their effects in humans and other animals.

Mycotoxins are regularly found in feed ingredients such as maize, sorghum grain, barley, wheat, rice meal, cottonseed meal, groundnuts and other legumes. Most are relatively stable compounds and are not destroyed by processing of feed and may even be concentrated in screenings. Different animal species metabolize mycotoxins in different ways. Mycotoxins, or their metabolites, can be detected in meat, visceral organs, milk and eggs. Their concentration in food is usually considerably lower than the levels present in the feed consumed by the animals and unlikely to cause acute intoxications in humans. However, residues of carcinogenic mycotoxins, such as aflatoxin B_1 and M_1, and ochratoxin A, when present in animal products pose a threat to human health, and their levels should be monitored and controlled. The extent of mycotoxin accumulation in fish tissues due to consumption of contaminated feed is poorly understood.

Veterinary drugs

Veterinary drugs may be administered in animal feeds for livestock and aquaculture. If good veterinary practices (GVP) are employed then Maximum Residue Limit (MRL) should not be exceeded, however, if GVP is not adhered to, residues in foods of animal origin may exceed MRL. Consumer concern about residue levels in foods has fuelled recent trends towards organic and other agricultural production systems that feature reduced reliance on artificial chemical inputs. Foods produced by such systems contain reduced or zero veterinary drug and chemical residues. Organic agriculture is one of the fastest growing sectors of agricultural production with an overall persistent growth rate; especially in Europe is around 25% per year for the last ten years. The need for the containment of anti-microbial resistance due to the use of anti-microbial in livestock, including their addition to feedstuffs, requires attention. Anti-microbial are used for therapeutic, prophylactic or growth purposes, and in the latter case they are added to feed and/or water. Even first-line antimicrobials (e.g. glycopeptides) are being used as feed additives for growth promotion. The assessment and containment of public health risks associated with the use of antimicrobials in livestock is a matter of priority.

Agricultural and other chemicals

Potential contaminants in feedstuffs include excessive residues of pesticides and fungicides, or other environmental contaminants such as the polychlorinated biphenyls (PCB), dioxins and heavy metals including mercury, lead, or cadmium. Dioxins and PCB are ubiquitously present as contaminants in the environment and dietary intake represents the most common route of human exposure. PCBs and dioxins have similar physical and chemical properties. Foods of animal origin are the greatest source of human exposure to these contaminants and animal feeds may be an important source of contamination for livestock. Plant materials growing in areas with high levels of other environmental pollutants such as radionuclides and heavy metals that are used as feed may also lead to unacceptably high levels of contamination in food products of animal origin. Similarly, fish oils used as animal feed ingredients may contain high levels of lipid-soluble contaminants if they are produced from fish grown in polluted areas.

Infectious agents

Animal feed may be the source of a limited number of infections for farm animals that could lead to human illness on consumption of foods of animal origin. However, the risks to human health from several other infectious agents, which may contaminate feed or forage, appear to be either negligible or non-existent. Heat treatments of varying severity are commonly used to ensure the microbiological quality of animal feed. Irradiation may be considered a potentially important control measure for certain microbial agents in the feed of food-producing animals. However, these measures are only for the knowledge neither practiced nor attention is given to practice in India.

Transmissible spongiform encephalopathies

The transmissible spongiform encephalopathies (TSE) in ruminants are non-febrile neurological disease of man and many animal species including ruminants. They have a long incubation period and are ultimately fatal. They are associated with incompletely defined agents, currently termed prions, which are resistant to normal heat treatment of feed and food. Sheep scrapie has been recognized for over 250 years, while bovine spongiform encephalopathy (BSE) was first recognized in the United Kingdom in 1986. For BSE, it has been postulated that the aetiological agent enters the feed primarily through rendered infected tissues (notably the tissue of the central nervous system and the reticuloendothelial system) under conditions of insufficient heat treatment to destroy or inactivate the infectious agent. At the present time, there is a strong presumption of a link between this new variant and the possible transmission of the infective agent from infected bovine tissue to humans. Though, all are related to animal feed, which required attention but it is still neglected to practice in the country.

Other infectious agents

Salmonellae are widely distributed in nature and animal feeds. Animal and plant origin feed ingredients are frequently contaminated with Salmonellae. Processed feed can be contaminated from these raw feed ingredients. There are over 2000 Salmonella serotypes and these can be divided arbitrarily into three unequally sized groups. These include the species-specific serotypes such as S. dublin (cattle); the invasive serotypes, which may cause septicaemic disease in several animal species (e.g. S. enteritidis and S. typhimurium); and the non-invasive serotypes, which tend not to result in septicaemia. Members of the first group are not recognized as food borne pathogens. The third group is by far the largest and may be associated with sub-clinical infections in farm livestock. For example, toxoplasma gondii, the protozoan is found in cats and based on serological surveys also in birds, other domesticated species including sheep, pigs, goats, and horses. The primary source of infection for animals is feedstuffs contaminated with faces of cats and possibly rodent tissues.

Genetically modified organisms

A new issue concerning animal feed relates to the inclusion of genetically modified organisms (GMO) in animal feeds given that approximately 60% of soybeans planted in the USA are GMO and this is a principal source for importing into Europe, it is inevitable that they are likely to be present in compound feeds made in the European Region. This also applies to maize products. While there are no proven harmful effects about the transfer of modified DNA along the food chain, there has been public concern and recent incidents involving protests against this practice. It addressed concerns over the transfer of foreign genes introduced into GM crops, especially antibiotic resistance marker genes to bacteria, which would render common infectious diseases untreatable. The main conclusions are that DNA is not degraded under most commercial processing conditions, or in silage, and that further studies on GM material should be undertaken. Notably, potential feed safety problems related to the use of antibiotic marker genes may be avoidable through the use of marker-free transgenic systems, which have now been, developed. The relevant future strategies and priorities for animal health are: (i) development of a new generation of vaccines and diagnostics, diagnostic technologies/ methodologies, and immunobiologicals against important livestock and poultry diseases, and (ii) development of centers of excellence for species-specific disease diagnosis and the establishment of better co-ordination and linkages at national and international level.

Code of practice for animal feed production, handling and storage

The FAO on animal feeding and food safety noted that quality assurance (QA) begins with the concept of what the feed product is to be, in terms of the species being fed and the results being sought. Ingredient specifications are important to quality assurance in defining the quality of the feedstuffs to be accepted by the processor when the raw materials are received for processing. The formulation of the finished feed, including any added medications, should meet the regulatory requirements of the government as well as satisfy the animal production objectives of the customer. Other QA factors involve the manufacture and distribution of the feed. Key elements in effective quality assurance at the feed production facility should include proper sampling, laboratory testing and microscopy, in-plant quality control, control of drug carry-over, plant sanitation and integrated pest management, plant cleanliness, the receiving area, and storage. Quality assurance procedures must be clearly documented and records maintained.

Clear guidance from governments to industry regarding required features of feed quality assurance programs constitutes an important preventative approach to feed safety. Such guidance can be provided through codes of practice for the animal feed and livestock sectors covering all stages of feed production, handling and storage to its ultimate use. The code could cover such issues as: production of feed materials; sourcing of feed materials; use of additives and veterinary medicinal substances; structures, equipment and materials used in feed manufacture, packaging, handling or storage; product and personnel flows through the facilities; processing operations; labeling requirements; transport and storage of finished product; tractability mechanisms; and, appropriate record-keeping.

All potential safety hazards are thoroughly analyzed and assessed, that critical limits are established for all points along the chain that must be controlled to avoid occurrence of safety hazards, that effective systems for monitoring the critical control points are in place, and that plans for corrective action are established in the event of problems within the production chain. Processors and handlers of animal feed must further ensure that adequate documentation is maintained to demonstrate their adherence to defined principles.

The work currently being undertaken at international level (by FAO) to establish a Code of Practice for animal feeding will be instrumental in ensuring a uniform approach to assuring the quality of feed.

Transportation system for fodder

There is an ambitious program of highways and roads ready for investment with a length hundred thousands kilometers. The Investment Act of Government encourages foreign investors to construct or improve infrastructure (roads, ports, and communication). Thus, infrastructures, especially for surface transport will be better in coming days. But the modes of transport for fodders and feed still require great attention for improvement and practices. Other forms of fodder transport requiring investment include: (i) design of wagons for transport of fodders in the railway system, (ii) rehabilitation of the way river transport, and (iii) rehabilitation of the sea transport.

Important modes of transport of fodders practiced in India are: (1) bullock cart, (ii) tractor trolley, (iii) trucks and (iv) rails. Use of sea and air for fodder transport is beyond the accommodation in Indian transport systems, especially at present rate of development in this sector (feed and fodder). Scientifically designed transport system for the fodders are not available, neither receiving attention in the country. Fodders in tractor trolley and trucks are transported in haphazard way causing several accidents on road. Sometimes, there is huge loss of fodder in monetary value and losses of human life are unrecoverable. Very badly designed ballooned type canvas bags are used which disturbs to driver in controlling of vehicle and does not leave space to other vehicles to pass on. This system is widely used throughout the country. Drivers driving the vehicles are also not trained to run it on the road.

The other major transport in rural areas is bullock cart. Its capacity for fodder is very small but loaded like pumping air in balloon. It makes the driver always conscious and disturbs his mind while driving. Though it is of small types not dangerous to human life, but accident with this transport is very common. Sometimes, it is dangerous to life during accident.

The fodders are prepared as per application and use to animal for feeding. In general, feeding systems and fodder used patterns throughout the country are:

1. Dry fodders from wheat and paddy straws, sorghum and millet stover is a major source of fodder fed to types, dairy and draught animals.
2. Dry fodders and sorghum stovers are the most valuable fodder in the ruminant diet in the summer months.
3. Animals prefer rice straw in the villages during the Kharif season.
4. Grasses and weeds complement crop residues in the animal diet during the monsoon period.

5. Sorghum stover is an important fodder for those who do not have access to irrigation, as they have fewer alternative sources.
6. Dairy and draft animals are given special care in the intensive-dairy village, and fed with a good mixture of feeds that include green fodders and concentrates.
7. Green fodder from forage sorghum varieties and other cultivated grasses are used for a much longer period in the intensive-dairy village.

Recommendations for improving system

Based on the studies and further discussions, the following measures are suggested for adoption to improve the systems:

1. There is needs for classification of fodders, like natural green grass fodder, cultivated green fodder, dry fodder from straws, dry fodder from jungle grasses, processed fodders, etc.
2. Define the storage life of fodders and feeds.
3. Identification of storage system for the fodders and feeds and construction of properly designed structures at suitable place.
4. Define the modes of transport of particular type fodder and practice the same.
5. Design of transport systems both for local uses (like bullock carts, tractor trolley, manual bags, etc) and for long distance (truck, railway wagons, ship, etc); design should be on the basis of scientific information and standard of code of practices.
6. Standard code of practices for maintenance of transport systems as well as storage of fodders should be followed strictly.
7. There is need of regulation for transporting fodders by roads or rails.
8. Review existing food safety and quality legislation so as to ensure that it provides an adequate basis for the control of feed-related hazards with the potential to cause public health risks.
9. Participate to the fullest extent possible in the work being undertaken by international organizations involved in developing standards, guidelines and recommendations relating to feed-borne hazards.
10. Support research into the public health implications of animal feeds to support risk analysis of feed borne hazards.
11. Develop suitable analytical and diagnostic methods for rapid screening and confirmation of feed borne hazards in national surveillance and monitoring programs as well as in routine regulatory testing.

12. Establish mechanisms to ensure multidisciplinary scientific input, involving human medicine, veterinary medicine, animal science, crop science, food/feed technology, environmental science and toxicology, into policy and program decisions relating to the control of feed borne hazards.
13. Encourage dialogue among producers of fodder and feed or feed ingredients, livestock and aquaculture industries and government as an essential part of the process of elaborating codes of practice for the feed industry, which will ensure that eventual regulations and guidelines are practicable and widely supported.
14. Establish, as needed, coordination and cooperation among several government agencies and departments to ensure successful and efficient implementation of feed control programs.
15. Establish policies and programs for the evaluation, authorization and labeling of feed and fodders; these programs should be no more restrictive than required to meet legitimate objectives.

Certain chemical substances and biological agents incorporated into feed at any stage of production up to the point of feeding, either intentionally or unintentionally, can result in hazards in food of animal origin. National food safety programs should therefore include the control food borne hazards that originate in feedstuffs and fodders. Effective control of animal feedstuffs requires multidisciplinary input. There is a need for collaboration between all parties involved in feed and animal production, especially those in a position to provide veterinary clinical and epidemiological information, to establish the linkage between any identified or potential hazard and the level of risk. Such information is essential for the development and maintenance of appropriate risk management options and safe feeding practices. Transportation of fodder has not received attention in the country. Safe storage of fodder is equally importance as food storage. Scientifically designed transport and storage systems for fodders and feeds are the need of time. It will save the losses, improve the health of animals and will thereby improve the national economy.

Bibliography

FAO, 2000. Agriculture: Towards 2015/30. The Technical Interim Report, Global Perspective Studies Unit. Rome.

Delgado, C., Rosegrant, M., Steinfeld, H., Ehui, S. and Courbois C. 1999 Livestock to 2020: The Next Food Revolution. International Food Policy Research Institute, Food, Agriculture, and the Environment Discussion Paper No. 28, 72 pp

ACC/SCN, 2000. 4th Report on the World Nutrition Situation, Geneva.

4.FAO, 1997. Animal feeding and food safety. FAO Food and Nutrition Paper no. 69 / Rome (Italy), FAO, 1997, 48 p.

5.Gardner, B. 1993. The EC animal feed sector 1993-1998. Agra-Europe-Special-Report. 1993, No. 68, 74pp.

6.FEFAC, 1997. Feed & Food. Statistical Yearbook, 1997. European Feed Manufacturers Federation, Brussels, 38 p.

7.FAO/NACA/WHO 1998. Report of a Technical Consultation on Food Safety Issues Associated with Products from Aquaculture. WHO Technical Series 883, Geneva.

8.FAO/WHO 1999. Discussion Paper on Dioxins. Codex Committee on Food Additives and Contaminants, 32ND Session, Beijing, People's Republic of China, 20-24 March 2000. CX/FAC 00/26.

9.WHO 1999. Surveillance programme for control of food borne infections and intoxications in Europe. 8th Report 1993-1997. Federal Institute for Health Protection of consumers and veterinary medicine.

10.FAO/IAEA/WHO, 1999. Report of the Joint Study Group on High Dose Irradiation, Geneva 15-20 September 1997. Technical Report Series 890, WHO.

11.FAO/WHO 1999. Codex Alimentarius Commission Guidelines for the Production, Processing, Labelling and Marketing of organically produced foods cac/gl 32-1999.

12.Ryan, Angela and Mae-Wan Ho. 2000. Transgenic DNA in Animal Feed Critique of MAFF Report CS0116 "Effect of feed processing conditions on DNA fragmentation". Institute of Science in Society and Biology Department Open University Walton Hall, Milton Keynes MK7 6 AA, UK.

13.Spillane, Charles 1999. Recent developments in biotechnology as they relate to plant genetic resources for food and agriculture. Background Study Paper No. 9. Commission on Genetic Resources for Food and Agriculture, 8th Session 19-23 April 1999.

14.FAO/WHO 2000. Report of the Codex Committee on Food Additives and Contaminants, 32nd Session, Beijing China, 20-24 March 2000. Alinorm 01/12.

15.FAO/WHO 2000. Report of the First Session of the Codex Ad hoc Inter-Governmental Task Force on Foods Derived from Biotechnology. Alinorm 01/34

CHAPTER - 21
FEED PROCESSING FOR MEAT PRODUCTION IN SHEEP

S.A. Karim

Animal husbandry contribute 27% Ag GDP of the country and in that small ruminants have sizable contribution to the total livestock production. The importance of economic contribution by sheep as a livestock species has higher significance as it is one of main source of economic sustenance by the rural poor, upliftment of whose economic status is a commitment by government. Sheep as a livestock species are predominantly reared by weaker segment of the society under extensive range management in regions inhospitable for conventional agricultural production. The species contribute mutton, wool, hide and manure for trading by the farmers. Earlier sheep were reared for wool as the main produce whereas with advent of man made quality synthetic fiber, mutton has replaced wool as the main produce from sheep. More than 75 % of Indian consumers are non vegetarian by choice and beef/pork are not included in their diet due to religious reasons while sheep meat is uniformly consumed by all segments of the society irrespective of their religious and ethnic background The country possesses 59 m sheep ranking fifth in world sheep population (50%). Considering that about 32 % of sheep are slaughtered annually, '\ the production potential of mutton in the country is 188.8 m kg (Karim, 2004).

Although cost of feeding alone accounts for more than 70 % investment in livestock production, the expenditure of sheep farmers on this account is virtually negligible. Some resource sufficient and progressive farmers in western part of the country use top feed resources as supplement during lean period to sustain their stock. However, the available feed resources remain inadequate during critical summer months resulting in migration of the stock to neighboring states in source of greener pasture. The production trait of Indian sheep is one of the lowest in the world because of unorganized breeding, feeding and health coverage. Past efforts to augment the production traits of Indian sheep by crossbreeding with improved breeds failed because the genetically improved crossbred having higher nutrient requirement could not sustain the rigors of field existence and its perennial feed scarcity and low plane of nutrition for most part of the year.

Small ruminant production potentia

Sheep earlier were reared for wool as the primary produce; however, under the changed scenario of sheep production, mutton has replaced the wool while the economic realization is almost similar from wool and manure. Under farmers field condition the surplus male lambs are disposed off around 9- 12 months of age weighing about 20- 22 and 18- 20 kg, respectively, with average carcass yield of 10 kg. Moreover all spent ewes and rams are also slaughtered for mutton production. As per the recent estimates the two species contribute 158 million kg mutton amounting to Rs.15800 million to ODP.

Majority of small ruminants presently are maintained under extensive range management with top feed supplementation during lean season. However, it is envisaged that with emerging demand for quality meat, in future, lamb and kid rearing for mutton/chevon production will be taken up under intensive and semi intensive system of feeding management opening avenues for processed feed. Maximization of gain by nutritional manipulation can only be possible by least cost feed formulation using local feed ingredients/resources. The least cost feed formulation envisage use of cheaper ingredients and agro-industrial byproducts, which invariably contain some anti nutritional factors requiring masking/dilution/annulment by suitable feed processing.. Although under in vogue system of small ruminant production, the cost of feeding is negligible still for intensive mutton/chevon production the feed industry has a definite role to play.

The existing production traits

The average body weight of surplus sheep sold for meat purpose range from 22- 25 kg yielding a carcass of 10 kg as against the production level of .20- 25 kg carcass weight in developed countries. Under farmers in vogue feeding management without organized inputs, the lambs in active phase of growth hardly attain 40- 50 g average daily gain (ADO) and marketed for slaughter around 9- 12 months of age weighing about 20- 22 kg. Three decades of concentrated research under intensive feeding program at Central Sheep and Wool Research Institute improved the growth rate of native sheep by 300 % with 1820 % feed conversion efficiency. The developed technology has been demonstrated successfully on farmers' field under aegis of Jaivigyan Project (Singh et al., 2003) while large-scale adoption of the technology is not effective due to poor economic background of the sheep farmers.

The community rangeland in the country is grossly overstocked and as against the carrying capacity of 0.1 to 0.5 ACU, the stocking density ranges between 5- 50 ACU in different locations. All farmers in the locality utilize the rangeland resource whereas its maintenance and improvement is grossly neglected, as a result the land is undergoing progressive degradation. In semiarid and arid zones of the country the rangeland resources are important feed component of sheep and goat rearing. Although it is realized that in organized livestock enterprise, the cost of feeding is the major component of total investment, the expenditure of sheep farmers on this account is negligible as a result the production levels are low. However, for quality mutton production program, the level of production as well as quality of the carcass has to be improved to make the enterprise competitive thus envisaging role of feed processing industry. Improvements in sheep for mutton production envisage input of processed feed for which the industry has to be geared to meet the challenge.

Three decades of concentrated research under intensive feeding program for economic mutton production has improved the growth rate of the animals by 300 % and the technology is ready for commercial application and adoption by progressive farmers and entrepreneurs. It was evident from these studies that the extensive range management cannot support a growth rate higher than 50- 70 g ADG)hence in order to increase the rate of gain to 150- 180 g ADG for its commercial applicability, organized feeding has to taken up requiring input of concentrate/processed feed. The requirement of processed feed for organized lamb for mutton production, in turn, will open up avenues for feed processing thereby envisaging a role of feed industry in the venture.

Requirement of DM and nutrients of sheep and goats in the country:

The country is hosting 49.3 million sheep with a rich genetic diversity of 40 approved breeds. The 9.86 m ACU of sheep, accounts for about 3 % of total range managed livestock species consuming 10 % DM resources of the country (Table 1). It is established that range managed sheep on an average require 0.4 MT DM/head/year thus the total requirement of the two species will be 19.71 m MT. Ranjhan (1997) calculated the total feed availability of the country to be 387.9 m MT crop residues, 573.5 m MT green fodder and 35.7 m MT concentrate ingredients indicating deficit of 40, 25 and 47 %, respectively, for these items. Based on the livestock population of the year 2000 Ramachandra et al., (2001), calculated feed fodder requirement of the country was 950 m MT of dry fodder and 1135 m MT of green fodder as against the availability of 450 and 320 m MT, indicating a deficit of 53 and 72 %, respectively. Moreover the total concentrate feed production in organized feed manufacturing sector is 5 m MT as against the demand of 42 m MT.

Table 1: Livestock population of India

Livestock species	Population (Millions)	Population (ACU)	Requirement (m MT)
Cattle	209.49	209.49	418.98
Buffalo	91.78	91.78	183.56
Sheep	49.28	9.86	19.72
Goat	128.48	25.70	51.40
Camel	1.03	6.18	12.36
Horse	0.99	0.99	1.98
Total	-	**344.00**	**688.00**

Since the livestock population of the country is not only surviving but also providing sustenance level of production for meeting the farmer's requirements, it is apparent that while calculating feed resource availability, the rangeland and top feed resources were not given due weight age. The fact remains that in critical semiarid and arid region of the country these two feed resources serve as backbone for sheep feeding and rearing.

Role of feed industry in small ruminant production

The total feed production in organized sector is around 5 m MT as against the total requirement of 42 m MT. The CLFMA member's feed production is about 5 % of total cattle feed and 20 % of total poultry feed requirement of the country. However at the present contribution of CLFMA in small ruminant feeding is negligible while it is envisaged that with the changing scenario of land utilization pattern and progressive shrinkage of rangeland resources due to intense land reclamation for conventional agriculture, majority of livestock will be reared on stall-feeding necessitating intensification of processed feed for sheep rearing. However, the feed industry at the present is geared to prepare mash and pellets using concentrate ingredients while for intensive feeding program of sheep a complete feed in pellet form with roughage content ranging from 30- 50 % has to be prepared. Suitable technological renovation in the existing plant structure for effective mixing and pelleting of feed ingredients having wide density gradient has to be devised.

Two categories of sheep are marketed for mutton production: the male lambslkids surplus to the breeding requirement and spent ewes/does, which have completed their productive phase in life. The male lambs and kids surplus to the breeding requirement have commercial viability under organized feeding management. Under in vogue extensive range management system, the male lambs are marketed for meat production around 6- 9 months of age weighing about 20- 22 kg with average yield of 10 kg low-grade carcass.

Under organized feeding management the weaner lambs and kids can achieve 27 and 32 kg at 6 months of age (Karim et al., 2004) wherein the role of feed industry is envisaged.

Farmers lack in resources to maintain lambs for higher gain under intensive feeding beyond weaning hence are disposed off at the earliest. Moreover the continuous drought in the western part of the country has changes the existing production system. Earlier the male lambs use to be disposed off at 9- 12 months of age weighing 22- 23 kg while under the changed scenario the farmers are disposing off lambs at 3- 4 months of age weighing 1617 kg which is colossal waste of animal resources. It is suggested that some resource sufficient farmer or entrepreneur may purchase the surplus males from the farmers and rear them under intensive feeding/organized feeding for commercial mutton/chevon production. Unfortunately the regions rich in sheep and goat population are deficient energy and protein supplements and vice versa. However, for commercial viability of the program weaner lambs purchased from the native tracts are to be intensively fed in situ for finishing and marketing/slaughter.

Feed processing on animal performance

The physical advantage of pelleted feed is the reduction in segregation of high density feed ingredients and restricts selection by the animals in feed consumption. Advantages of feed pelleting:

1. Increased feed density,
2. Reduced feed selectivity by the animals,
3. Reduced feed wastage by the animals,
4. Reduced microbial load in the processed feed,
5. Inactivation of thermo labile toxins,
6. Easier transportation and reduced storage space,
7. Improved nutrient utilization and production traits.

However, the impact of processed feed in nutrient utilization profile of the animals is controversial: some reports indicate improvement while others show no such effect. The uses of pelleted feed generally show improvement in DMI, more so in low-grade roughage feeding, while the relationship between DMI and nutrient digestibility remains unaffected i.e. within limits, higher the DMI lower is the nutrient utilization. The pelleted feed does have definite advantage in reducing dustiness and bulk and improvement in storage, transportation and marketing of feed. The processing cost of concentrate ingredients to a mash form of feed is the lower but increases when converted to pellet form. The cost further increase when the mash for pelleting has roughage component for development of complete feed. The pelleting cost of Rs. 18- 20/1 00 kg concentrate feed has to be viewed in light of the possible improvement. Pelleting of mash ration increase the processing cost by 57- 130 %. Moreover pelleting has definite advantage in masking the obnoxious character of those feed ingredients, which are not readily accepted by the animals and usually avoided in feed selection, if fed in mash form. Pelleting roughages generally improves the gain, particularly when animals are fed on high roughage levels.

Table 2: Availability of green fodder and rangeland resources in India

Resource	Area (m ha)	Yield (MT/ha)	Availability (m MT)
Cultivated fodder	6.6	40.0	264.00
Forest vegetation	66.7	3.0	200.10
Other lands	49.5	1.0	49.50
Total	-	-	**513.60**

Table 3: Dry crop residue availability in India

Category	Crop	Grain production (m MT)	Crop production (m MT)	Grain to roughage ratio
Straws	Wheat	55.09	55.09	1:1
	Rice	73.66	95.76	1:1.3
	Barley	1.65	2.14	1:1.3
Stover	Maize	7.98	23.94	1:3
	Millet	0.95	3.80	1:4
	Sorghum	8.36	33.44	1:4
	Bajra	4.64	18.56	1:4
	Ragi	2.68	10.72	1:4
Tuber crops	Potato	15.50	3.10	1:0.2
	Cassava	5.63	1.12	1:0.2
Legumes	Tur	2.19	8.76	1:4
	Gram	4.15	16.60	1:4
	Other pulses	5.71	22.84	1:4
Sugarcane	Cane	239.00	59.75	1:0.25
Oil seeds	Groundnut	7.06	14.14	1:2
	Sesame	0.67	1.34	1:2
	Niger seed	0.16	0.32	1:2
	Rape seed	5.84	11.68	1:2
	Safflower	0.19	0.38	1:2
	Sunflower	1.18	2.36	1:2
	Soybean	2.27	4.54	1:2
	Linseed	2.30	0.60	1:2
Total			**390.98**	

Source: Ranjhan, S. K. (1997).

Bibliography

Deoghare, P. R. and Kumar, S. (2001). Economics of sheep and goat meat production and marketing potential. In: Scope and strategies for increasing meat production from small ruminants and rabbits in new millennium. Proceedings IV National Seminar of ISSGPU, Bharathiar University, November 9- 11, Coimbatore, Tamil Nadu.

Karim S. A. 2004. Nutrition and feeding of sheep for mutton production. Souvenir National Seminar on Opportunities and challenges in nutrition and feeding management of sheep, goat and rabbit for sustainable production, Central Sheep and Wool Research Institute, Avikanagar, February 10- 12, pp.88- 100.

Karim, S. A., Santra, A. and Singh, V. K. 2004. Lamb feeding for mutton production. Bulletin, Central Sheep and Wool Research Institute, Avikanagar.

Ranjhan, S. K. 1997. Problems and prospective of animal nutrition research in India in twenty first century. Platinum Jubilee Lectures, Center of Advance Studies, IVRI, Izatnagar, India.

Ramachandra, K. S., Anandan, S. and Raju, S. S. 2001. Livestock feed resources availability in India and need for national data base. Emerging Nutritional Technologies for Sustainable Animal Production and Environmental Protection, Proceedings Xth Animal Nutrition Conference, NDRI, Kamal, 9- 11 November, pp. 134- 144.

Singh, V. K. Bhagwan, P. S. K., Arora, A. 1. and Karim, S. A. 2003. Annual Report Jaivigyan Project on “Improvement in migratory sheep production program for tribal farmers of North- West. Central Sheep and Wool Research Institute, Avikanagar, Rajasthan.

CHAPTER - 22
STATUS AND FUTURE OF DENSIFICATION TECHNOLOGY

Bharat Sharma and G.S.Singh

During the year 1994 with the help of world bank(RCTC) the first commercial unit for manufacturing 'Densified TMR Blocks' was started in the name of "M/S Anmol Feeds Limited at Bulandsahar in UP. The technical backup for product development and other aspects were taken care by National Dairy Reaserch Institute Karnal, Haryana while we as Advance Hydrau-tech Pvt. Ltd took initiative for manufacturing the entire plant and machines as per the need . The Project could not succeeded and came to closure in 2000 however the 'Product' caught attention of the Industry. During 2001-2002 another feed manufacturer M/S Poshak Feeds (India) Private Limited Karnal took initiative and started the project at Karnal in Haryana. This time project took heights and product was sold almost in all the states. The continuous product manufacturing provided the opportunity and atmosphere for continuous machine developments.

Over a decade working experience concluded some of the landmark machine developments which could benefit the Indian farmer as well as the concerned The machines made during the year 1994, were based on a primitive model for densifying the material , it was based on some mechanical machines. The machine was providing output from 0.3 to 0.5 MT in an hour with an input of 40 HP. Not only the output was of concern but the continuous commercial production was not viable at these quantities i.e 3 MT in 10 hrs. (One Shift).

The second generation machines were developed and operated in Karnal with a capacity of 0.7 MT/Hr (5.5 MT in 10 hrs or One shift)and the power was reduced to 30 HP as compared to 40 HP in first generation machine. The third generation machine came in operation in 2004 when the production capacity was increased upto 1.2-1.5 MT/h with the same power as in Generation second machine. This machine is not only capable of working 20 hrs at a stretch but with its semi-auto functions it also reduces the Manpower as compared to first and second generation machines to half. Now we are at the finishing stage of the Fourth Generation and Fifth Generation Fodder Machines which will be capable of producing 3 MT to 5 MT Densified TMR /Fodder Blocks in continuous operation in fully automatic mode with the same power inputs. With the guidelines/ suggestions of the researchs Institutions like NDRI, IGFRI etc. our special fodder machines will not only meet the requirements of the country but will compete the entire south east Asia in this specialized work.

Densified Fodder Technology "A Need of Time"
(GLIMPSES OF SOME CASE STUDIES)

Govt. of India Report .... A.K.Chaudhary)

Nutritive value of feed and fodder is directly linked to production and productivity of livestock

Three main sources of fodder are:

- Cultivated fodder
- Agriculture/crop residues
- Grass lands, pastures & forests.

ADVERSE EFFECTS DURING DROUGHT

Drought of 2002-03 affected 62% of bovine population (156 million) in 18 states.Due to inadequate nutritional support, milk production declined in many states. Average daily milk procurement reduced by 7% in Tamilnadu, 8% in Maharashtra and 22% in Rajasthan.Other adverse effects included decline in bovine population, draught energy, breeding cover and migration of populations.

SCARCITY SAMAVAT 2059 (Rajasthan Government)

As per the above report the government of Rajasthan was in need of approximately 93.33 lach metric tons of fodder for almost all of its districts except two /three which claimed to have enough fodder to meet the requirement during the year 2003 draught and a big amount was declared to support the transportation of the above required fodder as 'TRANSPORTATION SUBSIDY'. The transportation subsidy needs to be declared because of the low bulk density of fodder.

URBAN AREA FODDER REQUIREMENT

Beside the above there is regular transportation of fodder from rural to urban places and a huge amount is regularly drained as "Transportation Loss" . Delhi is having aproximately 3 lac number of animals and the animals subsist on crop residues only (Wheat Straw only). Every day 2000-3000 MT of fodder is transported to Delhi in ballooned trucks / trolleys every day to meet the above requirement. This movement of fodder is not only hazardous and accident prone but also causes 'Traffic Jams' and transportation loss to the end users.

AVAILABLE TECHNOLOGIES

These Technologies can facilitate the 'FODDER TRANSPORTATION'

- Fodder Baling Technology
- Fodder Densification Technology

FODDER BALING TECHNOLOGY

From pre-independence upto post independence period mechanical baling of grasses has been in use by Military for supply of grasses from surplus to difficult terrains for horses.However even after so many years baling could not proved as a complete solution as in case of loose wheat straw or when the material is chopped.

In developing countries with their situations it is still the only solution.

A Glimplse .... In most developing countries 'Harvesting hay' includes the use of a .... Mower, Rake, and Baler.

A crop with a water content of 80 to 90 percent. First, the MOWER or mower-conditioner cuts the standing crop and lays it in a windrow to allow the hay to dry in the field until it reaches safe storage water content levels. Next, rake and / or related equipment moves the windrow to aid the drying process, creates a narrower windrow and/or brings two or more windrows together for a more efficient baling operation. After the raked hay has dried to the proper water content, the baler gathers hay from the windrow and compresses the hay into a denser package (bale) for ease of handling, storing. Since the Agriculture land use is specific and big acreages are involved and good quality crop residues (lucern, grasses etc.) are available as stocks the above systems in those countries is vastly in use however in our case where small holdings are in practice and poor quality crop residues (Wheat Straw / Paddy straw) are available after threshing (in chopped form), is not as viable.

In Our Country ....

The above Mower Rake Baler concept becomes unviable except in the cases for Forest Grasses and other similar activities.

DENSIFICATION

Densification increases the density of crop residues(chopped & thressed) and decreases the volume, which brings down the transport cost and reduces the space requirement for storage. Wheat straw is transported through trucks, which can not carry more than 40 quintals of straw when it is loose. But after mixing/without mixing with concentrate it is densified to such an extent that the material of four trucks can be carried in one. The reduction in the volume-to-weight ratio result in improved per unit cost of transportation and handling.

WHY DENSIFICATION

'DENSIFICATION' seems to be the ultimate and only solution for the effective utilization of crop residues like wheat straw and paddy straw etc

For Transporting the fodder from surplus to scarcity areas

(Many parts of the country such as Gujarat and Rajasthan suffer from the shortages of fodder while other parts of the country like Punjab and Haryana are surplus in fodder to the extent that it is burnt in absence of any demand and difficulties involved in its Transportation). IGFRI jhansi .....

The Crop Residues-(Straws, Stovers, Sugarcane Tops) are high in volume but low in Value so cannot be economically transported to distant areas.

Economic Movement of Dry Fodder can be facilitated with DENSIFICATION PROCESS or by making 'Densified Fodder Blocks' with fully automatic plants.

UTILITY OF DENSIFICATION TECHNOLOGY

With Densification Process Loose Material Density can be converted in finished product in a ratio of 1: 6 effectively. A roughage as wheat straw etc. with loose material density from 60 kg /m^3 to 70 kg/m^3 can be densified in a range between 300 to 400 kg /m^3 However if the loose material density ranges from 120 to 150 kg/m^3 (as in the case of TMR). The finished product density level goes up in the ratio of 1:4 . Thus the finished product density level will be between 450 to 550 kg /m^3

This Densification Process has proved commercial utility and viability for manufacturing.

a) Densified Fodder Blocks

b) Densified Enriched Fodder Blocks

c) Densified Complete Balanced Feed 'Densified TMR Blocks'.

Future Prospects

- The available fodder resources may be effectively utilized as the transportation is facilitated with the current technology.
- Farmers in the Surplus Area who at present are burning the fodder may start supplying fodder to the scarcity states. This situation not only provides an opportunity for getting handsome remunerative price for the fodder but also provides simultaneous opportunity of Animal Husbandry in the Fodder Scarcity areas.
- The Government of India every year releases huge funds as 'Transport Subsidy' for supplying fodder from surplus areas to scarcity areas can be utilized for other developmental work.
- Each state can establish their own 'Fodder Banks'.
- The agriculture land which otherwise is not in current use for cereal and other valuable crops can be effectively utilized for growing fodder.
- In case of further surplus position fodder can be exported to the countries who imports fodder i.e. Japan, Middle east , Korea. This situation can turn fodder in a valuable cash crop and the country gets handsome foreign currency.

PROCESSING

The finely chopped fodder is weighed and put into the mixer followed by different feed ingredients and Mineral/Vitamin premix. The TMR- mixer may be run for 10 minutes so that concentrate and roughage(low density material with high density material) are evenly mixed. Then allow molasses to trickle down to the mixer slowly and entangle each particle of roughage & concentrate. Allow the mixer to run for another 5 mins. Then mixed material is taken out manually or automatically to the feeding unit.

A fixed amount of feed mix is emptied into the receiver of the densifying machine. The material is densified through Block Maker. The block is taken from the exit point and packed immediately in the bag manually or automatically. More is the time taken for packing, more it will be difficult to pack since block expands after soon.

The following steps are involved in processing:

- Preparation of concentrate pre mix
- Threshed crop residues (wheat straw)/chopping of roughage
- Mixing of chopped fodder and concentrate in especially designed TMR mixer.

- Transfering the loose TMR mix in the densification machine after weighing.
- Getting the final product from densification machine before packaging
- Packaging the finished product in desirable packing material.

8. Essential Machines Required

- High Speed fodder Shredder/Chaff cutter
- Concentrate Pre Mix section
- TMR Mixers
- Densification Machines
- Other Ancillary machines compatible to the per hour production rate

Machines Available

Different models of the commercial machines are available in the capacity from 1 MT/ h to 3 MT/ h with single machine in manual / semi-automatic/fully automatic modes.

- Turn-key projects from 15 MT/day capacity upto 100 MT/Day capacity in fully manual /semi-automatic/fully automatic mode.
- Individual /customizedmachinerysolution in isolation as per the need of the industry.

DENSIFIED FODDER BLOCKS

This Machine is designed so as to make Densified fodder Blocks without adding any additional raw materials/binders etc like concentrates or Molasses etc.

It is a continuous system operated in manual / semi-automatic/fully automatic.

The chopped roughage is unloaded and stacked in Godowns / Silos and than it comes to the Densification Machines . The machine Densifies the material and converts into Blocks. Immediate packaging solution stops the final block expansion.

Machinery available

Commercial machines are available in manual /semi-automatic/fully automatic mode in the capacity of 1 MT/ h to 3 MT / h with single machine.

CHAPTER - 23

LIVESTOCK FEEDING IN BUNDELKHAND

P. Sharma, B.S. Meena, S. Pandey, M. Singh and R.N. Dwivedi

In the economy of world's highest milk producing nation i.e. India, livestock contribute 6.7 per cent to agricultural GDP in various ways viz., milk, meat, draft power, dung and by-products. In India several livestock species are reared under crop livestock integrated farming system. The average number of livestock/ ha. of land is 0.3 and 2.63 for holding above 10 ha and below 1 ha respectively, which shows the importance attached to livestock by small and marginal farmers (Mangurkar, 2001). Indian dairying is characterized by small, scattered milk production. Dairying has come to be India's largest self-sustaining rural employment program. In 1998-99, the value of output of milk and milk products was Rs. 822640 million and it accounted for 21 per cent of the value of agricultural output that year. It is estimated that 70 million households' livelihood is linked to the livestock sector, which is critical for sustenance and supplementing the income of farmers though the average productivity is low. Dairying is emerging as independent agriculture enterprise with bovine population comprising of 204 m cow (9.7 m crossbred cows), 78 m buffaloes making a milch herd of 103 m generating estimated revenue of Rs. 680000 million with an annual growth of 4.5 per cent. Therefore the development agencies are also encouraging livestock development for the welfare of small and marginal farmers and agricultural laborers (Mudgal, 2000).

India has already become highest milk producer in 1998-99. The present milk production of India is 84 million tones, which is about 14 per cent of the world's milk production. The per capita availability in 1999-2000 was 220 g/ day (Varshney 2000). There is huge gap in amount of milk produced by different agro- climatic zones and so is in milk consumption. This is due to difference in production potential of cattle and buffaloes, status of animal husbandry practices, soil, climatic conditions, agricultural priorities and socio-economic importance of milk production. To improve this situation efforts are being made since independence through various projects such as Key village scheme, ICDP, IRDP and OF etc. But problem of unequal distribution still persists.

The Bundelkhand

The Bundelkhand, though not functional administratively, is of about 71618 km^2 area comprising of 7 districts of Uttar Pradesh viz., Jhansi, Jalaun, Lalitpur, Hamirpur, Mahoba, Banda and S.M. Nagar and 6 districts of Madhya Pradesh viz., Datia, Tikamgarh, Chhatarpur, Damoh, Sagar and Panna, states of India, inhabited by about 15.61 m (2001) people and more than 10 m cattle and buffaloes. The Bundelkhand is characterized by rugged topography, limited underground water resources, high proportion of wastelands, high summer temperature, erratic rainfall concentrated in a short monsoon period and infertile soil with about 67 per cent population dependant on agriculture. Human population is increasing by 2 per cent. The animal density is 1.16 + 0.056 animal / hectare. Fodder is grown on 1 per cent area (Gomez et al.,. 1998). They further concluded that free range grazing and managed feeding were not mutually exclusive practices but often coexisted, both at zonal, village and household/ herd level. However the relative importance of two systems varied between zones, villages in each zone and households in each village. In Bundelkhand the animal pressure is generally more than human pressure. The recommended maximum animal density for this region is 0.78 adult cattle unit/ ha (Tyagi, 1997). The carrying capacity of land is less. Locally adapted breeds are low producer. The average is 1-2 kg/ cow and 4-5 kg / buffaloes. The average density in MP part of Bundelkhand was 181 ACU/ 100 ha which was higher than UP part. The district of Tikamgarh (227), Chhattarpur (218) and Panna were having higher density thus bringing higher pressure of livestock on their cropped area. Cow dung is not utilized properly. Annapratha, the free grazing practice, is still followed (Tyagi 1997). The Murrahs' survival at farmers field need better management. Farmer's adoption of Jersey crossbred is less. The sustainable option is improving local breeds and adoption of improved management practices based on milk shed based approach.

Livestock population

In Bundelkhand the animals are reared as insurance against crop failure due to low productivity of soils and unreliable rains. The drought generally occurs once in four year. Animals are reared for milk, manure, and draft and as wealth. In year 1985 there were 5141456, 1490511, 1699874, 361833 cattle, buffalo, goat and sheep respectively. Buffalo and goats were more in U.P. part while cattle and sheep were more in M.P. part. In 2001 the population of cattle, buffalo, goat and sheep changed to 4327750, 1764373, 1837052 and 408269 respectively. Compared to 1985, there has been a decrease of 813706 in cattle population, while, buffaloes, goat and sheep, increased by 273862, 137178 and 46436 respectively. There is decrease in buffalo's population in Banda and

Chhatarpur while in all other districts there is increase from 1985. In Bundelkhand only about 0.25% cattle are crossbred. Both sheep and goat have increased in number. It is evident from the figure 1 that the numbers of livestock population have decreased during past 17 years. The increase in goat population is about three times as compared to sheep. During 2001, it was found that, the cattle population is more in Chhatarpur, Sagar, Lalitpur and Hamirpur districts where rangelands are generally more. The buffalo population is more in Banda, Jalaun and Chhatarpur. These are districts where DM from cultivated fodder and crop residues is available in sufficient quantity. The goat population is more in Chhatarpur, Banda, Hamirpur and Jhansi district. Incidentally these districts supply more DM from wastelands. The sheep population was more in Chhatarpur, Tikamgarh, Jhansi and Hamirpur probably due to more forest area in these districts.

Regarding feed resources, Saran et al.,. 2000 concluded that crop residues constituted the major feed resources for the domesticated livestock in the Bundelkhand region. Berseem (green fodder), wheat and pea's straw (dry fodder) were major source of livestock feeding during rabi. Sorghum (green) and crop residues of soybean and black gram were sources of livestock feed during kharif. More than 88 per cent of annual availability of dry fodder and 86 per cent of green fodder was obtained during rabi season. Annapratha, the system of leaving the livestock free for grazing was commonly practiced during lean season as also during rainy season. Soaked oil cakes and concentrate mixture with dry fodder (wheat + pea straw) were also offered to the milch animals during the periods of green fodder scarcity. They further observed that livestock: land ratio of Adult Cattle Unit (ACU) / ha was 3.45 for small (<2 ha land) 1.77 medium (2-4 ha land) and 0.88 for large (>4 ha land) in Bundelkhand region (Table 1).

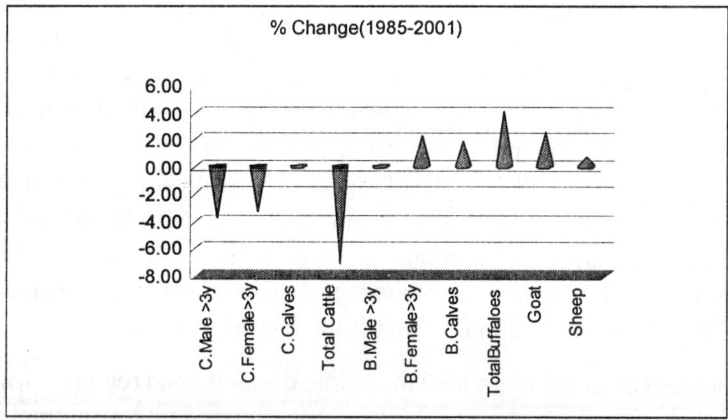

Fig. 1. Change in Livestock Population in Bundelkhand region

Table 1: Land and livestock holding pattern in Bundelkhand

Category	N	Land holding size (ha)	Average size (ha)	Proportion of holding (per cent)	Livestock holding (ACU)	Livestock: Land (ACU/ha)	Average family size (No.s)	Man: land ratio (No/ha)
Small	23	<2	1.18 (0.22)	69.20	4.07	3.45	5.76 (0.34)	4.88
Medium	25	2-4	3.02 (0.32)	23.84	5.35	1.77	8.12 (0.77)	2.69
Large	33	>4	7.42 (1.54)	6.96	6.53	0.88	9.06 (0.61)	1.22
Overall	99	-	2.05	100.00	4.55	2.22	6.55	1.08

1 ACU = 1 cow=0.75 buffalo=6 sheep/goat=1 ox=0.75 bull=4 calves; Figure in parenthesis are SE, mean.

Being a pastoral economy, the livestock population is very high but milk productivity is generally very low. The rural migration rate is high and Industrial growth is almost stagnant. There has been no promotion of trade and industry. There are limited sources of livelihood and the land productivity is less. Thus, basic needs are barely met. Rampant under-employment and unemployment, loss of skills and skill-based livelihoods has further complicated the entire issue. With suitable interventions at appropriate stage dairy farming can be profitable and source of sustainable livelihood. With better management higher net margin can be obtained. Several hectares of land is available in the form of permanent pastures, forest grazing lands and cultivable wastelands which can be profitably used.

Livestock density

Gomez et al.,. (1998) observed that a huge animal population was living in Bundelkhand (a mean of 1.16 ± 0.056 animals/ha). About 1% land is spared for cultivated fodder making the animals almost dependent on crop residues, which are in deficit. This compels farmers for free range grazing of forest areas, on wastelands or the residues, which remains in the field after harvesting of crops. The recommended maximum animal density for the type of land found in the region is 0.78 ACU/ha (Tyagi, 1997). High animal pressure against recommended results in deforestation and grassland degradation. In 2001, the average density (number of animals per ha of potential land) was 1.37 (Table-2). The districts having more than average density are Jhansi, Lalitpur, Banda, Datia, Tikamgarh and Damoh district is having least density.

Gomez et al.,. (1998) divided the whole Bundelkhand region in four zones. The Southern zones, zone 1 and 2, were characterized by low animal pressure (0.7 – 1.1 animals/ha) while the zones in the north, zone 3 and 4 had high ratio of population of animals to land area (1.4 – 2.4 animals/ha). They found that highest animal pressure was in Jhansi, Datia and Gwalior districts, it had been suggested that high human population would result in high animal population. This was confirmed by the statistics, which showed that the highest human population (220 people/ km^2) also occurred in areas of highest animal population. Further, the number of animals/person (0.44 – 0.91) varied little across whole region except for few but significant localized areas where it was as high as 1.8 or even 4.16 animals/person.

Table 2: Livestock Density in Bundelkhand region

Districts	**Potential area (ha) 2001**	**No of animals/ ha of potential land**				
		Cattle	**Buffalo**	**Sheep**	**Goat**	**Total**
Jhansi	427091	0.79	0.39	0.17	0.50	1.85
Lalitpur	458566	0.87	0.30	0.07	0.32	1.56
Jalaun	407907	0.59	0.53	0.09	0.48	1.69
Hamirpur	643126	0.61	0.26	0.07	0.35	1.28
Banda	628639	0.57	0.42	0.06	0.39	1.44
Sub. Total	2565329	0.68	0.37	0.09	0.40	1.53
Datia	241604	0.70	0.47	0.12	0.39	1.68
Tikamgarh	377857	0.91	0.33	0.17	0.41	1.80
Chhatarpur	747438	0.77	0.27	0.08	0.33	1.46
Panna	595265	0.82	0.23	0.03	0.18	1.25
Damoh	639935	0.71	0.15	0.01	0.16	1.03
Sagar	931178	0.60	0.16	0.01	0.12	0.89
Sub.Total	3533277	0.73	0.23	0.05	0.23	1.25
Grand Total	6098606	0.71	0.29	0.07	0.30	1.37

A brief reconnaissance of the area and interaction with the livestock farmers revealed following specific problems in Bundelkhand region.

(A) Scarcity of green forage

Milk production in Bundelkhand is roughage oriented utilizing mostly by-products of arable farming. Under such farming conditions where energy supply is likely to be restricted to 3000 mcal per year, only small to medium breeds or crossbred may have higher net efficiency. That is why there is more concentration of non-descript animals. The poor quality feed along with higher thermal stress conditions and host of other problems decreases productivity. The main problems causing poor performance are low amount of digestible nutrients, inadequate fodder supply, and coarseness of fodder, fiber and lignifications.

(B) Low milk yield of animals

The milk biosynthesis is function of both genetic constitution as well as external environmental factors. Low genetic potential is one of the important reasons. With this the environmental stress activates complex of nervous, endocrine and motor function to maintain homeostasis. Then there is poor feeding. Survival in these conditions makes milk yield the first causality.

(C) Severe mineral deficiency

Most of Bundelkhand soils are deficient in macro and microelements, which in turn makes the forages deficient in these elements. Among these trace elements are particularly neglected. Hardly there is any mineral mixture supplementation in diet. It reduces milk yield, makes reproductive cycle irregular thus reducing lifetime productivity and overall health is deteriorated.

(D) High internal and external parasitic load

The common parasitic infestation of cattle and buffaloes are ascaris, fasciola, babesiosis, cocciodiosis, Ticks and mange etc. They survive well in poor hygienic conditions causing heavy mortality & morbidity, diarrhoea, emaciation, loss of appetite, pot bellied appearance, stunted growth and death.

(E) Lower return on milk yield

The return to investment is low mainly due to defects in input and output delivery system. The profit margin goes to middlemen. The system should also be net efficient i.e. the animal maintained should be profitable. The broad pathways of livestock development i.e. breeding, feeding, management and disease control are shown to be part of the strategy. The major issues amongst these to be addressed are:

A. Breed improvement: The suitable indigenous breed should be selected. Its bulls should be widely used especially through AI. If inputs and management condition are good, crossbreed may also be tried. For particular tract a properly structured breeding policy is needed.

B. Focus on increased lactating period: The efforts be made to exploit full production potential of the animals through proper reproductive management.

C. Low cost feeding strategy: The balanced ration based on locally available feed resources with proper treatment will form major aspect of strategy. The strategic supplementations of mineral mixture will have several advantages in health, milk production and reproductive cycle restoration. Fodder such as stylo, perennial grasses and subabool etc. from wasteland and common property grazing resources have good potential. The straw treatment may be tried. Silage and hay making are also beneficial to conserve excess green fodder.

D. Disease control : For major diseases such as Foot & Mouth Disease (FMD), Hemorrhagic Septicemia (HS) and Black Quarter (BQ), timely vaccination is the best remedy. The mineral mixture supplementation will take care of several nutritional deficiency syndromes.

E. Sustainable input : output delivery system:- In sustainable input – output delivery system, actual cost and maximum price should be realized. Actual cost based payment will also encourage private service providers. In field condition most of veterinary services are paid, although through govt. appointed officials. The producers can be organised in cooperatives which will by then produce, market it collectively and will provide quality input to the users.

The main focus of action component will be on low cost feeding strategy, disease control, and sustainable input – output delivery system.

Forage Requirement, Availability and Balance Sheets

Based on Standards requirement of total forage (crop residue, roughages, forage from forest and other grazing lands and concentrate etc.) to feed the entire livestock population of the region is estimated as 14357931 DM (t/y), which is only 77.07% of the total estimated need. It is revealed from the table 5 that the situation in UP and MP part of the region is different. The availability in the UP part is 70.01% while in the MP part is 82.76%. The regional variation may also be noticed from the table that the availability is surplus in Chhatarpur and Sagar district (+54.39% and +24.09% respectively) and deficit in the rest part of the region but the situation is more alarming in Panna district (-34.43%). Roughages are major source of forage in this region and it is found surplus (+7.82% above the actual requirement). Crop residues of paddy, wheat, coarse cereals, pulses, and sugarcane and groundnut etc. have been estimated and found surplus in the UP part and deficit in the MP part of the region.

The average annual availability of forages in comparison to the forage requirement in the region seems to be satisfactory but the real story is not the same. It is found surplus from July to November and deficit from December to June. So, there is an urgent need to conserve the excess green forage production during monsoon season for lean period feeding.

Table 3: DM availability from roughages t/year (2000-2001)

	Total from crop residues	**DM from forest forest**	**DM from range-land**	**DM from waste-land**	**DM from CF CF**	**DM from rabi fodder**	**DM kharif fodder**	**DM from zaid fodder**	**DM from fodder**	**Total**
Jalaun	731456.2	89740	64.5	2046	62145	12285	25576.25	0	37861.25	961174
Jhansi	568144.7	119486.5	315	8557.5	99102.5	8067.5	1365	200.2	9632.7	814872
Lalitpur	402131	264684	1736	40592	44945	1628	88	162	1876.7	757841
Banda	779261	27237	230	16125	103180	6458	31859	31	38347.05	1002726
Hamirpur	670140	133308	430	11775	79923	2433	8374	185	10991.05	917557
Sub total t	**3151132**	**634456**	**2775**	**79095**	**389295**	**30870**	**67261**	**578**	**98708.75**	**4454170**
Sub total DM t	**2837999**	**634456**	**2775**	**79095**	**389295**	**30870**	**67261**	**578**	**98708.75**	**4141036**
Sagar	527203	1034765	0	33497	76768	0	0	0	246105	1918337
Datia	310445	82170	0	6348	15268	0	0	0	8190	422420
Chhattarpur	525628	748174	0	3867	171743	0	0	0	189608	1639019
Damoh	344513	934511	0	9420	36730	0	0	0	32993	1358165
Panna	272169	190813	0	31177	89043	0	0	0	0	583202
Tikamgarh	360747	94427	0	10994	77738	0	0	0	0	543905
Sub total t	**2340704**	**3084858**		**95302**	**467288**				**476895**	**6465047**
Sub total DM t	**2106722**	**3084858**	**0**	**95302**	**467288**	**0**	**0**	**0**	**476895**	**6231065**
Total	5491837	3719314	2775	174397	856583	30870	67261	578	575604	10919217
DM	4944721	3719314	2775	174397	856583	30870	67261	578	575604	10372101

Mahoba and SM Nagar districts of UP included in their parent districts. Data of Tikamgarh (1996) Panna (1996) Damoh (2001)

Table 4: Concentrate Production in Bulndelkhand-1999-2000

District	**Total cereals**	**Total pulses**	**Total oilseeds**	**Total**	**DM**
Jalaun	39384.95	55447.3	2201.2	97033.45	87330.11
Jhansi	21520.95	48845.4	7016.8	77383.15	69644.84
Lalitpur	23787.7	19168.55	2981.2	45937.45	41343.71
Banda	63648.584	49270.15	1227.154	114145.9	102731.3
Hamirpur	28213.082	54510.14	2349.058	85072.28	76565.05
SubTotal	176555.27	227241.5	15775.41	419572.2	377615
Sagar	17851.65	33791.9	22388	74031.55	66628.4
Datia	12110	19482	1700	33292	29962.8
Chhattarpur	23572.6	27556.2	8570.6	59699.4	53729.46
Damoh 01	25496.35	35349.8	14193	75039.15	67535.24
Panna 96	29338.75	22733.3	5330.4	57402.45	51662.21
Tikamgarh 96	19139.7	12439.9	19776.2	51355.8	46220.22
SubTotal	127509.05	151353.1	71958.2	350820.4	315738.3
Total	304064.32	378594.6	87733.61	770392.6	693353.3

Economic Analysis of Livestock Production Systems

Gomez et al.,, 1998 found four livestock production systems prevailing in Bundelkhand i.e. Free Range Grazing system, Mixed system, Extensive Stall fed system and Intensive Stall fed system. The economic analysis as presented in table 6. Revealed that the gross margins ranged from Rs.1147 (Free Range Grazing system) to Rs. 22560 (Intensive Stall fed system). They further observed that in real economic terms i.e. when all the costs (including cost that do not require cash payments, like imputed value of farm produced fodder and crop byproducts, imputed interest on capital, and family labour) are taken into account- Free Range Grazing would be the only viable system. All other system would produce negative real gross margins. The positive household cash incomes fro the mixed and stall fed systems would only be obtained when farmer do not calculate the costs of their own labour and capital involved nor in the case of mixed system of the crop residues from their own land. The fact that the stall fed only produced a positive household cash income if we assume maximum production at best estimate (medium) input levels indicate that these systems can only be viable if they use animals with a high conversion efficiency.

Livelihood Analysis

By livelihood analysis of Jhansi district, though not a comprehensive analysis, the share of different enterprises in farmers total income has been estimated. The various enterprises included were crop, dairy, labour, business and job. The landless farmers' livelihood was dependant mainly on labour and dairy. In case of marginal farmers, crop and dairy was major source of income. The share of crop in small farmers' livelihood increased to 67 per cent and that of dairying reduced to 26 per cent. In case of medium farmers', their 77 per cent income was from crops and about 20 per cent income was from dairying. About 86 per cent income of large farmers was from crop enterprise and remaining 14 per cent from dairying.

Knowledge and adoption of feed and fodder and their processing technologies

In a sample survey in four villages of Bundelkhand (2004) i.e. Chirula and Garera (Datia), Ronija and Baral (Jhansi) representing the four directions, the knowledge and adoption of feed and fodder and their processing technologies were estimated. The farmers were well aware of the different sources of dry and green fodder; and concentrate composition but the adoption was very low due to various reasons. The knowledge and adoption of the compound feed, dry fodder storage, enrichment, conservation was very low. A cursory look on table 7 revealed that farmers were having 41.16 percent knowledge on feed and

Table 5: **Balance sheet of DM t/year (2000-2001)**

District	DM required	DM required from roughage	DM required from conc.	DM available roughages	DM available concentrate	Total DM available	Total DM Surplus /deficit	% availability of Total DM	% availablity of Total DM from roughages	% availablity of Total DM from concentrate
U.P.										
Jalaun	1114574	746764.58	246432.3114	961174.25	87330.11	1048504	-66069.648	94.1	128.71	35.44
Jhansi	1237089	828849.63	273520.3779	814871.56	69644.84	884516.4	-352572.61	71.5	98.31	25.46
Lalitpur	1252012	838848.04	276819.8532	757840.63	41343.71	799184.3	-452827.67	63.8	90.34	14.94
Banda	1508350	1010594.5	333496.185	1002726.4	102731.3	1105458	-402892.34	73.3	99.22	30.80
Hamirpur	1335132	894538.44	295197.6852	917557.06	76565.05	994122.1	-341009.89	74.5	102.57	25.94
Mahoba and SM Nagar included in their parent district										
Sub total	6447156	4319594.52	1425466.192	4141036.4	377615	4518651	-1928504.6	70.1	95.87	26.49
M.P.										
Sagar	1600461	1072308.87	353861.9271	1918336.7	66628.4	1984965	384504.103	124	178.90	18.83
Datia	687847.5	460857.825	152083.0823	422419.5	29962.8	452382.3	-235465.2	65.8	91.66	19.70
Chhattarpur	1096422	734602.74	242418.9042	1639019.3	53729.46	1692749	596326.798	154.4	223.12	22.16
Damoh	1436702	962590.34	317654.8122	1358165.3	67535.24	1425701	-11001.455	99.2	141.09	21.26
Panna	1843850	1235379.5	407675.235	583201.53	51662.21	634863.7	-1208986.3	34.4	47.21	12.67
Tikamgarh	687847.5	460857.825	152083.0823	543904.54	46220.22	590124.8	-97722.743	85.8	118.02	30.39
Sub total	7910775	5300219.25	1749072.353	6231064.7	315738.3	6546803	-1363972	82.8	117.56	18.05
Total of MP &UP										
DM t/year	14357931	9619813.77	3174538.544	10372101	693353.3	11065454	-3292476.6	77.1	107.82	21.84

Table 6: Gross production value, costs and gross margins (in Rs./animal/Year) for various systems

Variable	Free Range Grazing system	Mixed system			Extensive Stall fed system		Intensive Stall fed system		
		Indigenous cow		Buffalo	Indigenous cow	Buffalo	Indigenous cow	Crossbred cow	Buffalo
		Poor	Good						
Fresh Milk	1260	1980	13020	27960	4080	16320	27440	41760	41400
Dung	150	300	300	300	500	500	600	600	600
Sale from calf (Imputed)	600	600	800	1000	800	1000	400	1000	1000
Gross production value	2010	2880	14120	29260	5380	17820	28440	43360	43000
Labour									
Hired	0	1200	1200	1200	0	0	1040	1040	1040
Family (Imputed)	390	640	640	640	360	360	440	440	440
Concentrates	0	900	1575	4275	1620	3240	1560	3120	3120
Other feeds	0	3240	2520	7020	2190	2670	1927	4150	3423
Veterinary services	0	250	250	250	400	400	250	250	250
Interst on capital (24%)	240	240	720	2880	720	4320	960	4800	6000
Depreciation	233	233	667	3000	667	500	1000	7000	9500
Transportation (Imputed)	0	72	96	120	240	300	0	0	0
Gross costs	863	6775	7668	19385	6197	16290	7177	20800	23773
Gross Margins	1147	3895	6452	9875	817	1530	21236	22560	19227
									Gomez *et al*, 1998

fodder and their processing technologies. The farmers were having adequate knowledge on dry fodder (75.35%), green fodder (71.48%) and concentrate composition (54.93%) whereas on enrichment of low graded roughages (2.11%), compound feed (12.68%) and forage conservation (24.65%), poor knowledge was observed. The over all adoption of these technologies was 28.77 percent. The concerned data revealed that the adoption of dry fodder, green fodder and concentrate composition was near about 50 percent while for other technologies the adoption was very poor.

Table 7: Knowledge and adoption of feed and fodder and their processing technologies

Sl.No	Parameters	Knowledge (%)	Adoption (%)
1	Dry fodder	75.35	54.23
2	Green Fodder	71.48	50.35
3	Dry fodder processing	44.72	23.94
4	Green Fodder process	45.77	26.06
5	Concentrate composition	54.93	46.48
6	Concentrate processing	44.72	30.28
7	Compound feed	12.68	3.52
8	Conservation	24.65	20.77
9	Dry fodder storage	35.21	30.63
10	Enrichment	2.11	1.41
	Pooled	41.16	28.77

Requirement of animal feed processing technologies

It is necessary to increase the knowledge and adoption of the feeding technologies of the farmers but before that it is equally important to ascertain the farmers' requirement regarding feeding technologies. The data presented in table-8 reveals that most of the farmers were not aware of feed processing technologies only progressive farmers were explained their requirement. The technologies mainly required by the farmers were silage making (25.35%), chaffing machine (21.13%), densification of straw (19.71%), storage for dry fodder (16.90%), grinding machine for concentrate (15.49%), and mineral supplementation (14.08%).

Table 8: Requirement of animal feed processing technologies

Sl.No	Dry fodder	Percentage	Rank
1	Storage	16.90	IV
2	Densification of fodder	19.71	III
3	Block making of straw	2.82	IX
4	Chaffing of fodder	21.13	II
5	Fresh fodder	7.04	VII
6	Market	4.23	VII
7	Grinding of Concentrates	15.49	V
8	Mineral Mixture	14.08	VI
9	Silage	25.35	I
10	Hay	7.04	VII

Problems in adoption of improved feed processing technologies

The farmers were asked to give their opinion regarding what problems they are facing in adoption of the improved feed and fodder processing technologies. The farmers were knowing and adopting the technologies such as haymaking, soaking, grinding, boiling and chaffing and make the Saani of the ingredients. Regarding the other technologies such as CRB Machine, Bailing / densification, Chemical treatment, Ensilage, Concentrate processing, Balanced rationing, Microbial treatment, Use of feed supplement, Compound feed etc.

The farmers of the IVLP adopted village were knowing these technologies mainly due to their exposure to IGFRI. For them the major problems were cost, easy availability of the technologies and poor quality of the compound feed available in the market. The farmers were of opinion that it is better to use home gron ingredients of concentrate in comparison to compound feed.

According to S.V. Vaidya (www.fao.org.DOCREP/ARTICLE/AGRIPPA/X9500E01.HTM) Cattle feeding practices are very traditional. Farmers choose their own ingredients and prepare their own formulations, believing that by these means they are able to pay more individual attention to their cattle. The productivity of the cattle is limited because of their poor genetic make-up, so high-quality compound feed (industry feed) may not necessarily generate a significant improvement in productivity and this has hampered growth of the cattle feed industry because most farmers are reluctant to use compound feed fully. Instead they compromise by using such feed in proportions of 5 to 60 percent, making up the balance with their own formulations. It is only in the case of highly productive animals that compound feed has been able to show its real potential and the importance of technology has been demonstrated.

The share of compound cattle feed manufactured by the industry, in relation to the overall potential, is low for the following reasons:

- The cattle population is fragmented and spread over large parts of the country. Farmers' low level of education and strong traditional beliefs mean that there is generally little awareness of compound cattle feed.
- More than 50 percent of the country's total milk production comes from a very large number of low-yielding cows and buffaloes. A further 25 percent of milk production comes from buffaloes and only the remaining 25 percent of the total is produced by cross-bred and improved cows.
- Industrially manufactured compound cattle feed has proved its value for cross-bred cows and buffaloes but not for low-yielding cattle because of their genetic limitations. Home-mixed feed is very frequently used for buffaloes and low-yielding cattle.

In the case of layer feed, cost is the main constraint in using compound feed. An innovative, high-value compound feed can result in increased numbers of eggs, but the risks are too high because of the birds' long life cycle. Compound feed has, however, made a major contribution to broiler feeding. This is an example of excellent coordination among instrument technology, formulations and use of feed additives and supplements. Cost is a less important factor because the performance improvements are greater than the cost increases and the birds' life cycle is short. Two types of poultry feed are prepared. One is ready-made and in the form of mash or pellets. The second is in concentrated form for mixing with an energy source. Concentrates are protein sources, balanced in amino acids and containing vitamins, minerals and feed additives. They are mixed with energy sources such as maize, sorghum or bajra to prepare poultry rations.

Singh and Mehra 1990 concluded that a complete feed block containing ammoniated wheat straw, molasses, minerals, salts, vitamins and a small amount of rumen bypass protein is suitable for buffalo experiencing drought and famine. Blocks occupy less space than uncompressed feed and are easy to handle and store. A 20-day digestibility trial using crossbred cattle 3 to 4 years old fed on baled or chaffed and dried hay (Trifolium alexandrinum) indicated that total digestible nutrients, digestible crude protein and DM intake was higher in the dried chaffed hay than in the baled hay (Gupta and Singh, 1992).

Gupta et al.,., 1988 reported that slowly released ammonia was entrapped efficiently by microbes, resulting in significantly higher digestibility of DM and crude fibre in vitro, indicating that Urea and rice straw or husk products (USP) can be utilized efficiently as a livestock feed. The use of fungi and bacteria to degrade the fibre components in straw to upgrade the nutritive value of straw for ruminants is reviewed (Manorama et al., 1991). Ramana et. al. 1989 concluded that 4% urea was effective in improving the nutritional quality of bajra stalks when incubated for 20 days at 30% moisture. Stiefel et al., 1991 indicated that under tropical conditions cereal straw treatment with urea 2 days instead of longer periods is worth considering by farmers with limited resources. Singh et al., 1993 Adoption of urea-ammonia treatment technology under village conditions depends on the effectiveness of the urea-treated straw as an animal feed. The overall economics of using urea-treated straw depends on the animal response, saving of concentrate and oil cakes and availability of green fodder. The high cost of urea, labour, water and straw. lack of awareness or low price of product and other socio-economic conditions are the constraints which result in the failure of technology of the farming systems. Virk and Kamphues 1995 concluded that feeding of urea ammoniated straw supplemented with barley, germinated barley or non-legume green forage as substitutes of protein, to growing buffaloes/cattle or milch animals was safe and profitable. With urea-treated paddy straw feeding under on-farm conditions, intake was improved with less wastage. The effect on milk yield was imperceptible. The cost of urea and the involvement of extra inputs (labour, time etc.) are major constraints in adoption of this technology. It is suggested that a series of extended demonstration trials in selected areas of North Bihar are needed to facilitate adoption of this system (Sohane 1994).

Rai et al.,., 1991 concluded that use of paddy straw reduced feed costs in villages and towns compared with wheat straw. The overall feed cost for maintenance plus milk production and income over feed costs was higher in urban than in rural areas. Rice bran was revealed as the cheapest energy source followed by cassava peelings, wheat bran, sorghum then maize. Cheapest protein sources were cottonseed oilmeal and sunflower oilmeal followed by groundnut oilmeal and soyabean oilmeal (Krishna and Ramana 1991). Thakur et al., 1997 reported that high-moisture green jowar fodder can be preserved satisfactorily after treating with 0.5% urea + 1% common salt. The benefits of using feed blocks in India instead of loose feeds or pelleted feeds include reduced cost of transportation, reduced cost of storage, easier to achieve a balanced diet with little wastage, cost, and the new technology may enable feed banks to be set up which will provide feed in times of floods or drought (Rathee et al., 1994). Misra et al., 2000 recommended for using plastic bags for ensiling grass in the field conditions specially for small farmers.

Causes of the generally poor quality of crop residues as feeds include their high contents of silica, lignin, oxalates and phenolic compounds, which inhibit intake, and their slow rate of digestion by microorganisms resulting in an increased amount of digesta retained in the rumen. Means of reducing these problems are considered under the following headings: physical treatments (chaffing, soaking, steam treatment, irradiation, grinding and pelleting), chemical treatment (including sodium hydroxide, calcium hydroxide, ammonia and urea treatment), supplementation (nitrogen as urea, oilcake and high rumen bypass protein such as cottonseed cake and fish meal; minerals; energy; urea-molasses; or forages). It is concluded that each method has its own advantages and limitations. The method to be adopted would depend on convenience, the facilities available at the field level and the economic benefits derived (Prasad and Rao, 1994).

Identification of Suitable Interventions in existing feeding system

S.No	Interventions
1.	Mineral mixture supplementation
2.	Availability of green fodder round the year specially green fodder in summer season.
3.	Balancing concentrate mixture
4.	The proportionate DM intake from different sources.
5.	Increasing awareness about treatment of poor quality roughages to improve their nutritive value
6.	Grazing resources improvement.
7.	Increase the knowledge on fodder crops
8.	Improving livestock feeding skills
9.	Easy to handle fodder conservation methods
10.	Adequate quantity of concentrates for advanced pregnant animals need to be fed.
11.	Balancing protein in ration of lactating cows in summer by all categories of farmers.
12.	The balanced ration for lactating buffaloes in terms of DM, DCP and TDN.
13.	Balanced ration of dry and pregnant animal
14.	Better feeding and management of heifers.
15.	Colostrum feeding to calves by all farmers.

List of feeding technologies for implementation

1. Mineral supplementation.
2. Summer fodder crops such as multi-cut sorghum, maize, bajra and cowpea should be grown.
3. Round the year fodder production involving the perennial grasses (H. Napier, Guinea grass, Setaria) on bunds and between bunds seasonal fodder crops can be grown.
4. Stylo can be grown on wasteland and grasslands.
5. Concentrate mixture preparation from locally available ingredients which have about 18% CP and about 70 % TDN.
6. Urea-Mollasses treatment of low grade roughages for improving their nutritive value.
7. Fodder conservation techniques such plastic bag silage, silo pit for large scale silage making, hay for maintaining supply during lean period.
8. Stylo leaf meal, leucena leaf meal for strategic supplementation during summers.
9. Managed / rotational grazing and promote stall feeding of high yielding animals.
10. Farmers can easily and conveniently follow thumb rule method of ration formulation.

Bibliography

Gomez, Norma.; Kombiok, James.;Mshana, Yahya.; Puskur, Ranjitha.; Soam, Sudhir Kumar. And Williams, Stephen.1998. Towards sustainable livestock development- opportunities and constraints for conversion free range grazing to managed feeding regime in Bundelkhand region. ICRA- India field study 1998.

Gupta-PC; Singh, K .1992. Chemical composition and nutritive value of baled hay vs. chaffed and dried hay of berseem. Livestock-Adviser. 17: 4, 25-27.

Gupta-RP; Bakshi-MPS; Punj-ML .1988. Preparation and evaluation of urea rice straw or husk product as livestock feed. Indian-Journal-of-Animal-Nutrition. 5: 4, 280-284.

Mangurkar, B.R. 2001. Livestock-environment interaction in mixed farming system. Indian Dairyman, 53(4): 50-54.

Manorama; Prajapati-JB; Neelakantan-S .1991. Bioconversion of straw for ruminant feeding - a review. Asian-Journal-of-Dairy-Research. 10: 1, 1-9.

Misra, A.K.; Samanta,A.K.; Verma,N.C. and Singh, A.P. 2000. On-farm trial on silage: effect of urea and molasses addition on the quality of garu (Coix lacryma jobi) grass silage ensiled in plastic bags. Indian-Journal-of-Animal-Sciences. 70: 6, 662-663.

Mudgal, V.D.2000. Sustainable milk production and productivity enhancement to match global competition. Indian Dairyman, 52 (12):33-37.

Prasad, C.S. and Rao, A.S. 1994. Methods to improve the intake and nutritive value of crop residues. Indian-Journal-of-Dairy-and-Biosciences. 5: 85-88.

Rai, S.N,; Schiere, J.B. and Gupta, B.N.1991. Cost of feeding (treated) straw vis-a-vis green fodder for milk production under different farming systems. Indian-Journal-of-Animal-Nutrition. 8: 2, 103-110.

Ramana, J.V. ; Krishna, N.; Parthasarathy, M.; Prasad ,J.R. and Prasad, D.A. 1989. Evaluation of urea treated bajra (Pennisetum typhoideum) stalks by chemical, in-vitro and in sacco techniques. Indian-Journal-of-Animal-Nutrition. 6: 3, 205-214.

Rathee, C.S.; Lohan, O.P.; Yadav, K.K. 1994. Densified feed blocks. Livestock-Adviser. 19: 3, 6-7.

Saran, Sandeep.; Singh, R.A., Singh, Rajveer.; Rani S. Issabella and Singh, K.K. 2000. Feed resources for rearing livestock in the Bulndelkhand region of Uttar Pradesh. Indian J. Anim. Sci. 70 (5): 526-529.

Singh, UB; Mehra, R. 1990. Utilization of ammoniated wheat straw given in a feed block and supplemented with varying quantities of fish meal and oil-extracted rice bran. Animal-Feed-Science-and-Technology. 28: 1-2, 129-134.

Sohane, R.K. 1994. On-farm research with urea treatment of paddy straw. Indian-Journal-of-Animal-Production-and-Management. 10: 4, 140-142.

Stiefel, J; Babu, Y.K.; Butchaiah, V. and Reddy, N.G. 1991. Short duration urea treatment of paddy straw as ruminant feed. Indian-Journal-of-Animal-Nutrition. 8: 3, 233-236.

Thakur, S.S.; Kundu, S.S.; Oberoi, P.S. and Sharma, D.D. 1997. Chemical characteristics and degradability of green jowar fodder preserved by urea plus common salt treatment. Indian-Journal-of-Animal-Nutrition. 14: 2, 65-69.

Tyagi, Rajkumar. 1997. Grassland and fodder atlas of Bulndelkhand. Bulletin published by IGFRI, Jhansi (UP)-284003, India.

Varshney, N.N.2000. Compatibility of Indian Dairy Industry with global market. Indian Dairyman, 52 (12): 129-137.

Virk-AS; Kamphues-J 1995Urea ammoniation of crop residues - nutritional enrichment vis-a-vis environmental pollution. Journal-of-Ecotoxicology-and-Environmental-Monitoring. 5: 2, 89-98.

Assumption and constant values

1. DM requirement

Category	B. Weight	DM required
Growing Cattle	200 kg	5.02 kg
Growing Buffalo	220 kg	5.30 kg
Cattle	350 kg	6.5 kg
Buffalo	400 kg	7.5 kg

2. Grain to straw ratio

Wheat	1:1.17	DM	Urd	1:0.5	DM
Rice	1:1	90%	Moong	1:0.5	90%
Barley	1:1		Masur	1:0.5	
Sorghum	1:4		Guar	1:0.6	
Bajra	1:4		Pea	1:0.6	
Maize	1:2.75		Arhar	1:0.6	
Other millets	1:1		Maloi seed	1:1	
Sugarcane	1:0.25	35%			

3. DM Source and yield of DM

Forest	3.5 t/year
Rangeland	0.5 t/year
Wasteland	0.5 t/year
Current Fallow	2.5 t/year
DM in concentrate	90%

4. Yield from

Rabi fodder	70 t/ha	DM 25%
Kharif	25 t/ha	DM 35%
Zaid	22 t/ha	DM 35%

5. Grain and by-products availability as concentrate

Arhar	30%	Soybean	15%
Gram	30%	Barley	70%
Pea	30%	Maize	30%
Masur	5%	Bajra	20%
Urd	10%	Sorghum	20%
Moong	10%	Wheat	5%
Til	5%	Rice	35%
Lentil	55%	Millets	20%
G.nut	15%	Mal oilseed	20%

CHAPTER - 24
CODEX ALIMENTARIUS AND LIVESTOCK

A. Kumar, R. Singh, S.S. Kundu and G.H. Pailan

The World Trade Organization (WTO) came into existence on 1 January 1995, after the protracted Uruguay Round negotiations spanning 1986-94. It brought into its preview the trade in services, knowledge (Intellectual Property Rights) and trade related investments. The Uruguay Round negotiations also resulted in a separate agreement on agricultural products. The Government of India ratified the WTO agreements after taking into account the perceived balance of rights and obligations conferred by these agreements. The WTO provides a rule based, transparent and predictable multilateral trading system which protects the member countries from the pressures of stronger trading partners. WTO rules envisage non-discrimination in the form of National Treatment and Most Favoured Nation (MFN) treatment to Member countries. The WTO rules also ensures that India's products once imported into the territory of other WTO members would not be discriminated vis-à-vis the domestic products in those countries. The MFN treatment ensures that members do not discriminate among various WTO members not only in their tariff regimes but also in respect of various other rules, regulations, incentives etc. The three main functions of WTO are (i) make international trade as free as possible by ensuring that the ground rules are made predictable and transparent, (ii) act as a forum for trade negotiations among countries, and (iii) work as a neutral agency for settlement of trade disputes.

The highest decision making body of the WTO- the Ministerial Conference meets every two years starting from 1996. The latest in the series was the Cancun meet (Sept.10-14, 2003) which was preceded by Doha (Nov.9-14, 2001). After the Cancun Ministerial Conference ended in deadlock, the WTO Members agreed to keep the Doha Round trade negotiations alive. The Doha round of Ministerial Declaration on agricultural sector mainly aimed at: (i) substantial improvement in market access, (ii) reduction of, with a view to phasing out , all forms of export subsidies, (iii) substantial reduction in trade-distorting domestic support and (iv) more effective special and differential treatment for developing countries by taking into account their developmental needs including food security and rural development.

The long term objective of the Agreement on Agriculture was to establish a fair and market oriented trading system through a programme of fundamental reform encompassing strengthened rules and specific commitments on domestic support, export subsidies and protection in order to correct and prevent restrictions and distortions in world agricultural markets. The agriculture sector in many developed countries continued to be characterized by high level of support and protection. Simultaneously, the market access to developing countries is being resisted by the developed countries and barriers on export of agricultural commodities including livestock products is being erected in the disguise of Sanitary and phytosanitary Measures (SPS) and Technical Barriers to Trade (TBT), both of which are in essence, non-tariff barriers (NBT).

The application of Sanitary and Phytosanitary Measures (SPS) which came into force with the establishment of WTO, confers the right of WTO Member countries to apply measures to protect human, animal and plant life and health. Member countries were asked to apply only those measures for protection that are based on scientific principles, only to the extent necessary and not in a manner which may constitute a disguised restriction on international trade.

A Technical barrier to trade (TBT) is a specific type of NTB that inhibits or prevents trade through the use of regulations or standards. These measures generally focus on ensuring that imported products satisfy domestic tastes, preferences and requirements with respect to quality or appropriate consideration of environmental concerns during manufacture, packaging and/or shipment of the products. A technical regulation can be defined as any document stating product characteristics or their related processes and production methods with which an imported product must conform. Safety regulations and environmental protection regulations are just a few examples. As compliance is mandatory, any import that does not meet the stated requirements of a technical regulation is not permitted entry into the country. The TBT is generally perceived to encômpass issues related to product weight, size of packaging material, ingredient or identity standards, mandatory nutritional labeling and shelf life restrictions. However, in future the TBT may also cover the aspects related to ethical issues in animal production, which are broadly covered under organic animal production.

The increase in volume of international trade in products of animal origin and the prospects of India making its presence felt may be seriously jeopardize if our products do not conform to the SPS measures and TBT. One way to beat these NTB is to get concessions in the multilateral negotiations held under the banner of WTO, but in the long run we need to have standards in place which conforms to international practice. The Codex standards serve as a benchmark for comparison of national sanitary and phytosanitary measures. While it is not compulsory for member states to apply Codex standards, it is in their best interests to harmonize their national food standards with those elaborated by Codex. The Codex Alimentarius Commission (CAC) is an intergovernmental body under FAO/ WHO that coordinates food standards at the international level. Its main objective are to protect the health of consumers and ensure fair practices in food trade. The CAC has proved to be most successful in achieving international harmonization in food quality and safety requirements. It has formulated international standards for a wide range of food products and specific requirements covering pesticide residues, food additives, veterinary drug residues, hygiene, food contaminants, labeling etc. Codex work has created worldwide awareness of food safety, quality and consumer protection issues and has achieved international consensus on how to deal with them scientifically through a risk based approach. As a result there has been a continuous appraisal of the principles of food safety and quality at the international level.

The Codex Alimentarius covering subjects relevant to livestock and livestock products are covered under following volume:

Vol. 1 : General requirements

Vol 1B: General requirements (food hygiene)

Vol. 2A : Pesticide residues in food (general texts)

Vol. 2B : Pesticide residues in food (maximum reside limits)

Vol. 3 : Residues of veterinary drugs in foods

Vol. 10 : Meat and meat products; soups and broths

Vol. 12 : Milk and milk products

Vol. 13: Methods of analysis and sampling

Importance of Sanitary and Phytosanitary Measures (SPS) and Technical barriers to Trade (TBT) for Livestock sector in India

The Indian Livestock industry is poised to take off in a big way in coming decades. The livestock revolution unfolding on Indian horizon has set the alarm bells ringing among some of the developed countries. The country which once was importing huge quantity of milk and milk products has now become the leading producer of milk in the world. Very soon the country may have a substantial quantity of milk and milk products ready for export. But the countries which already have surplus milk and milk products may resist it by raising the non-tariff barriers by bringing into the issues related to quality aspects of the products. They have also in the past tried to beat up the non issues like berating the milk from India as being produced by strange black animals (buffalo). Therefore, we cannot let ourselves be caught unawareness on the emerging issues where we can hold a significant share in the world market in products of animal origin. Similarly, the export potential of buffalo meat in middle east countries is yet to be realized where again the quality aspects will come to haunt us if we continue to ignore the hygiene and quality control aspects.

Therefore, the implementation of Codex Alimentarius needs attention, not only for production of food for human beings but also for animals. In the endeavour to produce livestock products for International trade, the attention has to be paid at the beginning of food chain which concerns with way the animals are fed and raised. The adoption of Codex Alimentarius for 'Good Animal Feeding Practices' (Annexure-I) will determine the quality of livestock product, which will allow us to catch a significant share in the world market. Even otherwise, the domestic consumers stand to get an assured high quality of products of animal origin. Moreover, the health of our livestock will see a marked improvement with fewer veterinary problems, simultaneously breaking the chain of several microorganisms and contaminants in the food chain.

HACCP in feed industry

The Hazard Analysis and Critical Control Point (HACCP) which forms an important component in the implementation of 'Good Animal Feeding Practices' and has a wider ramifications for the feed industry, has been described in detail. Application of HACCP in the Feed industry will ensure quality livestock feed available to animals to give animal product of International standards. Simultaneously, it will also help explore the opportunities for the export of their product. The products used in country's market will affect the health and well being of the people positively. HACCP is a system that identifies, evaluates and controls hazards, which make a food unsafe for human consumption. HACCP is a systematic approach for the control of food safety throughout the commodity system of food chain. Before HACCP, most of the food systems were based on end product testing which did not assure 100 per cent safety. It requires understanding of the relationship between cause and effect in order to be more proactive.

The Development of HACCP system was started in 1959 when Pittsburgh Company was asked to develop in cooperation with National Aeronautics Space Agency (NASA), US Army and US Air Force, a food that could be used in the space capsules. The demand was for such an advanced food, that would not be contaminated with bacterial/viral pathogen, toxins, chemicals or physical hazards, that could cause an illness or injury. Such a hazard might result in catastrophic space mission.

The existing technique of quality control at that time could not assure that there would not be a problem. While looking for a new system, Bauman and his colleagues developed a new preventive system that enabled the management control over raw materials, process, environment, personnel, storage and distribution as early in the system as they could. If a system could be established together with record keeping, one should be able to give a product with high degree of assurance that it would be safe. For practical purpose, if this system were implemented correctly, there would be no testing requireds of finished products other than for monitoring purposes (Pierson and Corlett, 1996). The basic principle of HACCP signaled a shift in emphasis from end product testing to preventive control of critical aspects of food production. The HACCP concept has been refined in and International Life Science Institute (ILSI) Europe gave the definition:

HACCP is a concept which permits a systematic approach to the identification and assessment of hazards and risks associated with the manufacture, distributions and use of food product, as well as the definition of preventive measures for their control. The application of HACCP in food production has been well known in developed countries. However, application of HACCP in feed production is not that known. Although the hazard for animals may not be as critical as for human, the growing concern on the animal product contamination in food chain- which may be caused by contamination of feed, make the concept of HACCP quite relevant for the livestock feed industry. Nutritional quality of feed should be guaranteed in order to achieve maximum and consistent animal production. This has relevance concerning not only the health of the population consuming the products of animal origin but also the International trade will be increasingly governed by safety of food products as defined by the Codex guidelines.

The HACCP system, which is science based and systematic, identifies specific hazards and measures for their control to ensure the safety of food. HACCP is a tool to assess hazards and establish control systems that focus on prevention rather than relying mainly on end product testing. Any HACCP system is capable of accommodating change, such as advances in equipment design, processing procedures or technological developments. The HACCP system consist of the following seven element:

1. Conduct hazard analysis
2. Determine the Critical Control Points (CCPs)
3. Establish critical limits
4. Establish system to monitor control of Critical Control Points
5. Executions of corrective actions when critical limits are not met
6. Establish procedures for verification that HACCP system is working
7. Record keeping

Conducting hazard analyses

It involves a systematic study of the ingredients, the final ration or products, condition of processing, handling, packaging, storage, distribution and consumer use. The study will cover identification of the hazard, assessment of the severity and the risk. From this information, it will be to determine the critical control points in the system that has to be monitored. In conducting the hazard analysis, wherever possible the following should be included:

- the likely occurrence of hazards and severity of their adverse health effects;
- the qualitative and/or quantitative evaluation of the presence of hazards;
- survival or multiplication of microorganisms of concern;
- production or persistence in foods of toxins, chemicals or physical agents; and,
- conditions leading to the above.

Biological, chemical, physical and nutritional hazard

In the HACCP for feed, the term of hazard refers to any aspect of the feed production chain that is unacceptable because it is a potential cause of feed safety problem. National Advisory Committee on Microbiological Criteria for Food defines: "Hazard as any biological, chemical and physical properties that may cause an unacceptable consumer risk"

Similar principle may be applied to animal feed that any hazard can be classified to biological, chemical and physical however additional hazard is also important to animal feed, which is nutritional hazard.

Biological hazard

Animal feed can also be a source of biological hazard causing food borne illness to human beings. The source of biological hazard in animal feed could be fungi, bacteria and insects. Salmonella, which cause illness in human beings through infected egg or broiler meat, affects an estimated 2-4 million people annually, however, only 40,000 cases are reported each year. Growth of fungi or mould is a major problem in tropics because of high temperature and humidity. Poor harvesting of cereals encourages growth of fungi/ mould on raw materials like corn, rice bran, oil cakes etc., which together with poor pelleting conditions may encourage fungi to grow in final product. Fungi not only reduces the nutritive value of feed but also cause a severe problem when mycotoxin is produced. High mortalities in turkeys have been reported in the beginning of aflatoxin cases. The toxin, however, is classified as chemical hazard. Aspergillus sp. causes aspergillosis to the bird and several cases has been reported to occur in broiler.

Another biological hazard is infestation of insect due to poor storage condition or sanitation. Infestation of Tribollium or Sitophyllus are common problem during storage of feed. Many farmers may reject feed that contain insect. Although insects are not dangerous to animals, they cause a reduction in the nutritive value of feed.

Table 1: Source of biological hazard and their control measures

Pathogen	Control measures
Bacillus cereus	Proper holding and cooling temperatures of foods; thermal processing of shelf-stable canned food.
Campylobacter jejuni	Proper pasteurization or cooking; avoiding cross contamination of utensils, equipment; freezing; atmospheric packaging.
Clostridium botulinum	Thermal processing of shelf-stable canned food; addition of nitrate and salt to cured processed meats; refrigeration of perishable vacuum packaged meats; acidification below pH 4.6; reduction of moisture below water activity of 0.93.
Clostridium perfringens	Proper holding and cooling temperatures of foods; proper cooking times and temperatures; adequate cooking and avoidance of cross-contamination by unsanitary equipment.
Listeria monocytogenes	Proper heat treatments; rigid environmental sanitation programme; separation of raw and ready – to – eat production areas and product.
Salmonella spp.	Proper heat treatments; separation of raw and cooked product; proper employee hygiene; fermentation controls; decreased water activity; withdrawing feed from animals before slaughter; avoiding exterior of hide from contacting carcass during skinning; antimicrobial rinses; scalding procedures; disinfecting knives.
Staphylococcus aureus	Employee hygiene; proper fermentation and pH control; proper heat treatment and post-process product handling practices; reduced water activity.
Yersinia enterocolitica	Proper refrigeration; heat treatments; control of salt and acidity; prevention of cross- contamination.
Mycobacterium tuberculosis	Heat treatment and proper employee hygiene
Eschericia coli	Proper heat treatment and prevention of contamination

Chemical hazard

The chemical hazard in animal feed can occur due to i) Naturally occurring chemical and/ or ii) added chemicals. Naturally occurring chemicals in feed include toxicants are found in feed ingredients or secondary by-product, produced during metabolism of feed.

The growth of microorganisms produces several toxins, which include different mycotoxins. Biochemical reactions due to enzyme or oxidation may results in biogenic amine in fish product and peroxide in fat containing materials. Chemicals in feed may be added intentionally or contaminated during handling of raw materials. Several feed additives have been used to improve the quality of feed such as minerals, vitamins, antibiotic, feed preservative, growth promoters, processing modifier, enzyme etc. Most of these chemicals may improve animal production when used at recommended level. However, excessive or improper use may cause hazard not only to animal but also to human. Contaminated chemicals that can be detected in feed include: insecticide, pesticide, toxic element etc. They may be detected in fumigated feedstuff or the minerals that contain toxic chemicals (Vanadium or Pb in Rock phosphate etc).

Control of chemical hazard includes:

- control before receiving
- control before use
- control storage and handling condition
- making inventory all chemicals in facility

Physical hazard

Physical hazard is caused by extraneous matter not normally found in feed. They may cause illness or injury to animals or health risk. Materials which pose physical hazard include sand, wood, stone, plastic, metal, insect, rubber, paper. Methods involved in controlling physical hazard include raw material specification and inspection along with vendor certification and guarantee. Various preventive measures are available to find and remove physical hazard including metal removal, feed cleaner screen etc. Effective pest control and foreign object removal from plant environment are also essential. Proper receiving, distribution, storage procedures as well as packaging materials, handling practice may be evaluated for potential hazard. Employee education and practice should involve knowledge and prevention of physical hazard.

Determining Critical Control Points

There may be more than one CCP at which control should be applied to address the same hazard. The determination of a CCP in the HACCP system can be facilitated by the application of a decision tree, which indicates a logic reasoning approach.

Establish critical limits for each CCP

Critical limits must be specified and validated if possible for each Critical Control Point. In some cases more than one critical limit will be elaborated at a particular step. Criteria often used include measurements of temperature, time, moisture level, pH, pressure etc.

Establish a monitoring system for each CCP

The monitoring procedures should be able to detect loss of control at the CCP. Where possible, process adjustments should be made when monitoring results indicate a trend towards loss of control at a CCP. The adjustments should be taken before a deviation occurs. The monitoring procedures for CCPs need to be done rapidly because they relate to on-line processes and there will not be time for lengthy analytical testing. Physical and chemical measurements are often preferred to microbiological testing because they may be done rapidly and can often indicate the microbiological control of the product.

Establish corrective actions

Specific corrective actions must be developed for each CCP in the HACCP system in order to deal with deviations when they occur. The actions must ensure that the CCP has been brought under control. Actions taken must also include proper disposition of the affected product.

Establish verification procedures

Verification and auditing methods, procedures and tests, including random sampling and analysis, can be used to determine if the HACCP system is working correctly. The frequency of verification should be sufficient to confirm that the HACCP system is working effectively. Examples of verification activities include:

Review of the HACCP system and its records;

Review of deviations and product dispositions;

Confirmation that CCPs are kept under control.

Establish Documentation and Record Keeping

Efficient and accurate record keeping is essential to the application of a HACCP system. HACCP procedures should be documented. Documentation and record keeping should be appropriate to the nature and size of the operation.

Documentation examples are:

Hazard analysis;

CCP determination;

Critical limit determination.

Record examples are:

CCP monitoring activities;

Deviations and associated corrective actions;

Modifications to the HACCP system.

Harmonizing Indian laws along Codex Alimentarius

The Government of India is seized upon the need for harmonizing laws with the International practice so that our livestock products stand the stringent test in the International market. To regulate and control the standards in food a 'Food Safety & Standard Authority of India' (FSSAI) is proposed to be established involving the Ministries of Agriculture, Commerce, Consumer's Affairs, Food Processing, Health and Law& Justice. The scientific committee and scientific panels to be set up under FSSAI include panels on Food Additives, Plant Health, GMOs, Dietetic Products, Biological Hazards, Contaminants in Food Chain and a Panel on Labeling.

Adoption of Codex Alimentarius by the livestock sector in India by adopting 'Practices on Good Animal Feeding, and 'Practices for Organic Animal Production' will address to all the issues related to the food and feed safety of consumers. It will open up the opportunities for the livestock products from India to enter into the world market creating more employment opportunities and income to people. The setting up of 'Food Safety & Standard Authority of India' (FSSAI) by the Government will be a positive step in this direction to regulate this emerging sector.

Bibliography

Boring, S.A. (1995). Soybean processing quality control. In Practical handbook of soybean processing and utilization (Ericson, D.R., ed). AOCS press and USB St. Louis, MO.

Codex Alimentarius (2005). Practices on Good Animal Feeding. www.codexalimentarius.net. Accessed January 2005.

FAO/ WHO (2001). Codex Alimentarius –Food Hygiene: Basic Texts. Joint FAO/WHO Food standards Programme, Codex Alimentarius Commission, Rome.

FAO /WHO (2001). Codex Alimentarius – Organically Produced Foods. Joint FAO/WHO Food standards Programme, Codex Alimentarius Commission, Rome.

Farnum, N.R. (1975). Modern statistical quality control and improvement. Outbury Press, Belmount Calfornia.

ILSI (1997). A simple guide to understanding and applying the hazard analysis critical control point concept. ILSI Europe Sci. Com. And Microbiology.

Leslie, A. (1997). Quality control in feedmill. Procedures for an effective programm. Proc. 5th Regional ASA Feed Tech. & Nutr. Workshop 25-29 May 1997, Chiangmai, Thailand.

Pierson, M.D. and Corlett, D.A. (1996). HACCP Principles and Application. Chapman & Hall, N.V.

Tangendjaja, B. (2003) The application of Hazard Analysis and Critical Control Point Principle in Feed Manufacturing. American Soybean Association, New Delhi.

ANNEXURE-I

Codex Alimentarius on Good Animal Feeding Practice

The code on Good Animal Feeding tries to establish a feed safety system for food producing animals which covers the whole food chain, taking into account relevant aspects of animal health and the environment in order to minimize risks to consumers' health. This Code applies in addition to the principles of food hygiene already established by the Codex Alimentarius Commission taking into account the special aspects of animal feeding. The objective of this Code is to help ensure the safety of food for human consumption through adherence to good animal feeding practice at the farm level and good manufacturing practices (GMPs) during the procurement, handling, storage, processing and distribution of animal feed and feed ingredients for food producing animals. This Code of Practice applies to the production and use of all materials destined for animal feed and feed ingredients at all levels whether produced industrially or on farm. It also includes grazing or free-range feeding, forage crop production and aquaculture.

Those issues of animal welfare other than food safety related animal health are not covered. Environmental contaminants should be considered where the level of such substances in the feed and feed ingredients could present a risk to consumers' health from the consumption of foods of animal origin. While recognizing that, in its totality, a feed safety system would address animal health and environmental issues, in addition to consumers' health, this Code of Practice, in fulfilling the Codex mandate of consumer protection, only addresses food safety. Notwithstanding this, best efforts have been made to ensure that the recommendations and practices in this Code of Practice will not be detrimental to the more general animal health and environmental aspects of animal feeding.

General Principles and Requirements

Feed and feed ingredients should be obtained and maintained in a stable condition so as to protect feed and feed ingredients from contamination by pests, or by chemical, physical or microbiological contaminants or other objectionable substances during production, handling, storage and transport. Feed should be in good condition and meet generally accepted quality standards. Where appropriate, good agricultural practices, good manufacturing practices (GMPs) and, where applicable, Hazard Analysis and Critical Control Point (HACCP) principles' should be followed to control hazards that may occur in food. Potential sources of contamination from the environment should be considered.

Parties that produce feed or feed ingredients, those that rear animals for use as food and those that produce such animal products need to collaborate to identify potential hazards and their levels of risk to consumers' health. Such collaboration will enable the development and maintenance of appropriate risk management options and safe feeding practices.

Feed Ingredients

Feed ingredients should be obtained from safe sources and be subject to a risk analysis where the ingredients are derived from processes or technologies not hithelio evaluated from a food safety point of view. The procedure used should be consistent with the Working Principles for Risk Analysis for Application in the Framework of the Codex Alimentarius. Manufacturers of feed additives in particular should provide cleat' information to the user to permit correct and safe use. Monitoring of feed ingredients should include inspection and sampling and analysis for undesirable substances using risk-based protocols. Feed ingredients should meet acceptable and. if applicable, statutory standards for levels of pathogens, mycotoxins, pesticides and undesirable substances that may give rise to consumers' health hazards.

Labelling

Labelling should be clear and informative as to how the user should handle, store and use feed and feed ingredients. Labelling should be consistent with any statutory requirements and should describe the feed and provide instructions for use. Labelling or the accompanying documents should contain, where appropriate:

- information about the species or category of animals for which the feed is intended; the purpose for which the feed is intended;
- a list of feed ingredients, including appropriate reference to additives. in descending order 01 proportion:
- contact information of manufacturer or registrant;
- registration number if available; directions and precautions for use;
- lot identification:
- manufacturing date; and
- use before or expiry date

Traceability/Product Tracing and Record Keeping of Feed and Feed Ingredients

Traceability/product tracing of feed and feed ingredients, including additives, should be enabled by proper record keeping for timely and effective withdrawal or recall of products if known or probable adverse effects on consumers' health are identified. Records should be maintained and readily available regarding the production. distribution and use of feed and feed ingredients to facilitate the prompt trace-back of feed and feed ingredients to the immediate previous source and trace-forward to the next subsequent recipients if known or probable adverse effects on consumers' health are identified.

Special Conditions Applicable to Emergency Situations

Operators should, as soon as reasonable, inform the competent authorities in the country if they consider that a feed or feed ingredient does not satisfy the feed safety requirements established in this Code. The information should be as detailed as possible and should at least contain a description of the nature of the problem. a description of the feed or feed ingredients, the species for which it is intended. the lot identifier. the name of the manufacturer and the place of origin. The competent authorities and operators should immediately take effective measures to ensure that those feed or feed ingredients do not pose any danger to consumers' health. As soon as it becomes likely that a particular feed or feed ingredient is to be traded internationally and may pose a danger to consumers' health. the competent authorities of the exp01l ing countries should notify, at least, the competent authorities of the relevant importing countries. The notification should be as detailed as possible and should at least contain the par l iculars indicated in the previous paragraph.

Inspection and Control Procedures

Feed and feed ingredients manufacturers and other relevant parts of industry should practice self-regulation/auto-control to secure compliance with required standards for production, storage and transpol1. It will also be necessary for risk-based official regulatory programmes to be established to check that feed and feed ingredients are produced, distributed and used in such a way that foods of animal origin for human consumption are both safe and suitable. Inspection and control procedures should be used to verify that feed and feed ingredients meet requirements in order to protect consumers against food-borne hazards. Inspection systems should be designed and operated on the basis of objective risk assessment appropriate to the circumstances. Preferably the risk assessment methodology employed should be consistent with internationally accepted approaches. Risk assessment should be based on current available scientific evidence.

Inspection And Control Procedures

* Monitoring of feed and feed ingredients, whether by industry or official inspection bodies, should include inspection and sampling and analysis to detect unacceptable levels of undesirable substances.

Health Hazards Associated With Animal Feed

* All feed and feed ingredients should meet minimum safety standards. It is essential that levels of undesirable substances are sufficiently low in feed and feed ingredients that their concentration in food for human consumption is consistently below the level of concern. Codex Maximum Residue Limits and Extraneous Maximum Residue Levels set for feed should be applied. Maximum residue limits set for food, such as those established by the Codex Alimentarius Commission, may be useful in determining minimum safety standards for feed.

Feed Additives And Veterinary Drugs Used In Medicated Feed

* Feed additives and veterinary drugs used in medicated feed should be assessed for safety and used under stated conditions of use as pre-approved by the competent authorities.

* Veterinary drugs used in medicated feed should comply with the provisions of the Codex Recommended International Code of Practice for the Control of the Use of Veterinary Drugs 10 .

* Borderlines between feed additives and veterinary drugs used in medicated feed may be set to avoid misuse.

* Feed additives should be received, handled and stored to maintain their integrity and to minimise misuse or unsafe contamination. Feed containing them should be used in strict accordance with clearly defined instructions for use.

* Antibiotics should not be used in feed for growth promoting purposes in the absence of a public health safety assessment.11

Feed And Feed Ingredients

* Feed and feed ingredients should only be produced, marketed, stored and used if they are safe and suitable, and, when used as intended, should not represent in any way an unacceptable risk to consumers' health. In particular, feed and feed ingredients contaminated with unacceptable levels of undesirable substances should be clearly identified as unsuitable for animal feed and not be marketed or used.

* Feed and feed ingredients should not be presented or marketed in a manner liable to mislead the user.

Undesirable Substances

* The presence in feed and feed ingredients of undesirable substances such as industrial and environmental contaminants, pesticides, radionuclides, persistent organic pollutants, pathogenic agents and toxins such as mycotoxins should be identified, controlled and minimised. Animal products that could be a source of the Bovine Spongiform Encephalopathy (BSE) agent 12 should not be used for feeding directly to, or for feed manufacturing for, ruminants. Control measures applied to reduce unacceptable level of undesirable substances should be assessed in terms of their impact on food safety.

* The risks of each undesirable substance to consumers' health should be assessed and such assessment may lead to the setting of maximum limits for feed and feed ingredients or the prohibition of certain materials from animal feeding.

Production, Processing, Storage, Transport and Distribution of Feed and Feed Ingredients

The production, processing, storage, transport and distribution of safe and suitable feed and feed ingredients is the responsibility of all participants in the feed chain, including farmers, feed ingredient manufacturers, feed compounders, truckers, etc. Each participant in the feed chain is responsible for all activities that are under their direct control, including compliance with any applicable statutory requirements. Feed and feed ingredients should not be produced, processed, stored, transported or distributed in facilities or using equipment where incompatible operations may affect their safety and lead to adverse effects on consumers' health. Due to the unique characteristics of aquaculture, the application of these general principles must consider the differences between aquaculture and terrestrial-based production.

Where appropriate, operators should follow GMPs and, where applicable, HACCP principles to control hazards that may affect food safety. The aim is to ensure feed safety and in particular to prevent contamination of animal feed and food of animal origin as far as this is reasonably achievable, recognising that total elimination of hazards is often not possible. The effective implementation of GMPs and, where applicable, HACCP-based approaches should ensure, in particular, that the following areas are addressed.

Premises

* Buildings and equipment used to process feed and feed ingredients should be constructed in a manner that permits ease of operation, maintenance and cleaning and minimises feed contamination. Process flow within the manufacturing facility should also be designed to minimise feed contamination.

* Water used in feed manufacture should meet hygienic standards and be of suitable quality for animals. Tanks, pipes and other equipment used to store and convey water should be of appropriate materials which do not produce unsafe levels of contamination.

* Sewage, waste and rain water should be disposed of in a manner which avoids contamination of equipment, feed and feed ingredients.

Receiving, Storage And Transportation

* Chemical fertilizers, pesticides and other materials not intended for use in feed and feed ingredients should be stored separately from feed and feed ingredients to avoid the potential for manufacturing errors and contamination of feed and feed ingredients.

* Processed feed and feed ingredients should be stored separately from unprocessed feed ingredients and appropriate packaging materials should be used. Feed and feed ingredients should be received, stored and transported in such a way so as to minimize the potential for any cross-contamination to occur at a level likely to have a negative impact on food safety.

* The presence of undesirable substances in feed and feed ingredients should be monitored and controlled.

* Feed and feed ingredients should be delivered and used as soon as possible. All feed and feed ingredients should be stored and transported in a manner which minimizes deterioration and contamination and enables the correct feed to be sent to the right animal group.

* Care should be taken to minimize deterioration and spoilage at all stages of handling, storage and transport of feed and feed ingredients. Special precautions should be taken to limit fungal and bacterial growth in moist and semi-moist feed. Condensation should be minimized in feed and feed ingredient manufacturing and processing facilities. Dry feed and feed ingredients should be kept dry in order to limit fungal and bacterial growth.

* Waste feed and feed ingredients and other material containing unsafe levels of undesirable substances or any other hazards should not be used as

feed, but, should be disposed of in an appropriate manner including compliance with any applicable statutory requirements.

Personnel Training

* All personnel involved in the manufacture, storage and handling of feed and feed ingredients should be adequately trained and aware of their role and responsibility in protecting food safety.

Sanitation And Pest Control

* Feed and feed ingredients, processing plants, storage facilities and their immediate surroundings should be kept clean and effective pest control programmes should be implemented.

* Containers and equipment used for manufacturing, processing, transport, storage, conveying, handling and weighing should be kept clean. Cleaning programmes should be effective and minimise residues of detergents and disinfectants.

* Machinery coming into contact with dry feed or feed ingredients should be dried following any wet cleaning process.

* Special precautions should be taken when cleaning machinery used for moist and semi-moist feed and feed ingredients to avoid fungal and bacterial growth.

Equipment Performance And Maintenance

* All scales and metering devices used in the manufacture of feed and feed ingredients should be appropriate for the range of weights and volumes to be measured, and be tested regularly for accuracy.

* All mixers used in the manufacture of feed and feed ingredients should be appropriate for the range of weights or volumes being mixed and be capable of manufacturing suitable homogeneous mixtures and homogeneous dilutions, and be tested regularly to verify their performance.

* All other equipment used in the manufacture of feed and feed ingredients should be appropriate for the range of weights or volumes being processed, and be monitored regularly.

Manufacturing Controls

* Manufacturing procedures should be used to avoid cross-contamination (for example flushing, sequencing and physical clean-out) between batches of

feed and feed ingredients containing restricted or otherwise potentially harmful materials (such as certain animal by-product meals, veterinary drugs). These procedures should also be used to minimise cross-contamination between medicated and non-medicated feed and other incompatible feed. In cases where the food safety risk associated with cross-contamination is high and the use of proper flushing and cleaning methods is deemed insufficient, consideration should be given to the use of completely separate production lines, transfer, storage and delivery equipment.

* Pathogen control procedures, such as heat treatment or the addition of authorised chemicals, should be used where appropriate, and monitored at the applicable steps in the manufacturing process.

Recalls

*Records and other information should be maintained to include the identity and distribution of feed and feed ingredients so that any feed or feed ingredient considered to pose a threat to consumers' health can be rapidly removed from the market and that animals exposed to the relevant feed can be identified.

On–Farm Production and Use of Feed and Feed Ingredients

* This section provides guidance on the cultivation, manufacture, management and use of feed and feed ingredients on farms and in aquaculture.

* This section should be used in conjunction with the applicable requirements of Sections 4 and 5 of this Code.

* To help ensure the safety of food used for human consumption, good agricultural practices 13 should be applied during all stages of on-farm production of pastures, cereal grain and forage crops used as feed or feed ingredients for food producing animals. For aquaculture the same principles should apply, where applicable.

Three types of contamination represent hazards at most stages of on-farm production of feed and feed ingredients, namely :

- Biological, such as bacteria, fungi and other microbial pathogens;
- Chemical, such as residues of medication, pesticides, fertilizer or other agricultural substances; and
- Physical, such as broken needles, machinery and other foreign material.

Agricultural Production of Feed

* Adherence to good agricultural practices is encouraged in the production of natural, improved and cultivated pastures and in the production of forage and cereal grain crops used as feed or feed ingredients for food producing animals. Following good agricultural practice standards will minimize the risk of biological, chemical and physical contaminants entering the food chain. If crop residuals and stubbles are grazed after harvest, or otherwise enter the food chain, they should also be considered as livestock feed. Most livestock will consume a portion of their bedding. Crops that produce bedding material or bedding materials such as straw or wood shavings should also be managed in the same manner as animal feed ingredients. Good pasture management practices, such as rotational grazing and dispersion of manure droppings, should be used to reduce cross-contamination between groups of animals.

Site Selection

* Land used for production of animal feed and feed ingredients should not be located in close proximity to industrial operations where industrial pollutants from air, ground water or runoff from adjacent land would be expected to result in the production of foods of animal origin that may present a food safety risk. Contaminants present in runoff from adjacent land and irrigation water should be below levels that present a food safety risk.

Fertilizers

* Where manure fertilization of crops or pastures is practised, an appropriate handling and storage system should be in place and maintained to minimize environmental contamination, which could negatively impact on the safety of foods of animal origin. There should be adequate time between applying the manure and grazing or forage harvesting (silage and hay making) to allow the manure to decompose and to minimize contamination.

* Manure, compost and other plant nutrients should be properly used and applied to minimize biological, chemical and physical contamination of foods of animal origin which could adversely affect food safety.

* Chemical fertilizers should be handled, stored and applied in a manner such that they do not have a negative impact on the safety of foods of animal origin.

Pesticides and other Agricultural Chemicals

* Pesticides and other agricultural chemicals should be obtained from safe sources. Where a regulatory system is in place, any chemical used must comply with the requirements of that system.

* Pesticides should be stored according to the manufacturer's instructions and used in accordance with Good Agricultural Practice in the Use of Pesticides (GAP)14 . It is important that farmers carefully follow the manufacturer's instructions for use for all agricultural chemicals.

* Pesticides and other agricultural chemicals should be disposed of responsibly in a manner that will not lead to contamination of any body of water, soil, feed or feed ingredients that may lead to the contamination of foods of animal origin which could adversely affect food safety.

Manufacturing of Feed on-Farm

Feed ingredients

* On-farm feed manufacturers should follow the applicable guidelines established in sub-section 4.1 of this Code when sourcing feed ingredients off the farm.

* Feed ingredients produced on the farm should meet the requirements established for feed ingredients sourced off the farm. For example, seed treated for planting should not be fed.

Mixing

* On-farm feed manufacturers should follow the applicable guidelines established earlier.

* In particular, feed should be mixed in a manner that will minimize the potential for cross-contamination between feed or feed ingredients that may have an effect on the safety or withholding period for the feed or feed ingredients.

Monitoring Records

* Appropriate records of feed manufacturing procedures followed by on-farm feed manufacturers should be maintained to assist in the investigations of possible feed-related contamination or disease events.

* Records should be kept of incoming feed ingredients, date of receipt and batches of feed produced in addition to other applicable records set out in sub-section 4.3 of the Code.

Good Animal Feeding Practice

* Good animal feeding practices include those practices that help to ensure the proper use of feed and feed ingredients on-farm while minimising biological, chemical and physical risks to consumers of foods of animal origin.

* Water

* Water for drinking or for aquaculture should be of appropriate quality for the animals being produced. Where there is reason to be concerned about contamination of animals from the water, measures should be taken to evaluate and minimise the hazards.

Pasture grazing

* The grazing of pastures and crop lands should be managed in a way that minimises the avoidable contamination of foods of animal origin by biological, chemical and physical food safety hazards.

* Where appropriate, an adequate period should be observed before allowing livestock to graze on pasture, crops and crop residuals and between grazing rotations to minimise biological cross-contamination from manure.

* Where agricultural chemicals are used, operators should ensure that the required withholding periods are observed.

Feeding

* It is important that the correct feed is fed to the right animal group and that the directions for use are followed. Contamination should be minimised during feeding. Information should be available of what is fed to animals and when, to ensure that food safety risks are managed.

* Animals receiving medicated feed should be identified and managed appropriately until the correct withholding period (if any) has been reached and records of these procedures must be maintained. Procedures to ensure that medicated feed are transported to the correct location and are fed to animals that require the medication should be followed. Feed transport vehicles and feeding equipment used to deliver and distribute medicated feed should be cleaned after use, if a different medicated feed or non-medicated feed or feed ingredient is to be transported next.

Stable Feeding and Lot/Intensive Feeding Units

* The animal production unit should be located in an area that does not result in the production of food of animal origin that poses a risk to food safety.

Care should be taken to avoid animal access to contaminated land, and to facilities with potential sources of toxicity.

Hygiene

* The animal production unit should be designed so that it can be adequately cleaned. The animal production unit and feeding equipment should be thoroughly cleaned regularly to prevent potential hazards to food safety. Chemicals used should be appropriate for cleaning and sanitising feed manufacturing equipment and should be used according to instructions. These products should be properly labelled and stored away from feed manufacturing, feed storage and feeding areas.

* A pest control system should be put in place to control the access of pests to the animal production unit to minimise potential hazards to food safety.

* Operators and employees working in the animal production unit should observe appropriate hygiene requirements to minimise potential hazards to food safety from feed.

* To ensure food safety, necessary precautions should be taken regarding culturing methods, culturing sites, technologies, materials and feed used to minimize contamination in order to reduce food hazards.

Methods of Sampling and Analysis

Sampling

* Sampling protocols should meet scientifically recognized principles and procedures.

Analysis

* Laboratory methods developed and validated using scientifically recognized principles and procedures should be used.16 When selecting methods, consideration should also be given to practicability, with preference given to those methods which are reliable and applicable for routine use. Laboratories conducting routine analyses of feed and feed ingredients should ensure their analytical competency with each method used and maintain appropriate documentation.

Index